Revolutionizing Youth Mental Health with Ethical AI

Transform Youth Mental Well-being by Exploring Responsible AI and GenAI Applications

Sharmistha Chatterjee
Azadeh Dindarian
Usha Rengaraju

Apress®

Revolutionizing Youth Mental Health with Ethical AI: Transform Youth Mental Well-being by Exploring Responsible AI and GenAI Applications

Sharmistha Chatterjee
Bangalore, Karnataka, India

Azadeh Dindarian
Berlin, Germany

Usha Rengaraju
Bangalore, India

ISBN-13 (pbk): 979-8-8688-1185-2
https://doi.org/10.1007/979-8-8688-1186-9

ISBN-13 (electronic): 979-8-8688-1186-9

Managing Director, Apress Media LLC: Welmoed Spahr
Acquisitions Editor: Celestin Suresh John
Desk Editor: Laura Berendson
Editorial Project Manager: Gryffin Winkler

Cover designed by eStudioCalamar

Cover image designed by Luke Jones

Distributed to the book trade worldwide by Springer Science+Business Media New York, 1 New York Plaza, New York, NY 10004. Phone 1-800-SPRINGER, fax (201) 348-4505, e-mail orders-ny@springer-sbm.com, or visit www.springeronline.com. Apress Media, LLC is a Delaware LLC and the sole member (owner) is Springer Science + Business Media Finance Inc (SSBM Finance Inc). SSBM Finance Inc is a **Delaware** corporation.

For information on translations, please e-mail booktranslations@springernature.com; for reprint, paperback, or audio rights, please e-mail bookpermissions@springernature.com.

Apress titles may be purchased in bulk for academic, corporate, or promotional use. eBook versions and licenses are also available for most titles. For more information, reference our Print and eBook Bulk Sales web page at http://www.apress.com/bulk-sales.

Any source code or other supplementary material referenced by the author in this book is available to readers on GitHub. For more detailed information, please visit https://www.apress.com/gp/services/source-code.

If disposing of this product, please recycle the paper

This book is dedicated to my mother, Anjali Chatterjee, and my late father, Subhas Chatterjee, who, as parents, provided me with immense encouragement to pursue a career in science, technology, engineering, and mathematics (STEM) and supported me in all the decisions that made me successful. This book reflects my better half, Abhisek Bakshi, who was the first to make me believe that I could contribute to societal problems like mental health and author a book to reshape this field by applying AI. The tremendous support, encouragement, mentorship, and vision he has laid out in front of me in my journey in the field of AI deserve special mention. Your intense support and enlightenment have been immensely helpful and shaped my research in AI in mental health.

Thank you for guiding my thoughts and making it possible to accomplish this with two little daughters, Aarya and Adrika, aged 7 and 5, who have made multiple sacrifices to allow me to dedicate myself to writing this book. Last but not least, I would like to thank the little ones for allowing me the time and space to complete this feat.

—Sharmistha Chatterjee

This book is dedicated to my daughter and my husband. To my daughter, thank you for guiding me back to myself. You showed me how to let go, to truly listen, and to move with the flow instead of trying to control every step. Thank you for being you.

To my husband, thank you for always believing in me and giving me the space to grow. You've supported me quietly and consistently, allowing me to explore, to make mistakes, and to find my own path. Your trust and patience have made all the difference. That kind of support is rare, and I don't take it for granted. This book would not exist without either of you.

—Azadeh Dindarian

I dedicate this book to my parents, Chellamuthu Rengaraju and Egambal, whose unwavering love and steadfast encouragement have been the cornerstone of every step I've taken. Your faith in me pushed me beyond the confines of my comfort zone and instilled in me the confidence to pursue my dreams fearlessly.

Thank you for celebrating each victory and guiding me through every challenge with patience and wisdom. This work stands as a testament to your endless support, and I hope it makes you as proud as I am to be your child.

—Usha Rengaraju

Table of Contents

About the Authors

 **Sharmistha Chatterjee** is an AI and cloud applications evangelist, who has worked across organizations like Commonwealth Bank of Australia, Publicis Sapient, SAP Labs, and Nokia. A graduate of Aalto University with a background in Computer Science and Engineering, she has established and led the practice, focusing on building scalable AI platforms. Sharmistha has published research papers, filed patents, and holds proven expertise in leading teams to develop scalable AI solutions across various industries. She is recognized as a Google Developer Expert (GDE), award-winning data scientist, and a nominee for several accolades, including 40 under 40 Data Scientists, Women to Watch in AI and Analytics, and global AI leader finalist. She actively contributes to the community through blogging, speaking at international conferences, mentoring startups, and serving as a guest speaker in universities. Sharmistha is also an advocate for diversity, equity, and inclusion, actively participating in DEI hiring initiatives and addressing gaps in AI tools. She is also an advocate for neurodiversity and championed around creating awareness in different organizations. She is also coauthor of the book *Platform and Model Design for Responsible AI*.

Prof. Dr. Azadeh Dindarian is a scholar and leader in digitalization, AI, and emerging technologies, with a unique career spanning engineering, entrepreneurship, and academia. She holds a PhD in Electrical and Electronic Engineering and a BEng in Computing and Communication Systems from the University of Manchester. Azadeh is the founder and CEO of Younivers app, an AI-powered mental fitness app designed to support teenagers' emotional well-being through psychoeducation. As a professor at SRH Berlin University of Applied Sciences, she specializes in the intersection of digital transformation, ethical AI, and technological applications across industries such as supply chain management. Azadeh is also a passionate advocate for diversity in technology, evidenced by her founding of Women AI Academy and Resili Tech Venture. She regularly contributes to academic research on topics including digitalization, smart cities, and sustainability, with numerous publications in scientific journals. Her leadership and vision continue to inspire the next generation of engineers, entrepreneurs, and digital innovators.

Usha Rengaraju the Head of Data Science Research at Exa Protocol, holds the distinction of being the world's first woman triple Kaggle Grandmaster. Specializing in deep learning and probabilistic graphical models, she served as a judge for prestigious challenges like TigerGraph's "Graph for all - Million Dollar Challenge" and the Intel oneAPI Hackathon in 2022. She is recognized as one of the top-ten data scientists in India (2020) by *Analytics*

India Magazine, a leading woman data scientist, and one of the top-150 AI leaders and influencers by *3AI* magazine, Usha is a trailblazer in her field. Her initiatives include organizing NeuroAI, India's first research symposium at the intersection of Neuroscience and Data Science, and the Neurodiversity India Summit, the country's inaugural neurodiversity conference. A champion of autism (neurodiversity), Usha focuses on leveraging AI to address the challenges faced by the autistic community. She led her team to victory in the WiDS 2022 Datathon, won the Kaggle ML Research Spotlight for 2022, and earned the TensorFlow Community Spotlight in 2023.

Passionate about applying AI to healthcare, Usha believes in its transformative potential for India's healthcare crisis. With expertise in deep learning for computer vision and NLP, she aims to revolutionize mental healthcare by enhancing public health surveillance, enabling early disease diagnosis, and advancing precision medicine.

About the Technical Reviewer

 **Siddhant Agarwal** is a seasoned DevRel professional with over a decade of experience cultivating innovation and scaling developer ecosystems globally. Currently leading Developer Relations across APAC at Neo4j and recognized as a Google Developer Expert in Gen-AI, Sid transforms local developer initiatives into global success stories with his signature "Local to Global" approach. Previously at Google managing flagship developer programs, he has shared his technical expertise at diverse forums worldwide, fueling inspiration and innovation. Know more about him at meetsid.dev.

Acknowledgments

I would like to express my heartfelt gratitude to several people whose constant support and mentorship have led to the coauthoring of this book. In the first place, I would like to extend my wholehearted thankfulness to my mentor and husband, Abhisek Bakshi, who has instilled an interest in researching the latest cutting-edge AI tools and technologies that can help common people when a serious societal problem like mental health worries adolescents, students, and parents.

In addition, I am honored and immensely grateful to my coauthors, Prof. Dr. Azadeh Dindarian and Usha Rengaraju, who have allowed me to collaborate with them, listened to my ideas patiently, given timely feedback, and reshaped several ideas presented in the book. I would also like to extend my gratitude to my mentor, Roopa Hungund, for her constant support and encouragement.

The book is a manifestation of the unwavering support I have received from the entire Apress team and the reviewers of the book, who have worked tirelessly to stitch the pieces together to bring a coherent story to this book. Special mention goes to Siddhant Agarwal, Celestin Suresh John, Gryffin Winkler, and Shobana Srinivasan, whose support, editorial expertise, and hard work have been essential in refining and polishing this book.

Without the collaborative efforts, encouragement, and guidance of these remarkable individuals, this book would not have been possible. Thank you so much for being a part of this wonderful journey.

—Sharmistha Chatterjee

ACKNOWLEDGMENTS

Writing this book has been a meaningful and, at times, challenging journey. It has been filled with reflection, deep learning, and personal growth. The final outcome is the result of many conversations, moments of clarity, and long hours of focused, often quiet, work.

The seed for this project was planted during the first year of the COVID-19 pandemic. I was deeply affected by the visible impact it had on the mental health and well-being of young people. As a mother and as an engineer, I couldn't look away. I saw too many teenagers and families struggling silently, often without access to professional help. That experience stayed with me.

It pushed me to explore how technology, combined with expert support, could provide help that is both more affordable and more accessible. That idea became Younivers app, a platform designed to offer personalized psychoeducational resources and connect teens and their families with professional guidance. My intention was simple: to help.

Along the way, I've had the privilege of meeting many inspiring individuals. I now call them my inner circle and dear friends. To Rilana and Yara, thank you for being constant sources of support, insight, and honest reflection. Your curiosity and openness have shaped how I think, build, teach, and write. This book carries your spirit throughout.

To my colleagues and the academic community, thank you for the rich conversations, helpful critique, and your generosity of thought. To my coauthors, Sharmistha Chatterjee and Usha Rengaraju, thank you for helping shape and develop these ideas. Our continuous exchange and feedback gave this work its depth and clarity.

To my husband and daughter, mum and dad, sister and brother, thank you for your patience, your quiet encouragement, and for giving me the space to focus. Your belief in me and in this work has carried me through more than you know.

Finally, to the editorial team at Apress and reviewers, thank you for your time, your thoughtful suggestions, and your trust. This book is stronger because of you.

With sincere gratitude, thank you all.

—Azadeh Dindarian

My heartfelt thanks go to my coauthors, Sharmistha Chatterjee and Prof. Dr. Azadeh Dindarian, for the invaluable opportunity to join forces with you on this book. Your collaboration and insights have enriched this work in countless ways.

I've long been a passionate advocate for autism, dedicating myself to raising awareness through initiatives like the Neurodiversity India Summit—India's first conference on neurodiversity and NeuroAI—and a research symposium at the intersection of neuroscience and data science. With mental health still widely misunderstood in India and public awareness sorely lacking, I hope this book sparks new conversations and demonstrates how AI-driven solutions can creatively address the challenges we face in promoting mental well-being.

—Usha Rengaraju

Introduction

This book is a comprehensive guide that delves into the dynamic intersection of artificial intelligence (AI) and youth mental health. It aims to bridge the gap between cutting-edge AI technology and its transformative potential in addressing youth mental health challenges.

The book's content is structured into three key sections, each focusing on different facets of AI applications in youth mental health. The first section provides a comprehensive background on the current state of youth mental health, analyzing the prevalence of mental health issues and identifying the unique challenges faced by the digital first generation. In the second section, we explore the foundational principles of AI and its potential for revolutionizing mental healthcare, including natural language processing, machine learning, and predictive analytics. In this section, readers will find in-depth case studies and real-world applications that showcase how AI-driven interventions have already transformed mental healthcare for youth across diverse contexts. Finally, the third section delves into ethical considerations, fairness, privacy concerns, and the responsible integration of AI in youth mental healthcare to design long-term sustainable solutions.

This book offers a unique and holistic perspective, making it an indispensable resource for anyone passionate about leveraging AI for the betterment of youth mental health. Through this book, readers will gain the knowledge and tools needed to design and implement effective AI-driven solutions, in a unified framework that has the potential to transform the mental health landscape for the benefit of future generations.

What You Will Learn

- Understand the current state of youth mental health, exploring the prevalence of mental health issues among the digital first generation.

- Understand natural language processing, machine learning, and predictive analytics.

- How AI interventions are already transforming mental healthcare for youth in diverse contexts.

- What are fairness, privacy concerns, and the responsible integration of AI in youth mental healthcare.

- Learn about the impact and usefulness of generative AI solutions and how agentic AI can transform mental health.

- Explore next-generation mental health solutions with synthetic data testing frameworks, recommender systems, and digital twin solutions.

Who This Book Is For

The primary goal of this book is to equip academics and researchers in the AI, computer science, and digital mental health domain as well as AI application developers with a deeper understanding of how AI-powered innovations can enhance the well-being of youth. In addition, innovation managers and policymakers who are interested in exploring the AI use cases would gain the benefit of the book content. This book can also add a lot of value to hospitals, clinics, and mental health diagnostics centers, which wish to implement an AI-enabled software for the benefit of patients, doctors, and care providers for providing proactive servicing and recommendations. In addition, mental health-based software companies can also benefit from the book as it gives them an understanding of how to design an AI-enabled framework.

SECTION 1

An Overview Of Mental Well Being Of Youth

CHAPTER 1

Current State of Mental Health of Youth

This chapter provides an overview of youth mental well-being, examining prevalent mental health challenges faced by young individuals. It analyzes relevant data and statistics to shed light on the prevalence of mental health disorders among young people. Additionally, it explores the impact of modern societal pressures, academic stress, and the influence of digital technologies on youth mental health.

In this chapter, these topics will be covered in the following sections:

- Introduction to youth mental health

- Prevalent mental health challenges faced by youth

- The role of digital technologies in youth mental health

Introduction to Youth Mental Health

Adolescence is a pivotal stage of growth, marked by significant physical, emotional, and behavioral transformations that shape an individual's transition into adulthood. According to the World Health Organization (2024), one in six people worldwide falls within the 10 to 19-year-old age range, making up a significant proportion of the global population. During

© Sharmistha Chatterjee, Azadeh Dindarian, Usha Rengaraju 2025
S. Chatterjee et al., *Revolutionizing Youth Mental Health with Ethical AI*,
https://doi.org/10.1007/979-8-8688-1186-9_1

this period, young people must navigate complex social relationships, build their identity, and develop independence (Lukoševičiūtė-Barauskienė et al., 2023).

However, adolescence is not only a time of self-discovery but also a period of immense pressure. Young people often face high expectations from family, school, and society, in addition to the influence of peer relationships and digital technology (The U.S. Surgeon General's Advisory, 2023). For instance, the pressure to perform well in school, meet high parental and societal expectations, and secure future career opportunities can create overwhelming stress for adolescents (Deng et al., 2022). These factors play a significant role in shaping their mental and emotional well-being. Because this stage lays the foundation for lifelong emotional health, equipping adolescents with resilience, coping strategies, and a strong support system is crucial for their long-term well-being. Support systems such as family, friends, schools, and mental health services play a pivotal role in helping young people manage these challenges.

In recent years, youth mental health has become a growing concern, drawing attention from educators, policymakers, researchers, and young people themselves. Reports indicate that mental health among adolescents has been steadily declining since 2000 (Coyne et al., 2025a). Research by Kessler et al. (2005) revealed that half of all lifetime mental health disorders begin by age 14, and three-quarters emerge before the age of 24. These findings emphasize the importance of early intervention and prevention, ensuring that young people receive support before mental health challenges escalate into long-term conditions.

The World Health Organization (WHO) (WHO, 2022b) defines mental health as:

> *a state of mental well-being that enables people to cope with the stresses of life, realize their abilities, learn well and work well, and contribute to their community. It is an integral component of health and well-being that underpins our individual*

and collective abilities to make decisions, build relationships and shape the world we live in. Mental health is a basic human right. And it is crucial to personal, community and socio-economic development.

This definition underscores the fundamental role of mental health in shaping personal growth and societal well-being. Beyond individual concerns, adolescent mental health is a key factor in social stability, economic productivity, and public health.

Building on WHO's definition, Ross et al. (2020) provide a broader framework for adolescent well-being. They define it as:

Adolescents have the support, confidence, and resources to thrive in contexts of secure and healthy relationships, realizing their full potential and rights.

Their framework identifies five key domains that contribute to overall adolescent well-being:

1. *Good health and optimum nutrition*

2. *Connectedness, positive values, and contribution to society*

3. *Safety and a supportive environment*

4. *Learning, competence, education, skills, and employability*

5. *Agency and resilience*

Mental and emotional well-being fall under the *"Good health and optimum nutrition"* domain, highlighting the deep connection between mental and physical health. Adolescents who experience good health and positive relationships are better equipped to handle stress, build resilience, and make healthy life choices (Ross et al., 2020).

A recent study by Lukoševičiūtė-Barauskienė et al. (2023) concluded that

> *most adolescents define mental health as the ability to control their emotions and adjust their behavior according to societal and age norms, which is reflected in the quality of their relationships with other people.*

This suggests that emotional regulation and social connectedness are core aspects of how young people perceive mental well-being.

The Alarming Reality of Youth Mental Health Disorders

Despite the importance of mental well-being, many adolescents face significant mental health challenges. Reports indicate that one in seven young people worldwide experiences a mental health disorder (WHO, 2024). The most prevalent conditions include depression, anxiety, and behavioral disorders, which are among the leading causes of illness and disability in adolescents (WHO, 2024). Yet, access to mental health support remains limited. Gulliver et al. (2010) found that only 18% to 34% of young people with severe depression or anxiety symptoms seek professional help. The barriers to seeking care include stigma, lack of awareness, financial constraints, and insufficient mental health services.

Mental health disorders have far-reaching consequences, affecting individuals, families, and entire societies. For young people, these conditions disrupt daily life, education, and social interactions. Families, in turn, bear the emotional and financial burden, which can strain relationships and create additional stress within the household (WHO, 2003; Bapuji et al., 2024). This can ultimately affect the overall well-being if left untreated; poor mental health in adolescence can lead to long-term consequences, affecting employment opportunities, economic stability, and overall quality of life.

The Role of Family in Youth Mental Well-Being

The family environment plays a pivotal role in adolescent mental health. Research by Rothenberg et al. (2020) indicates that teenagers who have strong, positive relationships with their families experience better mental health outcomes. A nurturing home environment provides emotional security, helping young people develop coping skills, resilience, and self-confidence (Pine et al., 2024).

On the other hand, conversely, family conflict, harsh criticism, or unsupportive parenting styles can contribute to mental health struggles during adolescence (Johnco & Rapee, 2018). In addition, the stress of high parental expectations, emotional neglect, or constant tension can lead to anxiety, depression, and low self-esteem.

Beyond emotional support, parents also play a vital role in recognizing early warning signs of mental health issues. The ability to identify these signs and seek professional help can prevent long-term consequences and provide adolescents with the care they need before issues escalate (Fearing, 2024). Therefore, involving families in mental health education and awareness can significantly enhance youth mental well-being.

Mental health challenges among young people do not just affect families, they have wider societal and economic consequences. Adolescents experiencing severe mental health conditions often face social isolation, which makes it harder to maintain relationships and seek support (WHO, 2022c).

Economically, mental health disorders place a significant financial strain on both individuals and society. Research by Doran and Kinchin (2017) found that the economic burden of mental health problems is substantial, leading to higher healthcare costs, reduced workplace productivity, and increased reliance on social welfare systems. Additionally, individuals with untreated mental health disorders may experience higher rates of absenteeism and lower work performance (Hilton et al., 2008).

The increasing demand for mental health services also puts pressure on social and healthcare systems (Marmot, 2010). Addressing youth mental health is not just a healthcare issue; in fact, it is an economic imperative. By investing in early intervention, expanding access to care, and fostering awareness, societies can reduce these economic burdens while improving overall well-being.

Climate Change and Its Impact on Youth Mental Health

Climate change significantly impacts the mental health of young people, both directly and indirectly, affecting their well-being in the short and long term (Majeed & Lee, 2017; Cianconi et al., 2020; Vergunst & Berry, 2022).

Hickman et al. (2021) surveyed 10,000 young people (16–25 years) across ten countries and found that "59% were extremely worried about climate change" and "84% felt at least moderately concerned." More than half of the respondents reported feeling "sad, anxious, angry, powerless, or guilty," with "45% saying their worries affected their daily lives." Additionally, many reported struggling with persistent negative thoughts, highlighting the urgent need for mental health support and climate action.

For many adolescents, these disasters bring "fear, anxiety, and emotional distress, often leading to post-traumatic stress disorder (PTSD), depression, and ongoing anxiety" (Majeed & Lee, 2017; Cianconi et al., 2020; Vergunst & Berry, 2022).

Heinz and Brandt (2024) highlight that climate change affects mental health in three interconnected ways as follows: directly (e.g., exposure to extreme heat, floods, and other natural disasters), indirectly (e.g., forced migration and displacement), and intersectional effects (e.g., social exclusion and discrimination).

A comprehensive review by Cianconi et al. (2020) highlights how these experiences shake a young person's sense of security, making them feel uncertain about the future and powerless in the face of a changing climate. Whether it's losing a home, witnessing environmental destruction, or simply living with the growing fear of what's next, the mental toll is real. These experiences shake a young person's sense of security, fostering uncertainty and powerlessness in the face of climate change.

Prevalent Mental Health Challenges Faced by Youth

Adolescence is a period of significant emotional and psychological development, but it is also a time when mental health issues begin to emerge, potentially shaping long-term well-being. A study conducted in Germany between 2014 and 2017 found that 16.9% of individuals under 18 were affected by mental health conditions (Klipker et al., 2018). The suicide rate in the United States has increased by one-third, and 94% of adults believe suicide is preventable (American Foundation for Suicide Prevention, 2024). Further, in 2022, 45% of LGBTQ+ youth reported contemplating suicide (The Trevor Project, 2022). These numbers reflect a growing global concern about youth mental health, emphasizing the pressing need for proactive interventions and tailored support systems.

External crises such as the COVID-19 pandemic have exacerbated existing mental health challenges (Hossain et al., 2022; Prichett et al., 2024). During lockdowns, school closures and social restrictions significantly disrupted the daily lives of young people, leading to increased stress, anxiety, and depression (UNICEF, 2023). According to UNICEF (2023), recent developments have led to increased isolation from peers, heightened academic uncertainty, and growing household stress. These factors have created an environment where many adolescents struggle to adapt to sudden and profound changes, impacting their mental well-being and resilience.

Stress and Mental Well-Being

Chronic stress has been identified as a key factor affecting adolescent well-being. In a study of 422 adolescents aged 16–17 in Norway, Østerås et al. (2015) found that 22% reported moderate to severe stress, and among them, 79.6% also experienced significant physical pain. The link between stress, emotional distress, and physical symptoms suggests that mental health challenges do not exist in isolation, and they have tangible effects on overall health and quality of life.

Anxiety and Depression: The Most Common Emotional Disorders

Anxiety and depression are among the most frequently diagnosed mental health conditions in adolescents, affecting 4.4% to 5.5% of older adolescents for anxiety and 1.4% to 3.5% for depression (WHO, 2024). Further, research by Lu et al. (2024) suggests that the prevalence of depression and depressive symptoms among youth has been steadily increasing. In the United States, parent-reported data shows a rise in childhood anxiety and depression diagnoses. According to Bitsko et al. (2018) based on parent reports in the United States, the lifetime diagnosis of anxiety or depression among children aged 6 to 17 increased from 5.4% in 2003 to 8.4% in 2011–2012.

Behavioral Disorders and ADHD

Behavioral disorders are more common in younger adolescents than in older ones (WHO, 2024). According to the World Health Organization (WHO, 2024), attention deficit hyperactivity disorder (ADHD) affects 2.9% of 10- to 14-year-olds and 2.2% of 15- to 19-year-olds worldwide, causing difficulties in focusing, hyperactivity, and impulsive behavior. In addition, conduct disorder, characterized by persistent rule breaking and aggressive behavior, affects 3.5% of 10- to 14-year-olds and 1.9% of 15- to 19-year-olds (WHO, 2024). Recent study indicates that approximately 11.4% of US children aged 3–17 years, equating to about 7.1 million children, have been diagnosed with ADHD (Danielson et al., 2024). Notably, boys are more frequently diagnosed with ADHD than girls (Danielson et al., 2024). However, research suggests that girls with ADHD are underidentified and underdiagnosed, leading to fewer referrals for treatment compared to boys (Nøvik et al., 2006). Further, research indicates that ADHD often coexists with other psychiatric disorders, such as depression, anxiety, autism, and eating disorders; this overlap complicates diagnosis and may require multifaceted treatment approaches (Faraone et al., 2021).

A study in Germany between 2014 and 2017 found that 4.4% of children and adolescents were diagnosed with ADHD (Göbel et al., 2018). Additionally, young people with ADHD are significantly more likely to experience depression compared to their peers without ADHD (Daviss, 2008). These conditions, when left untreated, can significantly impact academic performance, social interactions, and long-term career opportunities.

Autism Spectrum Disorders (ASD)

Autism (or autism spectrum disorders, ASD):

> *characterized not only by persistent impairments in reciprocal social communication and social interactions, but is also manifested by restricted, repetitive patterns of behavior, interests, or activities.*
>
> —Joon et al. (2021)

According to Joon et al. (2021), the global incidence of ASD has increased, influenced by genetic predispositions, environmental factors, and medical interventions.

A large-scale analysis by Zeidan et al. (2022) examined 71 studies published since 2012 and found that the global prevalence of ASD varies significantly. The median prevalence was 100 per 10,000 people, with a median male-to-female ratio of 4.2:1. Additionally, 33% of autism cases involve cooccurring intellectual disabilities, which can further complicate daily functioning and social integration.

Research suggests that many children diagnosed with ADHD also show traits commonly associated with autism spectrum disorder (ASD) (Green et al., 2015). Understanding this connection is crucial in ensuring that children receive the right support and interventions tailored to their specific needs, helping them navigate both their learning and social environments more effectively.

Eating Disorders

Eating disorders such as anorexia nervosa, bulimia nervosa, and binge-eating disorder[1] typically emerge during adolescence. These disorders are more common among girls than boys and are strongly associated with anxiety, depression, and substance use disorders (Ram & Shelke, 2023).

According to the World Health Organization (WHO, 2024), an estimated 0.1% of adolescents aged 10–14 and 0.4% of those aged 15–19 experience eating disorders, with severe cases increasing the risk of suicide. Early diagnosis and intervention are crucial in preventing long-term physical and psychological complications. Germany's Federal

---

[1] Eating disorders encompass a range of severe psychiatric conditions, including **anorexia nervosa**, characterized by self-imposed starvation and an intense fear of weight gain; **bulimia nervosa**, marked by recurrent episodes of binge eating followed by compensatory behaviors such as self-induced vomiting, fasting, or excessive exercise; and **binge-eating disorder (BED)**, which involves recurrent binge-eating episodes similar to bulimia, but without the regular use of compensatory strategies to prevent weight gain (e.g., purging, excessive exercise, or laxative misuse) (source: www.hopkinsmedicine.org).

Statistical Office (Destatis) (2025) reported that 190 suicides occurred in 2023 among individuals under 19 years old.

Beyond academics, mental health disorders can significantly impact social interactions. Adolescents experiencing emotional distress may withdraw from their peers due to feelings of isolation, shame, or anxiety (Heinrich & Gullone, 2006). This withdrawal can worsen their mental health, intensifying feelings of loneliness and disconnection, which are crucial factors in overall well-being (Turner et al., 2024). Research by Leigh-Hunt et al. (2017) found a strong link between social isolation and an increased risk of mortality, as well as a significant association with cardiovascular disease. These findings highlight the profound impact of social isolation on both mental and physical health, underscoring the importance of fostering supportive social environments for adolescents.

In severe cases, untreated depression can become overwhelming, leading to thoughts of self-harm or even suicide. Alarmingly, suicide is now one of the leading causes of death among teenagers, with the risk being significantly higher for those who do not receive timely mental healthcare (Prichett et al., 2024). Many adolescents struggling with emotional pain feel trapped and hopeless, particularly when they lack support systems. Furthermore, stigma surrounding mental health issues often discourages young people from seeking help, worsening their struggles and increasing their sense of isolation (Johnco & Rapee, 2018).

If left undetected, unmanaged, or untreated, mental health challenges can have long-term consequences that extend well beyond adolescence. Chronic anxiety, persistent depression, and difficulty building healthy relationships can continue into adulthood, affecting career prospects, personal relationships, and overall life satisfaction. Additionally, untreated mental health conditions can contribute to harmful coping mechanisms such as substance abuse, self-harm, and declining physical health, as adolescents may attempt to manage their distress through unhealthy behaviors.

Recognizing the early signs of mental health disorders is essential in developing effective treatment and intervention strategies. Early intervention plays a crucial role in managing symptoms, preventing mental health conditions from escalating, and improving overall well-being. It is important that schools, families, and healthcare providers work together to ensure that adolescents have access to mental health resources, supportive environments, and tools to build emotional resilience and self-regulation. By prioritizing mental health support during this critical developmental stage, we can foster healthier outcomes, empowering adolescents with the skills and coping mechanisms they need to thrive—both now and in the future.

The Role of Digital Technologies in Youth Mental Health

The widespread use of digital technologies, particularly social media, gaming, and online communication platforms, has transformed the way young people interact, learn, and perceive the world. However, these advancements come with both opportunities and risks, raising concerns among researchers, educators, and policymakers about their effects on adolescent mental well-being (American Academy of Pediatrics, 2021; Office of the Surgeon General, 2024; Weigle & Shafi, 2024). A 2023 survey of US teenagers found that 95% have access to a smartphone and 60% reported using TikTok or Instagram (Anderson et al., 2024). In 2024, about 93% of German teenager are owning a smartphone (Statista, 2024).

Research shows that while digital platforms provide valuable mental health resources, educational tools, and social connections, they also expose young people to cyberbullying, social comparison, unrealistic beauty standards, and academic pressure (Anderson et al., 2024; Bozzola et al., 2022; Nixon, 2014). Additionally, with the rise of Internet accessibility, children and adolescents are spending more time online, increasing their risk of cyber harassment, addiction, and digital fatigue. The 2022 study by DAK-Gesundheit and the University Medical Center Hamburg-Eppendorf (UKE) found that over 600,000 children and adolescents in Germany display pathological media usage behavior (Klicksafe, 2023). Additionally, the study highlighted that many surveyed children and adolescents experience physical discomfort, including neck pain and dry or itchy eyes.

To gain a comprehensive understanding of how digital technologies influence youth mental health, it is crucial to examine both their advantages and the challenges they present. While these technologies offer opportunities for social connectivity, learning, and emotional support, they also pose risks such as increased stress, social comparison, and potential overuse. However, it is important to emphasize that research in this field is still in its early stages, leaving significant room for further exploration to fully grasp both the benefits and the risks involved.

This section delves into the impact of digital tools on adolescent mental well-being, focusing on their role in shaping social interactions, academic pressure, emotional resilience, and self-perception. By analyzing these factors, we can better understand how digital engagement affects young people's mental health and identify strategies to foster a balanced and healthy digital environment.

The Positive Influence of Digital Technologies on Youth Mental Health

Digital technologies have reshaped how young people interact, learn, and express themselves, bringing significant benefits to their mental well-being. From fostering support, these tools play a crucial role in shaping adolescent experiences. The following sections briefly highlight these positive aspects, offering insights into how digital engagement can enhance emotional resilience, social belonging, and overall mental health.

Digital Access to Mental Health Resources and Support

One of the most significant advantages of digital technology is its ability to bridge the gap in mental healthcare, offering online therapy, self-help applications, and peer support networks. Adolescents who struggle with stigma, financial barriers, or lack of access to professional help can turn to mental health apps and virtual counseling platforms for guidance (Goldberg et al., 2022; Taba et al., 2022).

For instance, platforms like Headspace (Headspace, 2025), BetterHelp (BetterHelp, 2025), Calm (Clam, 2025), and Wysa (Waysa, 2025) provide adolescents with mindfulness exercises, guided therapy, and emotional regulation tools, helping them manage stress, anxiety, and depressive symptoms. Online support communities, such as Facebook and Reddit, provide young people with a platform to share their experiences in a safe and judgment-free environment, fostering a sense of belonging and emotional support (Naslund et al., 2016). According to the research by Naslund et al. (2016), such digital spaces can be particularly beneficial for those who may struggle with social interactions in offline settings, offering peer-driven encouragement and shared coping strategies. However, researchers emphasize the importance of critically evaluating the safety, credibility, and professional reliability of the advice shared within these platforms. While online communities can provide meaningful emotional support, they also present risks, including the spread of misinformation, unverified guidance, and potentially harmful content (Lavis & Winter, 2020; Naslund et al., 2016).

Social Connectivity and Online Communities

For many adolescents, social media and digital communication platforms serve as a lifeline for social interaction, especially for those experiencing social anxiety, geographical isolation, or identity struggles (Wojtowicz et al., 2024).

In addition, online communities provide a space for self-expression, where young people can connect with like-minded peers, engage in creative activities, and seek validation (Naslund et al., 2016).

Marginalized groups, including LGBTQ+ youth and individuals with disabilities, often turn to digital platforms as safe spaces to share their experiences, connect with supportive networks, and engage in advocacy (McAlister et al., 2024; McInroy & Craig, 2020). In these online communities, they can find solidarity, validation, and resources that may be inaccessible in their offline environments. These digital connections can be profoundly empowering, offering a sense of belonging that helps counteract loneliness and social exclusion. However, while these platforms provide critical support, the quality and safety of interactions must be considered, ensuring that they truly serve as inclusive and protective spaces given that negative online experiences, such as cyberbullying, online harassment, and toxic social environments, can have severe consequences on mental health (Twenge et al., 2020).

The Negative Impact of Digital Technologies on Youth Mental Health

Although digital platforms can serve as valuable sources of connection and support, the nature and quality of these interactions must be carefully evaluated.

Negative online experiences such as cyberbullying, harassment, and toxic digital environments can have profound consequences on mental health, potentially outweighing the benefits (Dorol-Beauroy-Eustache & Mishara, 2021; Nixon, 2014). Without adequate safeguards, a system designed for support can quickly become a source of stress, amplifying existing vulnerabilities rather than alleviating them. This reinforces the urgent need for responsible platform management, ethical oversight, and user empowerment, ensuring that digital spaces foster well-being rather than compromise it.

The following sections briefly examine these negative aspects, offering insights into how excessive or unregulated digital engagement can impact adolescent mental health, highlighting concerns such as cyberbullying, sleep disruption, and the pressure of online validation.

The Psychological Impact of Social Media

While social media enables connection and self-expression, it also promotes unhealthy comparisons, unrealistic beauty standards, and validation-seeking behaviors. Research in this area remains in its infancy, with growing calls for more in-depth, multifaceted studies to better understand its complex psychological and societal impacts (Coyne et al., 2025b; Marciano et al., 2025). A longitudinal study (2019-2023) investigating the effects of the COVID-19 pandemic on social media use among 3,697 Israeli children and adolescents aged 8 to 14 found that increased social media use was linked to both heightened psychiatric symptoms and negative emotions, as well as enhanced positive emotions and life satisfaction (Shoshani et al., 2024). Social support was a key factor in reducing the negative impact of excessive social media use observed in this longitudinal study.

The curated, idealized portrayals of life on platforms such as Instagram, TikTok, and Snapchat can contribute to body dissatisfaction (Hosokawa et al., 2023), anxiety, and depressive symptoms (The U.S. Surgeon General's Advisory, 2023). Adolescents who spend more than three hours a day on social media have been found to experience higher rates of emotional distress and self-esteem issues.

A longitudinal cohort study by Riehm et al. (2019) involving 6,595 US adolescents aged 12 to 15, which accounted for initial mental health status, found that "those engaging with social media for over three hours daily are at an increased risk for mental health issues, particularly internalizing problems."

Cyberbullying and Online Harassment

With the increase in online activity, the risk of cyberbullying, online harassment, and digital exploitation has also risen. A study found that approximately 9.5% of early adolescents in the United States reported experiencing cyberbullying victimization (Nagata et al., 2022). Unlike traditional bullying, cyberbullying is persistent, difficult to escape, and often anonymous, making it even more damaging for victims (Dorol-Beauroy-Eustache & Mishara, 2021). Research highlights the psychological toll of cyberbullying, with victims experiencing higher levels of anxiety, depression, and suicidal ideation; further, many adolescents also struggle to report cyberbullying (Kumar & Goldstein, 2020). However, stronger digital literacy programs and anti-cyberbullying policies are essential in mitigating these risks (Nagata et al., 2022).

Digital Addiction and Mental Fatigue

Excessive screen use has been associated with poorer sleep quality, heightened stress levels, and a decline in cognitive function (Riehm et al., 2019). Additionally, many adolescents become reliant on digital entertainment, including gaming and social media, which can contribute to difficulties in impulse control, diminished face-to-face communication abilities, and greater emotional instability (Ding & Li, 2023) as well as lack of sleep (Baiden et al., 2019).

As given above, the impact of digital technologies on youth mental health is complex, presenting both valuable opportunities and critical challenges. While these tools facilitate access to mental health support, education, and social interaction, they also heighten risks such as cyberbullying, digital dependency, and the pressure of unrealistic societal standards. The key challenge is not merely in their use but in managing their influence, ensuring that adolescents harness the benefits of technology while safeguarding their well-being against its potential harms.

Conclusion

In this chapter, we explored the mental well-being of young people, focusing on the growing challenges they face in today's world. Adolescence is a critical period of development, yet many young individuals struggle with mental health disorders such as anxiety, depression, and behavioral conditions. Research highlights that one in seven young people worldwide experience mental health issues, with half of all lifetime disorders beginning by age 14 (Kessler et al., 2005). Despite the increasing need for support, stigma, financial barriers, and limited access to mental health services often prevent adolescents from seeking help.

Family and social environments play a crucial role in shaping mental well-being. Supportive family relationships contribute to resilience, while conflict, high parental expectations, and emotional neglect can intensify mental health struggles. Beyond the home, society bears the consequences of untreated youth mental health conditions, affecting education, social interactions, and long-term economic stability.

Nowadays, digital technology is part of young people's lives, offering both opportunities and risks. While online platforms provide access to mental health resources and peer support, they also expose adolescents to cyberbullying, social comparison, and digital overuse. The psychological impact of social media is a growing concern, with studies linking excessive use to emotional distress, low self-esteem, and increased psychiatric symptoms. Cyberbullying, in particular, has emerged as a major issue, affecting nearly 9.5% of early adolescents in the United States (Nagata et al., 2022).

As we have seen, untreated mental health challenges can have lasting effects, shaping an individual's future relationships, career prospects, and overall quality of life. With one in seven adolescents facing mental health challenges (WHO, 2024), early recognition is key. Recognizing the early signs of mental health struggles is crucial, and families, schools, and healthcare providers must work together to provide the support young people need.

References

American Academy of Pediatrics. (2021). *AAP-AACAP-CHA Declaration of a National Emergency in Child and Adolescent Mental Health*. https://www.aap.org/en/advocacy/child-and-adolescent-healthy-mental-development/aap-aacap-cha-declaration-of-a-national-emergency-in-child-and-adolescent-mental-health/

American Foundation for Suicide Prevention. (2024). *Suicide statistics*. https://afsp.org/suicide-statistics/

Anderson, M., Faverio, M., & Park, E. (2024). How Teens and Parents Approach Screen Time. *Pew Research Center*. https://www.pewresearch.org/internet/2024/03/11/how-teens-and-parents-approach-screen-time/

Baiden, P., Tadeo, S. K., & Peters, K. E. (2019). The association between excessive screen-time behaviors and insufficient sleep among adolescents: Findings from the 2017 youth risk behavior surveillance system. *Psychiatry Research, 281*, 112586. https://doi.org/10.1016/j.psychres.2019.112586

Bapuji, S. B., Hansen, A., Marembo, M. H., Olivier, P., & Yap, M. B. H. (2024). Modifiable parental factors associated with the mental health of youth from immigrant families in high-income countries: A systematic review and meta-analysis. *Clinical Psychology Review, 110*, 102429. https://doi.org/10.1016/j.cpr.2024.102429

BetterHelp. (2025). *BetterHelp*. http://www.betterhelp.com

Bitsko, R. H., Holbrook, J. R., Ghandour, R. M., Blumberg, S. J., Visser, S. N., Perou, R., & Walkup, J. T. (2018). Epidemiology and Impact of Health Care Provider–Diagnosed Anxiety and Depression Among US Children. *Journal of Developmental and Behavioral Pediatrics : JDBP, 39*(5), 395–403. https://doi.org/10.1097/DBP.0000000000000571

Bozzola, E., Spina, G., Agostiniani, R., Barni, S., Russo, R., Scarpato, E., Di Mauro, A., Di Stefano, A. V., Caruso, C., Corsello, G., & Staiano, A. (2022). The Use of Social Media in Children and Adolescents: Scoping Review on the Potential Risks. *International Journal of Environmental Research and Public Health*, *19*(16), Article 16. https://doi.org/10.3390/ijerph19169960

Cianconi, P., Betrò, S., & Janiri, L. (2020). The Impact of Climate Change on Mental Health: A Systematic Descriptive Review. *Frontiers in Psychiatry*, *11*. https://doi.org/10.3389/fpsyt.2020.00074

Clam. (2025). *Calm.com*. https://www.calm.com/

Coyne, S. M., Escobar-Viera, C., Bekalu, M. A., Charmaraman, L., Primack, B., Shafi, R. M. A., Valkenburg, P. M., & Williams, K. D. A. (2025a). Social Media and Youth Mental Health: A Departure from the Status Quo. In D. A. Christakis & L. Hale (Eds.), *Handbook of Children and Screens: Digital Media, Development, and Well-Being from Birth Through Adolescence* (pp. 121–127). Springer Nature Switzerland. https://doi.org/10.1007/978-3-031-69362-5_17

Coyne, S. M., Escobar-Viera, C., Bekalu, M. A., Charmaraman, L., Primack, B., Shafi, R. M. A., Valkenburg, P. M., & Williams, K. D. A. (2025b). Social Media and Youth Mental Health: A Departure from the Status Quo. In D. A. Christakis & L. Hale (Eds.), *Handbook of Children and Screens: Digital Media, Development, and Well-Being from Birth Through Adolescence* (pp. 121–127). Springer Nature Switzerland. https://doi.org/10.1007/978-3-031-69362-5_17

Danielson, M. L., Claussen, A. H., Bitsko, R. H., Katz, S. M., Newsome, K., Blumberg, S. J., Kogan, M. D., & Ghandour, R. (2024). ADHD Prevalence Among U.S. Children and Adolescents in 2022: Diagnosis, Severity, Co-Occurring Disorders, and Treatment. *Journal of Clinical Child and Adolescent Psychology: The Official Journal for the Society of Clinical Child and Adolescent Psychology, American Psychological Association, Division 53*(3), 343–360. https://doi.org/10.1080/15374416.2024.2335625

Daviss, W. B. (2008). A Review of Co-Morbid Depression in Pediatric ADHD: Etiologies, Phenomenology, and Treatment. *Journal of Child and Adolescent Psychopharmacology, 18*(6), 565–571. https://doi.org/10.1089/cap.2008.032

Deng, Y., Cherian, J., Khan, N. U. N., Kumari, K., Sial, M. S., Comite, U., Gavurova, B., & Popp, J. (2022). Family and Academic Stress and Their Impact on Students' Depression Level and Academic Performance. *Frontiers in Psychiatry, 13.* https://doi.org/10.3389/fpsyt.2022.869337

Ding, K., & Li, H. (2023). Digital Addiction Intervention for Children and Adolescents: A Scoping Review. *International Journal of Environmental Research and Public Health, 20*(6), 4777. https://doi.org/10.3390/ijerph20064777

Doran, C. M., & Kinchin, I. (2017). A review of the economic impact of mental illness. *Australian Health Review, 43*(1), 43–48. https://doi.org/10.1071/AH16115

Dorol-Beauroy-Eustache, O., & Mishara, B. L. (2021). Systematic review of risk and protective factors for suicidal and self-harm behaviors among children and adolescents involved with cyberbullying. *Preventive Medicine, 152*(Pt 1), 106684. https://doi.org/10.1016/j.ypmed.2021.106684

Faraone, S. V., Banaschewski, T., Coghill, D., Zheng, Y., Biederman, J., Bellgrove, M. A., Newcorn, J. H., Gignac, M., Al Saud, N. M., Manor, I., Rohde, L. A., Yang, L., Cortese, S., Almagor, D., Stein, M. A., Albatti, T. H., Aljoudi, H. F., Alqahtani, M. M. J., Asherson, P., ... Wang, Y. (2021). The World Federation of ADHD International Consensus Statement: 208 Evidence-based conclusions about the disorder. *Neuroscience & Biobehavioral Reviews, 128*, 789–818. https://doi.org/10.1016/j.neubiorev.2021.01.022

Fearing, G. (2024). Caregivers' help-seeking for child and adolescent mental health: A look into their journey through the lens of mental health literacy. *Children and Youth Services Review, 158*, 107479. https://doi.org/10.1016/j.childyouth.2024.107479

Göbel, K., Baumgarten, F., Kuntz, B., Hölling, H., & Schlacks, Robert. (2018). ADHD in children and adolescents in Germany. Results of the cross-sectional KiGGS Wave 2 study and trend. *Journal of Health Monitoring, Robert Koch Institute, Berlin, 3*(3). https://doi.org/DOI 10.17886/RKI-GBE-2018-085

Goldberg, S. B., Lam, S. U., Simonsson, O., Torous, J., & Sun, S. (2022). Mobile phone-based interventions for mental health: A systematic meta-review of 14 meta-analyses of randomized controlled trials. *PLOS Digital Health, 1*(1), e0000002. https://doi.org/10.1371/journal.pdig.0000002

Green, J. L., Rinehart, N., Anderson, V., Nicholson, J. M., Jongeling, B., & Sciberras, E. (2015). Autism spectrum disorder symptoms in children with ADHD: A community-based study. *Research in Developmental Disabilities, 47*, 175–184. https://doi.org/10.1016/j.ridd.2015.09.016

Gulliver, A., Griffiths, K. M., & Christensen, H. (2010). Perceived barriers and facilitators to mental health help-seeking in young people: A systematic review. *BMC Psychiatry, 10*(1), 113. https://doi.org/10.118 6/1471-244X-10-113

Headspace. (2025). *Headspace.* Headspace. https://www.headspace.com/

Heinrich, L. M., & Gullone, E. (2006). The clinical significance of loneliness: A literature review. *Clinical Psychology Review, 26*(6), 695–718. https://doi.org/10.1016/j.cpr.2006.04.002

Heinz, A., & Brandt, L. (2024). Climate change and mental health: Direct, indirect, and intersectional effects. *The Lancet Regional Health – Europe, 43.* https://doi.org/10.1016/j.lanepe.2024.100969

Hickman, C., Marks, E., Pihkala, P., Clayton, S., Lewandowski, R. E., Mayall, E. E., Wray, B., Mellor, C., & Susteren, L. van. (2021). Climate anxiety in children and young people and their beliefs about government responses to climate change: A global survey. *The Lancet Planetary Health, 5*(12), e863–e873. https://doi.org/10.1016/S2542-5196(21)00278-3

Hilton, M. F., Scuffham, P. A., Sheridan, J., Cleary, C. M., & Whiteford, H. A. (2008). Mental Ill-Health and the Differential Effect of Employee Type on Absenteeism and Presenteeism. *Journal of Occupational and Environmental Medicine, 50*(11), 1228. https://doi.org/10.1097/JOM.0b013e31818c30a8

Hosokawa, R., Kawabe, K., Nakachi, K., Soga, J., Horiuchi, F., & Ueno, S.-I. (2023). Effects of social media on body dissatisfaction in junior high school girls in Japan. *Eating Behaviors, 48*, 101685. https://doi.org/10.1016/j.eatbeh.2022.101685

Hossain, M. M., Nesa, F., Das, J., Aggad, R., Tasnim, S., Bairwa, M., Ma, P., & Ramirez, G. (2022). Global burden of mental health problems among children and adolescents during COVID-19 pandemic: An umbrella review. *Psychiatry Research, 317*, 114814. https://doi.org/10.1016/j.psychres.2022.114814

Johnco, C., & Rapee, R. M. (2018). Depression literacy and stigma influence how parents perceive and respond to adolescent depressive symptoms. *Journal of Affective Disorders, 241*, 599–607. https://doi.org/10.1016/j.jad.2018.08.062

Joon, P., Kumar, A., & Parle, M. (2021). What is autism? *Pharmacological Reports, 73*(5), 1255–1264. https://doi.org/10.1007/s43440-021-00244-0

Kessler, R. C., Berglund, P., Demler, O., Jin, R., Merikangas, K. R., & Walters, E. E. (2005). Lifetime Prevalence and Age-of-Onset Distributions of DSM-IV Disorders in the National Comorbidity Survey Replication. *Archives of General Psychiatry, 62*(6), 593–602. https://doi.org/10.1001/archpsyc.62.6.593

Klicksafe. (2023). *According to the latest DAK study / Over 600,000 children in Germany are addicted to media.* Klicksafe.Eu. https://www.klicksafe.eu/en/news/ueber-600000-kinder-in-deutschland-sind-mediensuechtig

Klipker, K., Baumgarten, F., Göbel, K., Lampert, Thomas, & Hölling, H. (2018). Mental health problems in children and adolescents in Germany. Results of the cross-sectional KiGGS Wave 2 study and trends. *Journal of Health Monitoring, Robert Koch Institute, Berlin, 3*(3). https://doi.org/DOI 10.17886/RKI-GBE-2018-084

Kumar, V. L., & Goldstein, M. A. (2020). Cyberbullying and Adolescents. *Current Pediatrics Reports, 8*(3), 86–92. https://doi.org/10.1007/s40124-020-00217-6

Lavis, A., & Winter, R. (2020). #Online harms or benefits? An ethnographic analysis of the positives and negatives of peer-support around self-harm on social media. *Journal of Child Psychology and Psychiatry, and Allied Disciplines, 61*(8), 842–854. https://doi.org/10.1111/jcpp.13245

Leigh-Hunt, N., Bagguley, D., Bash, K., Turner, V., Turnbull, S., Valtorta, N., & Caan, W. (2017). An overview of systematic reviews on the public health consequences of social isolation and loneliness. *Public Health, 152,* 157–171. https://doi.org/10.1016/j.puhe.2017.07.035

Lu, B., Lin, L., & Su, X. (2024). Global burden of depression or depressive symptoms in children and adolescents: A systematic review and meta-analysis. *Journal of Affective Disorders, 354,* 553–562. https://doi.org/10.1016/j.jad.2024.03.074

Lukoševičiūtė-Barauskienė, J., Žemaitaitytė, M., Šūmakarienė, V., & Šmigelskas, K. (2023). Adolescent Perception of Mental Health: It's Not Only about Oneself, It's about Others Too. *Children, 10*(7), 1109. https://doi.org/10.3390/children10071109

Majeed, H., & Lee, J. (2017). The impact of climate change on youth depression and mental health. *The Lancet Planetary Health, 1*(3), e94–e95. https://doi.org/10.1016/S2542-5196(17)30045-1

Marciano, L., Dubicka, B., Magis-Weinberg, L., Morese, R., Viswanath, K., & Weber, R. (2025). Digital Media, Cognition, and Brain Development in Adolescence. In D. A. Christakis & L. Hale (Eds.), *Handbook of Children and Screens: Digital Media, Development, and Well-Being from Birth Through Adolescence* (pp. 21–29). Springer Nature Switzerland. https://doi.org/10.1007/978-3-031-69362-5_4

McAlister, K. L., Beatty, C. C., Smith-Caswell, J. E., Yourell, J. L., & Huberty, J. L. (2024). Social Media Use in Adolescents: Bans, Benefits, and Emotion Regulation Behaviors. *JMIR Mental Health, 11*(1), e64626. https://doi.org/10.2196/64626

McInroy, L. B., & Craig, S. L. (2020). "It's like a safe haven fantasy world": Online fandom communities and the identity development activities of sexual and gender minority youth. *Psychology of Popular Media, 9*(2), 236–246. https://doi.org/10.1037/ppm0000234

Nagata, J. M., Trompeter, N., Singh, G., Ganson, K. T., Testa, A., Jackson, D. B., Assari, S., Murray, S. B., Bibbins-Domingo, K., & Baker, F. C. (2022). Social Epidemiology of Early Adolescent Cyberbullying in the United States. *Academic Pediatrics, 22*(8), 1287–1293. https://doi.org/10.1016/j.acap.2022.07.003

Naslund, J. A., Aschbrenner, K. A., Marsch, L. A., & Bartels, S. J. (2016). The future of mental health care: Peer-to-peer support and social media. *Epidemiology and Psychiatric Sciences, 25*(2), 113–122. https://doi.org/10.1017/S2045796015001067

Nixon, C. L. (2014). Current perspectives: The impact of cyberbullying on adolescent health. *Adolescent Health, Medicine and Therapeutics, 5,* 143–158. https://doi.org/10.2147/AHMT.S36456

Nøvik, T. S., Hervas, A., Ralston, S. J., Dalsgaard, S., Rodrigues Pereira, R., Lorenzo, M. J., & ADORE Study Group. (2006). Influence of gender on attention-deficit/hyperactivity disorder in Europe – ADORE. *European Child & Adolescent Psychiatry, 15 Suppl 1,* I15-24. https://doi.org/10.1007/s00787-006-1003-z

Office of the Surgeon General. (2024). *Youth Mental Health* [Page]. https://www.hhs.gov/surgeongeneral/reports-and-publications/youth-mental-health/index.html

Østerås, B., Sigmundsson, H., & Haga, M. (2015). Perceived stress and musculoskeletal pain are prevalent and significantly associated in adolescents: An epidemiological cross-sectional study. *BMC Public Health, 15*(1), 1081. https://doi.org/10.1186/s12889-015-2414-x

Pine, A. E., Baumann, M. G., Modugno, G., & Compas, B. E. (2024). Parental Involvement in Adolescent Psychological Interventions: A Meta-analysis. *Clinical Child and Family Psychology Review, 27*(3), 1–20. https://doi.org/10.1007/s10567-024-00481-8

Prichett, L. M., Yolken, R. H., Severance, E. G., Carmichael, D., Zeng, Y., Lu, Y., Young, A. S., & Kumra, T. (2024). COVID-19 and Youth Mental Health Disparities: Intersectional Trends in Depression, Anxiety and Suicide Risk-Related Diagnoses. *Academic Pediatrics, 24*(5), 837–847. https://doi.org/10.1016/j.acap.2024.01.021

Ram, J. R., & Shelke, S. B. (2023). Understanding Eating Disorders in Children and Adolescent Population. *Journal of Indian Association for Child and Adolescent Mental Health, 19*(1), 60–69. https://doi.org/10.1177/09731342231179267

Riehm, K. E., Feder, K. A., Tormohlen, K. N., Crum, R. M., Young, A. S., Green, K. M., Pacek, L. R., La Flair, L. N., & Mojtabai, R. (2019). Associations Between Time Spent Using Social Media and Internalizing and Externalizing Problems Among US Youth. *JAMA Psychiatry, 76*(12), 1266–1273. https://doi.org/10.1001/jamapsychiatry.2019.2325

Ross, D. A., Hinton, R., Melles-Brewer, M., Engel, D., Zeck, W., Fagan, L., Herat, J., Phaladi, G., Imbago-Jácome, D., Anyona, P., Sanchez, A., Damji, N., Terki, F., Baltag, V., Patton, G., Silverman, A., Fogstad, H., Banerjee, A., & Mohan, A. (2020). Adolescent Well-Being: A Definition and Conceptual Framework. *The Journal of Adolescent Health, 67*(4), 472–476. https://doi.org/10.1016/j.jadohealth.2020.06.042

Rothenberg, W. A., Lansford, J. E., Bornstein, M. H., Chang, L., Deater-Deckard, K., Di Giunta, L., Dodge, K. A., Malone, P. S., Oburu, P., Pastorelli, C., Skinner, A. T., Sorbring, E., Steinberg, L., Tapanya, S., Uribe Tirado, L. M., Yotanyamaneewong, S., Alampay, L. P., Al-Hassan,

S. M., & Bacchini, D. (2020). Effects of Parental Warmth and Behavioral Control on Adolescent Externalizing and Internalizing Trajectories Across Cultures. *Journal of Research on Adolescence, 30*(4), 835–855. https://doi.org/10.1111/jora.12566

Shoshani, A., Kor, A., & Bar, S. (2024). The impact of social media use on psychiatric symptoms and well-being of children and adolescents in the Post-COVID-19 era: A four-year longitudinal study. *European Child & Adolescent Psychiatry, 33*(11), 4013–4027. https://doi.org/10.1007/s00787-024-02454-2

Statista. (2024). *Smartphone ownership teenagers Germany 2011-2024.* Statista. https://www.statista.com/statistics/828257/smartphone-ownership-teenagers-germany/

Statistisches Bundesamt. (2025). *Anzahl der Suizide 2023.* Statistisches Bundesamt. https://www.destatis.de/DE/Themen/Gesellschaft-Umwelt/Gesundheit/Todesursachen/Tabellen/suizide.html

Taba, M., Allen, T. B., Caldwell, P. H. Y., Skinner, S. R., Kang, M., McCaffery, K., & Scott, K. M. (2022). Adolescents' self-efficacy and digital health literacy: A cross-sectional mixed methods study. *BMC Public Health, 22*(1), 1223. https://doi.org/10.1186/s12889-022-13599-7

The Trevor Project. (2022). *2022 National Survey on LGBTQ Youth Mental Health.* The Trevor Project. https://www.thetrevorproject.org/

The U.S. Surgeon General's Advisory. (2023). Social media usage and anxiety and depression in youth. *Social Media and Youth Mental Health: The U.S. Surgeon General's Advisory.*

Twenge JM. Increases in Depression, Self-Harm, and Suicide Among U.S. Adolescents After 2012 and Links to Technology Use: Possible Mechanisms. Psychiatr Res Clin Pract. 2020 Sep 9;2(1):19-25. doi: 10.1176/appi.prcp.20190015. PMID: 36101887; PMCID: PMC9176070.

Turner, S., Fulop, A., & Woodcock, K. A. (2024). Loneliness: Adolescents' perspectives on what causes it, and ways youth services can prevent it. *Children and Youth Services Review, 157,* 107442. https://doi.org/10.1016/j.childyouth.2024.107442

UNICEF. (2023). *Belastungen durch Covid-19 nur „Spitze des Eisbergs".* https://www.unicef.de/informieren/aktuelles/presse/-/sowcr-2021-mentale-gesundheit/277324

Vergunst, F., & Berry, H. L. (2022). Climate Change and Children's Mental Health: A Developmental Perspective. *Clinical Psychological Science, 10*(4), 767–785. https://doi.org/10.1177/21677026211040787

Waysa. (2025). *Wysa.* Wysa - Everyday Mental Health. https://www.wysa.com/children-and-young-people

Weigle, P. E., & Shafi, R. M. A. (2024). Social Media and Youth Mental Health. *Current Psychiatry Reports, 26*(1), 1–8. https://doi.org/10.1007/s11920-023-01478-w

WHO. (2003). *Investing in mental health.* World Health Organization.

WHO. (2022a). *Mental disorders.* https://www.who.int/news-room/fact-sheets/detail/mental-disorders

WHO. (2022b). *Mental health.* https://www.who.int/news-room/fact-sheets/detail/mental-health-strengthening-our-response

WHO. (2022c). *World mental health report: Transforming mental health for all.* World Health Organization.

WHO. (2024). *Mental health of adolescents.* https://www.who.int/news-room/fact-sheets/detail/adolescent-mental-health

Wojtowicz, A., Buckley, G. J., & Galea, S. (2024). Potential Benefits of Social Media. In *Social Media and Adolescent Health* (Social Media and Adolescent Health.). National Academies Press (US). https://www.ncbi.nlm.nih.gov/books/NBK603438/

Zeidan, J., Fombonne, E., Scorah, J., Ibrahim, A., Durkin, M. S., Saxena, S., Yusuf, A., Shih, A., & Elsabbagh, M. (2022). Global prevalence of autism: A systematic review update. *Autism Research, 15*(5), 778–790. https://doi.org/10.1002/aur.2696

CHAPTER 2

Societal Implications of Youth Mental Health Challenges

This chapter explores the profound societal implications of youth mental health challenges, a critical yet often underestimated issue. It examines the multifaceted impact on key societal domains, including education, workforce productivity, public health, and economic stability. Beyond individual suffering, untreated mental health conditions among young people contribute to increased healthcare expenditures, diminished academic achievement, and social integration difficulties. This chapter emphasizes that prioritizing youth mental health is both a moral responsibility and a strategic investment in building stronger, more resilient communities.

Beyond the immediate effects, the chapter explores the long-term ripple effects on the economy and the overall well-being of communities. The youth mental health struggle has indirect costs to the society that demand urgent action. By bringing these implications to the forefront, the chapter advocates comprehensive solutions, such as early intervention programs, efforts to destigmatize mental health struggles, and the development of accessible and youth-friendly mental health services.

© Sharmistha Chatterjee, Azadeh Dindarian, Usha Rengaraju 2025
S. Chatterjee et al., *Revolutionizing Youth Mental Health with Ethical AI*,
https://doi.org/10.1007/979-8-8688-1186-9_2

Additionally, the chapter highlights the significant economic and societal benefits of investing in youth mental well-being. It argues that prioritizing this issue is not only a moral imperative but also a pathway to fostering healthier, more productive, and thriving communities for the future.

In this chapter, these topics will be covered in the following sections:

- Understanding the societal impact of youth mental health challenges

- The economic burden of youth mental health challenges

- Strategies addressing youth mental health challenges

- Economic benefits of prioritizing youth mental well-being

Understanding the Societal Impact of Youth Mental Health Challenges

Youth mental health is not solely an individual concern; research indicates that it poses a significant challenge for families, schools, and communities worldwide (McGorry et al., 2024). When a young person struggles with anxiety, depression, or other mental health challenges, the effects extend far beyond them. Parents and caregivers are under significant stress, teachers see a decline in performance, and healthcare and juvenile justice systems become increasingly strained. These challenges ripple outward, impacting families, schools, and communities, ultimately shaping the emotional and social fabric of society. In the previous chapter, we highlighted that today, more young people than ever are facing mental health difficulties. In particular, genetics, undiagnosed mental health conditions, academic pressure, social media comparisons, economic and political uncertainty, environmental issues, and personal struggles, among others, create a heavy burden that many adolescents struggle to navigate on their own (The U.S. Surgeon General's Advisory, 2023).

In this section, we will explore the significant societal impacts of poor youth mental health across various domains, including families, academic performance, the economy and job market, public health, and juvenile delinquency.

Family Dynamic and Youth Mental Health

The adverse effects of poor youth mental health often manifest first within the family unit. Challenges in communication, increased conflicts, and strained relationships become prevalent as families navigate these difficulties. Research indicates that parents and caregivers may experience a decline in productivity, which can adversely affect their mental health and, in some cases, lead to financial instability or job loss (Evernorth, 2024; Grodberg et al., 2022). In addition, many caregivers find themselves needing to take additional time off work to support their child, attend diagnostic appointments, or manage ongoing treatment, further exacerbating the emotional and economic strain on the household. Additionally, a study by Smith et al. (2010) found that childhood psychological problems can result in a lifetime cost of approximately $300,000 in lost family income, underscoring the profound economic burden on families. A 2024 study by the Evernorth Research Institute (2024) found that 54% of working parents with children facing mental health challenges have also been diagnosed with a mental health condition themselves. This indicates that these parents are 2.7 times more likely to experience mental health issues compared to those whose children do not have such challenges. Further, it is reported that many caregivers find themselves needing to take additional time off work to support their child, attend diagnostic appointments, or manage ongoing treatment, further exacerbating the emotional and economic strain on the household (Evernorth, 2024; Grodberg et al., 2022). The existing research consistently emphasizes the urgent need to prioritize support for families with children facing mental health challenges. Therefore, when caregivers receive the right resources, whether through accessible mental health services, financial assistance, or workplace flexibility, they are better equipped to provide the care their children need while also maintaining their own well-being.

Academic Performance and Youth Mental Health

The academic struggles faced by youth with mental health challenges are reported, and it is significant to the overall well-being of the youth (Agnafors et al., 2021; Deng et al., 2022). Mental health issues such as anxiety, depression, and stress can hinder concentration, motivation, and overall academic performance (Jaycox et al., 2009). Students facing mental health challenges often struggle with school attendance, keeping up with assignments, meeting deadlines, and engaging with their peers. This not only affects their educational outcomes but can also lower their self-esteem, future opportunities in the job market, and being part of the active society member. In this regard, adequate support from education institutes, teachers, and families is needed to help those who need educational opportunities as early as possible.

Beyond personal and academic struggles, youth mental health challenges take a toll on society in profound ways. Worldwide, societies see the effects in rising healthcare costs, increased demand for mental health services, and the strain on support systems (Cardoso & McHayle, 2024; Kirby, 2022; Patel et al., 2007). Unaddressed mental health challenges during adolescence can significantly hinder young individuals' ability to secure and maintain stable employment as they transition into adulthood (Gmitroski et al., 2018).

Workforce Challenges and the Impact of Youth Mental Health

In light of the current global labor shortages, it's crucial to recognize the potential long-term impact of adolescent mental health challenges on the future workforce (Buchholz, 2024). Reports indicate that with one in seven individuals aged 10 to 19 experiencing a mental health condition (WHO, 2024). If these issues remain unaddressed, a significant portion of the emerging workforce may face difficulties in securing and maintaining stable employment. Workplaces lose potential talent when young people are unable to transition into stable careers due to unresolved mental health issues (Brook et al., 2013; Porru et al., 2023). Researchers have highlighted the strong connection between unemployment and mental health, particularly its impact on anxiety, mood disorders, and suicidal behavior (Virgolino et al., 2022). Virgolino et al.'s (2022) study indicates that men and young adults are especially vulnerable, with higher rates of psychological distress linked to job loss and economic instability.

Public Health System and Youth Mental Health

The rising prevalence of mental health disorders among adolescents poses significant challenges for public health systems. Overburdened hospitals and clinicians are struggling to meet the growing demand, making this a critical public health concern (Feuer et al., 2023). The shortage of healthcare workers further exacerbates the issue, limiting access to timely and adequate mental healthcare for children and adolescents (Patel et al., 2007; Federal Statistical Office, 2021; Singer, 2024). In addition, the COVID-19 pandemic intensified the workload of healthcare workers, further straining an already burdened system and amplifying existing challenges (Feuer et al., 2023). A recent study in England (Ward et al., 2025) examined hospital admissions for children and young people (ages 5–18) over a ten-year period from 2012 to 2022. The findings highlight a concerning rise in mental health-related hospitalizations. In 2021–2022 alone, out of 342,511 total admissions, nearly 40,000 (11.7%) were due to mental health issues, a sharp 65% increase from a decade earlier. Meanwhile, overall hospital admissions rose by 10.1%. The increase was particularly alarming among girls aged 11–15, where cases more than doubled. Additionally, the study found out that mental health-related hospital stays were often prolonged, with 7.8% lasting more than a week, compared to 3.5% of all other hospital admissions.

The Impact of Poor Mental Health on Juvenile Delinquency

Juvenile delinquency, or the involvement of minors in criminal activities, is a growing concern with deep-rooted social and psychological implications (Gupta et al., 2022; Miao et al., 2025). While multiple factors contribute to delinquent behavior, research highlights a strong correlation between youth mental health issues and engagement in criminal activity (see, e.g., Goldstein et al., 2005; Overbeek et al., 2005; Siouti et al., 2025). Adolescents struggling with untreated mental health disorders, such as anxiety, depression, ADHD, and substance abuse, are at a higher risk of engaging in risky behaviors, including theft, vandalism, drug use, and violence (Fazel et al., 2008). For example, a meta-analysis by Fazel et al. (2008) on adolescents in detention and correctional facilities found that 11.7% of boys and 18.5% of girls were diagnosed with ADHD.

This highlights that without early diagnosis, intervention, and adequate support, these behaviors can escalate, resulting in legal consequences and long-term societal challenges.

Juvenile delinquency imposes substantial economic and social burdens on societies worldwide. In the United States, for instance, the average annual cost to incarcerate a young person is approximately $214,620, with some states spending over $500,000 per youth per year (Justice Policy Institute, 2020).

The Economic Burden of Youth Mental Health Challenges

In this section, we delve into the economic implications of youth mental health disorders, which extend beyond individual suffering to impose significant financial burdens on families and society at large. Mental health issues among children and adolescents lead to increased healthcare expenditures, educational challenges, and reduced productivity within the workforce.

The economic strain is evident in various sectors, such as healthcare, education, and lost parental income. According to WHO estimates, the global cost of mental illness is expected to more than double by 2030, reaching US$ 6.0 trillion overall, with direct and indirect costs projected at $2.0 trillion and $4.0 trillion, respectively (Bloom et al., 2011).

A 2017 study by Lizheng et al. (2019) examined the financial impact of mental health disorders in Chinese youth, revealing a staggering economic burden of $1.191 trillion. Direct costs accounted for $565 million, with inpatient and outpatient services comprising $351 million and $213 million, respectively. If every child in need had received professional therapy, these costs would have escalated to $49 billion. Beyond medical expenses, families faced substantial indirect costs, including $47 billion in lost income due to caregivers reducing work hours or leaving jobs. On average, each affected family experienced an income loss of $1,653, and the total indirect costs to society reached $1.181 trillion.

These numbers tell a deeper story than just financial costs; further, they reveal the everyday struggles of families caring for children with mental health challenges. As Lizheng et al. (2019) highlighted, the burden isn't just about hospital bills; it disrupts daily life, affects parents' ability to work, and takes an emotional toll on millions. This isn't just a personal struggle, but it's a growing challenge for societies everywhere, as countries face the rising costs and impact of mental health issues in children and adolescents.

The financial impact of childhood and adolescent mental health disorders is not confined to one country, but it is a global challenge. In the Netherlands, Bodden et al. (2018) found that clinically referred adolescents cost Dutch society nearly €38 million annually. However, when more recent international prevalence rates were applied, this figure increased to €59 million per year.

Looking beyond Europe, the economic toll is just as striking. In the United States, serious mental illness resulted in a staggering $193.2 billion loss in personal earnings across the entire population in 2002 (Kessler et al., 2008). Meanwhile, in England, the annual cost of mental ill health among children and young people under 20 is estimated at £18.8 billion (Cardoso & McHayle, 2024).

These statistics illustrate that mental health disorders in young people affect not only the individuals but also their families and the broader economy. The financial strain extends beyond medical expenses, as parents and caregivers often juggle work and caregiving responsibilities, leading to lost productivity and, in some cases, financial instability. Many families also rely on social support systems and special education programs, further adding to the overall economic impact.

An often-overlooked consequence is the significant emotional and financial strain on caregivers. The stress of caring for a child with mental health challenges can take a toll on their own well-being, sometimes leading to anxiety or depression (Zapata, 2025). This, in turn, can further impact their ability to work and provide financial stability, creating a cycle that's difficult to break. Considering these hidden costs alongside the long-term impact on national economies, it becomes evident that investing in early intervention and mental health support is not only a social responsibility but also an economic necessity. Furthermore, these private costs have ripple effects on broader society (Cardoso & McHayle, 2024). The increased demand for mental health services, special education resources, and caregiving support strains public health and welfare systems. Inadequate support for children and adolescents with mental health conditions may lead to long-term social costs, including reduced workforce participation, higher unemployment rates, and an increased reliance on social services as individuals struggle with the ongoing impact of untreated mental disorders.

Addressing the mental well-being of adolescents and children not only yields individual well-being but significant economic dividends for society in terms of avoided long-term healthcare, educational, and social welfare costs. Targeted early intervention, preventive interventions, and extended access to mental care can prevent these costs and make a positive impact in the lives of adolescents and children, their family, and society (Jackson-Morris et al., 2024). Providing mental health support for adolescents and children is not just an ethical responsibility but also a crucial investment in building a healthier and more productive future generation.

Access to psychosocial services, adherence to treatment, and achieving positive outcomes are vital for well-being. However, children and families in low-income situations often face significant barriers to obtaining high-quality mental healthcare (Hodgkinson et al., 2017). These challenges include limited availability of services, financial constraints, and systemic inequalities that hinder their ability to receive necessary support. Many adolescents struggle to find trustworthy, easy-to-understand information tailored to their needs for personal development (Taba et al., 2022). They need a positive, unbiased environment and knowledgeable support network to thrive. A 2024 survey (SUN, 2024) found that 40% of students aged 11 to 18 lack someone at school to discuss mental health concerns, and nearly half are hesitant to seek help. Given that schools are where youth spend most of their time, and mental health support can be cost-effective and easily accessible there, it is crucial to create a supportive environment, reduce stigma, and improve access to resources.

Access to regular therapy remains a major challenge for many individuals, largely due to the "ongoing shortage of mental health professionals" and "geographical barriers" that limit availability and accessibility of care. Globally, the median number of mental health workers is approximately 9 per 100,000 population, with substantial disparities between low- and high-income countries (Hanna et al., 2018).

Strategies Addressing Youth Mental Health Challenges

Addressing youth mental health challenges requires a comprehensive approach that includes early intervention and prevention, accessible mental health services for youth and their families, and caregivers. Globally, one in seven adolescents aged 10–19 experiences a mental disorder, contributing to 15% of the global disease burden in this age group (WHO, 2024). This section explores strategies for tackling these challenges by examining existing literature on the effectiveness of early intervention, improved access to mental health services, and the creation of supportive environments. By implementing these measures, we can mitigate the long-term consequences of untreated mental health issues and prevent the direct and indirect costs from escalating further.

Early Detection, Intervention, and Prevention

Detecting early signs of mental health challenges in children and adolescents can significantly improve their well-being. Early identification allows for timely intervention by professionals, which can alleviate immediate distress and reduce the risk of developing more severe issues, such as depression and long-term consequences (McGorry & Mei, 2018).

Early intervention and prevention are pivotal in addressing youth mental health challenges, aiming to identify and mitigate issues before they escalate into more severe conditions (Thapar et al., 2022). Research by Parry (1992) shows that "early intervention is clearly effective in offering parental support, fostering parent/child relationships and diminishing anxiety." Another research by Fisak et al. (2023) found that the early intervention to reduce anxiety in young children was effective in reducing the anxiety. Further, the evidence shows that *"Integrated and multidisciplinary services are needed to increase the range of possible interventions and limit the risk of poor long-term outcome, with also potential benefits in terms of healthcare system costs"* (Colizzi et al., 2020).

Prevention programs also play a vital role in fostering adolescent mental well-being, with school-based social and emotional learning (SEL) programs emerging as one of the most cost-effective strategies (Cipriano et al., 2023). These initiatives equip students with essential life skills, such as emotional regulation, effective communication, and self-awareness, helping them build resilience against mental health challenges (Jones et al., 2011). By addressing risk factors linked to mental health challenges, SEL programs create a supportive school environment that encourages early detection and intervention.

Global investments in the prevention of mental illnesses are still low (World Health Organization, 2019). Parents and caregivers play the most crucial role in helping and supporting a teenager with mental problems (Zapata, 2025). Preventive strategies for mental health can be highly effective, encompassing early support for mothers and young children, parenting workshops, school-based anti-bullying initiatives, and mindfulness programs for adults (Kirby, 2022). Investing in such interventions can yield significant long-term benefits, with studies indicating that "for every £1 spent on parental programs, there is a potential return of up to £15.80 in cost savings through improved mental well-being and reduced healthcare expenses" (Kirby, 2022).

It is therefore crucial to provide support and education to all stakeholders in an affordable and accessible format to create a positive environment and empower the younger generation through open prevention, diagnosis, treatment, medication management, and maintaining mental well-being.

Access to Mental Health Services

Ensuring that all adolescents have equitable access to mental health services is crucial for their well-being. Barriers such as stigma, lack of awareness, and limited availability of services can impede young people from seeking help. Yet, access to regular therapy is a significant challenge for many individuals, primarily due to the scarcity of mental health professionals and geographical limitations. Globally, in 2020, the median number of mental health workers is approximately 13 per 100,000 population, with substantial disparities between low- and high-income countries (World Health Organization, 2021).

In Germany, the situation reflects both progress and ongoing challenges. As of 2019, there were about 48,000 psychological psychotherapists and child and adolescent psychotherapists (Federal Statistical Office, 2021). Despite this growth, a significant number of clinics face difficulties in staffing. Approximately 80% of clinics report challenges in filling physician positions in psychiatry, psychotherapy, psychosomatic medicine, and child and adolescent psychiatry or psychotherapy (Deutsches Krankenhausinstitut e. V., 2020). These staffing shortages have tangible effects on patient care. In particular, patients often encounter long waiting times, with some regions reporting delays of several months before therapy can commence (Singer, 2024). This delay can exacerbate mental health conditions, making timely intervention critical.

In addition, geographical disparities create significant barriers to accessing mental health services, limiting availability for those who need them most (Saxena et al., 2007). As a result, individuals in underserved areas may struggle to receive timely and professional mental health support.

Digital technologies hold significant promise in enhancing access to mental health services, particularly for adolescents. By leveraging digital platforms, care can become more accessible, overcoming barriers such as geographical limitations and resource constraints. Implementing these solutions can provide timely support to younger populations, bridging existing gaps in mental healthcare delivery (Bantjes, 2022). (Note: While digital mental health services offer significant benefits, they also present challenges that will be discussed in detail in a later section of this book.)

Addressing these challenges requires a multifaceted approach, including increasing the number of trained mental health professionals, implementing policies to ensure equitable distribution across regions, and developing digital platforms to provide support and education, particularly for children and adolescents. Such measures are essential to bridge the gap in mental health service accessibility and to meet the growing demand for psychological support.

Economic Benefits of Prioritizing Youth Mental Health

Investing in adolescent mental health yields substantial economic benefits, both immediate and long term, including reducing disability and preventing premature death (WHO, World Health, 2003).

Research indicates that for every dollar invested in preventing and treating adolescent mental disorders, there is a return of $24 in health and economic benefits over an 80-year period (Stelmach et al., 2022). This return is particularly significant in low- and middle-income countries, where the impact of such interventions can be profound.

Addressing mental health issues during adolescence can significantly enhance individuals' lifetime earnings. A recent study in the UK estimates that untreated mental health problems among young people could lead to a £1.16 trillion loss in lifetime earnings, underscoring the importance of early mental health support (Bawden, 2025). Furthermore, persistent absence from school, often linked to mental health challenges, cost £1.17 billion in the 2023–2024 school year (Bawden, 2025). A 2016 study led by the World Health Organization (2016) estimated that "every US$1 invested in scaling up treatment for depression and anxiety leads to a return of US$4 in better health and increased ability to work," highlighting the significant health and economic benefits of investing in mental healthcare. Prioritizing mental well-being in educational settings leads to "better academic performance, higher retention rates, and lower dropout levels" (UNISEF, 2023).

Conclusion

This chapter examined the societal impact of youth mental health issues. We saw how untreated mental health issues in young people disrupt not only their personal lives but also affect education, work productivity, and public health. These struggles lead to higher healthcare costs, lower academic achievements, and difficulties in social connections.

We also discussed the long-term economic impacts, noting that poor mental health during adolescence can result in significant losses in lifetime earnings and increased societal costs. For instance, a recent study in the UK estimated that untreated mental health problems among young people could lead to a £1.16 trillion loss in lifetime earnings, underscoring the importance of early mental health support (Bawden, 2025).

To tackle these challenges, we emphasized the need for comprehensive solutions. This includes early intervention programs, efforts to remove the stigma around mental health struggles, and creating accessible, youth-friendly mental health services. Investing in the mental well-being of young people is not just the right thing to do; it's also a smart move for building healthier, more productive communities. In this chapter, we looked at how mental health issues affect families, school performance, the economy, public health, and youth behavior. We analyzed the financial implications, including increased healthcare spending, educational challenges, and reduced productivity. Further, we discussed strategies like early detection, intervention, prevention, and improving access to mental health services. And finally, we highlighted the long-term advantages of investing in mental health initiatives for young people.

References

Agnafors, S., Barmark, M., & Sydsjö, G. (2021). Mental health and academic performance: A study on selection and causation effects from childhood to early adulthood. Social Psychiatry and Psychiatric Epidemiology, 56(5), 857–866. https://doi.org/10.1007/s00127-020-01934-5

Bantjes, J. (2022). Digital solutions to promote adolescent mental health: Opportunities and challenges for research and practice. PLoS Medicine, 19(5), e1004008. https://doi.org/10.1371/journal.pmed.1004008

Bawden, A. (2025). UK childhood mental health crisis to cost. £1.1tn in lost pay, study finds. The Guardian. https://www.theguardian.com/society/2025/feb/05/uk-childhood-mental-health-crisis-to-cost-1tn-in-lost-pay-study-finds

Bloom, Cafiero, Jané-Llopis, Abrahams-Gessel, Bloom, Fathima, Feigl, & A.B., Gaziano, T., Mowafi, M., Pandya, A., Prettner, K., Rosenberg, L., Seligman, B., Stein, A.Z., & Weinstein, C. (2011). The Global Economic Burden of Noncommunicable Diseases. World Economic Forum.

Bodden, D. H. M., Stikkelbroek, Y., & Dirksen, C. D. (2018). Societal burden of adolescent depression, an overview and cost-of-illness study. Journal of Affective Disorders, 241, 256–262. https://doi.org/10.1016/j.jad.2018.06.015

Brook, J. S., Brook, D. W., Zhang, C., Seltzer, N., & Finch, S. J. (2013). Adolescent ADHD and Adult Physical and Mental Health, Work Performance, and Financial Stress. Pediatrics, 131(1), 5–13. https://doi.org/10.1542/peds.2012-1725

Buchholz, K. (2024, March 25). Infographic: Global Staff Shortages Balloon in Just Six Years. Statista Daily Data. https://www.statista.com/chart/4690/the-countries-facing-the-greatest-skill-shortages

Cardoso, F., & McHayle, Z. (2024). THE ECONOMIC AND SOCIAL COSTS OF MENTAL ILL HEALTH REVIEW OF METHODOLOGY AND UPDATE OF CALCULATIONS. Centre for Mental Health. https://www.centreformentalhealth.org.uk/wp-content/uploads/2024/03/CentreforMH_TheEconomicSocialCostsofMentalIllHealth-1.pdf?

Cipriano, C., Strambler, M. J., Naples, L. H., Ha, C., Kirk, M., Wood, M., Sehgal, K., Zieher, A. K., Eveleigh, A., McCarthy, M., Funaro, M., Ponnock, A., Chow, J. C., & Durlak, J. (2023). The state of evidence for social and emotional learning: A contemporary meta-analysis of universal school-based SEL interventions. Child Development, 94(5), 1181–1204. https://doi.org/10.1111/cdev.13968

Colizzi, M., Lasalvia, A., & Ruggeri, M. (2020). Prevention and early intervention in youth mental health: Is it time for a multidisciplinary and trans-diagnostic model for care? International Journal of Mental Health Systems, 14(1), 23. https://doi.org/10.1186/s13033-020-00356-9

Deng, Y., Cherian, J., Khan, N. U. N., Kumari, K., Sial, M. S., Comite, U., Gavurova, B., & Popp, J. (2022). Family and Academic Stress and Their Impact on Students' Depression Level and Academic Performance. Frontiers in Psychiatry, 13. https://doi.org/10.3389/fpsyt.2022.869337

Deutsches Krankenhausinstitut e. V. (2020). PSYCHIATRIE BAROMETER 2019/2020.

Evernorth. (2024). Research: Employers must strengthen their approach to family behavioral health. https://www.evernorth.com/articles/www.evernorth.com/articles/youth-mental-health-research-may-2024

Fazel, S., Doll, H., & Långström, N. (2008). Mental disorders among adolescents in juvenile detention and correctional facilities: A systematic review and metaregression analysis of 25 surveys. Journal of the American Academy of Child and Adolescent Psychiatry, 47(9), 1010–1019. https://doi.org/10.1097/CHI.0b013e31817eecf3

Federal Statistical Office. (2021). Number of psychotherapists up 19% between 2015 and 2019. Federal Statistical Office. https://www.destatis.de/EN/Press/2021/03/PE21_N022_23.html

Feuer, V., Mooneyham, G. C., Malas, N. M., & Pediatric Boarding Consensus Guidelines Panel. (2023). Addressing the Pediatric Mental Health Crisis in Emergency Departments in the US: Findings of a National Pediatric Boarding Consensus Panel. Journal of the Academy of Consultation-Liaison Psychiatry, 64(6), 501–511. https://doi.org/10.1016/j.jaclp.2023.06.003

Fisak, B., Penna, A., Mian, N. D., Lamoli, L., Margaris, A., & Cruz, S. A. M. F. D. (2023). The Effectiveness of Anxiety Interventions for Young Children: A Meta-Analytic Review. Journal of Child and Family Studies, 1–12. https://doi.org/10.1007/s10826-023-02596-y

Gmitroski, T., Bradley, C., Heinemann, L., Liu, G., Blanchard, P., Beck, C., Mathias, S., Leon, A., & Barbic, S. P. (2018). Barriers and facilitators to employment for young adults with mental illness: A scoping review. BMJ Open, 8(12), e024487. https://doi.org/10.1136/bmjopen-2018-024487

Goldstein, N., Olubadewo, O., Redding, R. E., & Lexcen, F. (2005). Mental Health Disorders: The Neglected Risk Factor in Juvenile Delinquency (SSRN Scholarly Paper No. 2398584). Social Science Research Network. https://papers.ssrn.com/abstract=2398584

Grodberg, D., Bridgewater, J., Loo, T., & Bravata, D. (2022). Examining the Relationship Between Pediatric Behavioral Health and Parent Productivity Through a Parent-Reported Survey in the Time of COVID-19: Exploratory Study. JMIR Formative Research, 6(8), e37285. https://doi.org/10.2196/37285

Gupta, M. K., Mohapatra, S., & Mahanta, P. K. (2022). Juvenile's Delinquent Behavior, Risk Factors, and Quantitative Assessment Approach: A Systematic Review. Indian Journal of Community Medicine: Official Publication of Indian Association of Preventive & Social Medicine, 47(4), 483–490. https://doi.org/10.4103/ijcm.ijcm_1061_21

Hanna, F., Barbui, C., Dua, T., Lora, A., van Regteren Altena, M., & Saxena, S. (2018). Global mental health: How are we doing? World Psychiatry, 17(3), 367–368. https://doi.org/10.1002/wps.20572

Hodgkinson, S., Godoy, L., Beers, L. S., & Lewin, A. (2017). Improving Mental Health Access for Low-Income Children and Families in the Primary Care Setting. Pediatrics, 139(1), e20151175. https://doi.org/10.1542/peds.2015-1175

Jackson-Morris, A., Meyer, C. L., Morgan, A., Stelmach, R., Jamison, L., & Currie, C. (2024). An investment case analysis for the prevention and treatment of adolescent mental disorders and suicide in England. European Journal of Public Health, 34(1), 107–113. https://doi.org/10.1093/eurpub/ckad193

Jaycox, L. H., Stein, B. D., Paddock, S., Miles, J. N. V., Chandra, A., Meredith, L. S., Tanielian, T., Hickey, S., & Burnam, M. A. (2009). Impact of Teen Depression on Academic, Social, and Physical Functioning. Pediatrics, 124(4), e596–e605. https://doi.org/10.1542/peds.2008-3348

Jones, S. M., Brown, J. L., & Lawrence Aber, J. (2011). Two-Year Impacts of a Universal School-Based Social-Emotional and Literacy Intervention: An Experiment in Translational Developmental Research. Child Development, 82(2), 533–554. https://doi.org/10.1111/j.1467-8624.2010.01560.x

Justice Policy Institute. (2020). Sticker Shock 2020: The Cost of Youth Incarceration. Justice Policy Institute. https://justicepolicy.org/research/policy-brief-2020-sticker-shock-the-cost-of-youth-incarceration/

Kessler, R. C., Heeringa, S., Lakoma, M. D., Petukhova, M., Rupp, A. E., Schoenbaum, M., Wang, P. S., & Zaslavsky, A. M. (2008). Individual and Societal Effects of Mental Disorders on Earnings in the United States: Results From the National Comorbidity Survey Replication. American Journal of Psychiatry, 165(6), 703–711. https://doi.org/10.1176/appi.ajp.2008.08010126

Kirby, J. (2022). Mental health problems 'cost the UK economy at least £118 billion a year'. The Standard. https://www.standard.co.uk/news/uk/gdp-mental-health-foundation-london-school-of-economics-and-political-science-b985721.html

Lizheng, X., Chaofan, L., Jiajing, L., Fan, Y., & Jian, W. (2019). The economic burden of mental disorders in children and adolescents in China: A cross-sectional study. The Lancet, 394, S48. https://doi.org/10.1016/S0140-6736(19)32384-0

McGorry, P. D., & Mei, C. (2018). Early intervention in youth mental health: Progress and future directions. Evidence-Based Mental Health, 21(4), 182–184. https://doi.org/10.1136/ebmental-2018-300060

McGorry, P. D., Mei, C., Dalal, N., Alvarez-Jimenez, M., Blakemore, S.-J., Browne, V., Dooley, B., Hickie, I. B., Jones, P. B., McDaid, D., Mihalopoulos, C., Wood, S. J., Azzouzi, F. A. E., Fazio, J., Gow, E., Hanjabam, S., Hayes, A., Morris, A., Pang, E., ... Killackey, E. (2024). The Lancet Psychiatry Commission on youth mental health. The Lancet Psychiatry, 11(9), 731–774. https://doi.org/10.1016/S2215-0366(24)00163-9

Miao, L., Jiang, J., & Wang, H. (2025). Psychological process and risk factors of juvenile delinquency: Evidence from a qualitative analysis. Child Abuse & Neglect, 161, 107259. https://doi.org/10.1016/j.chiabu.2025.107259

Overbeek, G., Vollebergh, W., Engels, R., & Meeus, W. (2005). Juvenile delinquency as acting out: Emotional disturbance mediating the effects of parental attachment and life events. European Journal of Developmental Psychology, 2(1), 39–46. https://doi.org/10.1080/17405620444000184

Parry, T. S. (1992). The effectiveness of early intervention: A critical review. Journal of Paediatrics and Child Health, 28(5), 343–346. https://doi.org/10.1111/j.1440-1754.1992.tb02688.x

Patel, V., Flisher, A. J., Hetrick, S., & McGorry, P. (2007). Mental health of young people: A global public-health challenge. The Lancet, 369(9569), 1302–1313. https://doi.org/10.1016/S0140-6736(07)60368-7

Porru, F., Schuring, M., Hoogendijk, W. J. G., Burdorf, A., & Robroek, S. J. W. (2023). Impact of mental disorders during education on work participation: A register-based longitudinal study on young adults with 10 years follow-up. Journal of Epidemiology and Community Health, 77(9), 549–557. https://doi.org/10.1136/jech-2022-219487

Saxena, S., Thornicroft, G., Knapp, M., & Whiteford, H. (2007). Resources for mental health: Scarcity, inequity, and inefficiency. The Lancet, 370(9590), 878–889. https://doi.org/10.1016/S0140-6736(07)61239-2

Singer, S. (2024). Almost no Change in Waiting Times for Outpatient Psychotherapy After an Amendment to the Law (17.05.2024). Aerzteblatt. https://www.aerzteblatt.de/int/archive/article?id=239104

Siouti, Z., Tolis, E., Theodoratou, M., Kaltsouda, A., Megari, K., & Kougioumtzis, G. A. (2025). Juvenile Delinquency: Risk Factors and Prevention Techniques. In Exploring Cognitive and Psychosocial Dynamics Across Childhood and Adolescence (pp. 309–322). IGI Global Scientific Publishing. https://doi.org/10.4018/979-8-3693-4022-6.ch016

Smith, J. P., Monica, S., & Smith, G. C. (2010). Long-Term Economic Costs of Psychological Problems During Childhood. Social Science & Medicine (1982), 71(1), 110–115. https://doi.org/10.1016/j.socscimed.2010.02.046

Stelmach, R., Kocher, E. L., Kataria, I., Jackson-Morris, A. M., Saxena, S., & Nugent, R. (2022). The global return on investment from preventing and treating adolescent mental disorders and suicide: A modelling study. BMJ Global Health, 7(6), e007759. https://doi.org/10.1136/bmjgh-2021-007759

SUN. (2024). Four in 10 schoolkids struggle to discuss mental health—As they battle mood swings and sleep issues | The Sun. SUN. https://www.thesun.co.uk/health/30120033/schoolkids-struggle-discuss-mental-health-mood-swings-sleep-issues/?utm_source=chatgpt.com

Taba, M., Allen, T. B., Caldwell, P. H. Y., Skinner, S. R., Kang, M., McCaffery, K., & Scott, K. M. (2022). Adolescents' self-efficacy and digital health literacy: A cross-sectional mixed methods study. BMC Public Health, 22(1), 1223. https://doi.org/10.1186/s12889-022-13599-7

Thapar, A., Eyre, O., Patel, V., & Brent, D. (2022). Depression in young people. The Lancet, 400(10352), 617–631. https://doi.org/10.1016/S0140-6736(22)01012-1

The U.S. Surgeon General's Advisory. (2023). Social media usage and anxiety and depression in youth. Social Media and Youth Mental Health: The U.S. Surgeon General's Advisory.

UNISEF. (2023). The benefits of investing in school-based mental health support. UNICEF. https://www.unicef.org/reports/benefits-investing-school-based-mental-health-support

Virgolino, A., Costa, J., Santos, O., Pereira, M. E., Antunes, R., Ambrósio, S., Heitor, M. J., & Vaz Carneiro, A. (2022). Lost in transition: A systematic review of the association between unemployment and mental health. Journal of Mental Health (Abingdon, England), 31(3), 432–444. https://doi.org/10.1080/09638237.2021.2022615

Ward, J. L., Vázquez-Vázquez, A., Phillips, K., Settle, K., Pilvar, H., Cornaglia, F., Gibson, F., Nicholls, D., Roland, D., Mathews, G., Roberts, H., Viner, R. M., & Hudson, L. D. (2025). Admission to acute medical wards for mental health concerns among children and young people in England from 2012 to 2022: A cohort study. The Lancet Child & Adolescent Health, 9(2), 112–120. https://doi.org/10.1016/S2352-4642(24)00333-X

WHO, W. H. (2024). Mental health of adolescents. https://www.who.int/news-room/fact-sheets/detail/adolescent-mental-health

WHO, World Health. (2003). Investing in mental health. World Health Organization.

World Health Organization. (2016). Investing in treatment for depression and anxiety leads to fourfold return. https://www.who.int/news/item/13-04-2016-investing-in-treatment-for-depression-and-anxiety-leads-to-fourfold-return

World Health Organization. (2019). Making the investment case for mental health (No. WHO/UHC/CD-NCD/19.97). https://iris.who.int/bitstream/handle/10665/325116/WHO-UHC-CD-NCD-19.97-eng.pdf

World Health Organization. (2021). WHO report highlights global shortfall in investment in mental health. https://www.who.int/news/item/08-10-2021-who-report-highlights-global-shortfall-in-investment-in-mental-health

Zapata, K. (2025). The Rising Cost of Mental Health Care Is Putting a Strain On Families. Parents. https://www.parents.com/the-doctor-is-out-how-financial-barriers-prevent-many-families-from-accessing-mental-health-care-8762774

CHAPTER 3

Digital Mental Health: Bridging Gaps in Adolescent Mental Health Services

Adolescence (11- to 19-year-olds) is a pivotal stage of growth, filled with rapid physical, emotional, and social changes. During this period, many young individuals experience mental health challenges, such as anxiety, depression, and behavioral disorders. Despite the increasing prevalence of these issues, access to timely and effective mental health services remains a significant barrier. Digital mental health, which encompasses digital tools and platforms designed to deliver mental healthcare, has emerged as a promising solution to bridge these gaps in recent years. This chapter explores the role of digital mental health in addressing the challenges within adolescent mental health services, examining its benefits, limitations, and future directions. Further, this chapter explores how digital mental health tools are making mental health support more accessible and affordable. While these innovations offer great potential, they also come with challenges, keeping users engaged, ensuring they are backed by scientific research and clinical trials, and addressing ethical concerns around privacy and AI-driven care.

© Sharmistha Chatterjee, Azadeh Dindarian, Usha Rengaraju 2025
S. Chatterjee et al., *Revolutionizing Youth Mental Health with Ethical AI*,
https://doi.org/10.1007/979-8-8688-1186-9_3

In this chapter, these topics will be covered in the following sections:

- Emergence of digital mental health solutions

- Benefits of digital mental health for adolescents

- Challenges and limitations of digital mental health for adolescents

- Examples of digital mental health solutions for adolescents

Emergence of Digital Mental Health Solutions

In recent years, advancements in information technology and global connectivity have heightened interest among stakeholders in applying these technologies within the mental health sector. Challenges such as the stigma surrounding child and adolescent mental health, a shortage of mental health professionals, and the high costs associated with traditional therapy have prompted the exploration of innovative solutions. Further, COVID-19 pandemic has enormously helped with the adoption of the digital mental health solutions (Sorkin et al., 2021). Digital mental health or also known as e-mental health leverages digital technologies such as machine learning (ML) and artificial intelligence (AI) to deliver mental health services and support, such as online therapy platforms, mental health apps, chatbots, and virtual reality interventions. Digital mental health solutions have revolutionized access to care, especially individuals living in remote areas, individuals with limited mobility, or anyone who values the privacy of virtual interactions (Lal & Adair, 2014).

Utilizing technologies like artificial intelligence and machine learning in online therapy and case management offers promising avenues to address these issues.

Research indicates that as Internet connectivity increases, a growing number of individuals are utilizing online resources to seek support for their mental health. For instance, a study by Wetterlin et al. (2014) found that 61.6% of participants aged 17 to 24 had used the Internet to seek information or help for their feelings. This trend underscores the importance of accessible and reliable digital mental health resources. A comprehensive review of e-mental health applications by Lal and Adair (2014) has identified four key areas in mental health service delivery that these digital tools address in particular for adult and adolescent population:

1. **Information provision**: Offering accessible mental health information to users

2. **Screening, assessment, and monitoring**: Facilitating the evaluation and tracking of mental health symptoms

3. **Intervention**: Providing therapeutic strategies and support to manage mental health conditions

4. **Social support**: Creating platforms for users to connect and share experiences

Mind.org.uk,[1] a leading UK-based platform, offers comprehensive mental health resources, including information on conditions, treatments, and expert advice tailored for young people. Similarly, NHS Every Mind Matters,[2] an initiative by the National Health Service (NHS), provides self-help guides, practical tips, and expert-backed articles to support mental well-being. These platforms serve as valuable digital tools, empowering individuals to take charge of their mental health with accessible, evidence-based support.

---

[1] www.mind.org.uk
[2] www.mind.org.uk

Currently, there are "between 10,000 to 20,000 mental health apps available on the market," offering a range of services from symptom tracking to cognitive behavioral therapy (CBT), mindfulness exercises, and tools for managing conditions like ADHD (Gross, 2024). Symptom-tracking mobile applications, like MindDoc,[3] a Germany-based mental health platform, showcase how information and communication technology (ICT) is transforming mental health monitoring. Designed to help individuals track their emotional well-being, MindDoc offers personalized insights, mood assessments, and self-reflection tools. As one of many innovative digital solutions, it demonstrates the increasing role of technology in mental healthcare, providing users with a data-driven approach to recognizing and managing their mental health needs.

Digital tools offering guided therapy, mindfulness exercises, and coping strategies are increasingly available to help adolescents manage their mental health. For instance, Woebot[4] is an "AI-powered chatbot that provides cognitive behavioral therapy (CBT) techniques" to assist users in navigating emotional distress. According to their website, "through conversations, Woebot helps users develop skills for emotional regulation and supports mood monitoring and management, with tools such as mood tracking, progress reflection, gratitude journaling, and mindfulness practice."

Online communities and forums play a vital role in offering social support, especially for young people navigating life's challenges. Many adolescents turn to these platforms not just to stay connected with friends but to share their struggles and seek advice from others who truly understand what they're going through (Naslund et al., 2016).

For example, The Mighty[5] is a US-based online community where young people can read, write, and share personal stories about mental

---

[3] www.minddoc.com

[4] www.woebothealth.com

[5] www.themighty.com

health. It's a space designed to foster openness, allowing individuals to express their feelings and find reassurance in knowing they're not alone.

Similarly, Reddit[6] hosts various support-focused communities, known as subreddits, where people openly discuss their mental health experiences. Subreddit spaces like "r/mentalhealth," "r/depression," and "r/anxiety" offer a sense of belonging and understanding, where individuals can exchange advice, encouragement, and personal stories. Digital mental health solutions have become increasingly prevalent, offering various benefits and presenting certain risks. In the following section, we will explore these advantages and potential drawbacks in detail.

Benefits of Digital Mental Health for Adolescents

Digital mental health solutions are changing the way adolescents access support, making mental healthcare more flexible, accessible, and stigma free. By leveraging technology, these services offer innovative ways to overcome traditional barriers in mental healthcare, such as stigma, clinician shortages, and high costs. In this section, we will explore the benefits of digital mental health for adolescents, focusing on accessibility, effectiveness, personalization, cost-effectiveness, and the potential to reduce stigma.

Accessibility

One of the primary benefits of digital mental health solutions is their enhanced accessibility. A recent systematic review by Wies et al. (2021) analyzed 36 relevant studies and identified key accessibility benefits

---

[6]www.reddit.com

of digital mental health interventions. These include expanded reach
and affordability, the ability to overcome geographical barriers, 24/7
availability and continuous support, promotion of health equity and
inclusion, and the empowerment of patient–therapist relationships.

Telemental services and online therapies have revolutionized mental
healthcare for adolescents by making support readily available, regardless
of geographical location. This is especially beneficial for people living in
rural or underserved areas, where finding a mental health professional can
be difficult (Serafin, 2020; Sorkin et al., 2021). By utilizing everyday devices
such as laptops, tablets, or smartphones, adolescents can conveniently
engage in therapy sessions without the need for extensive travel. Research
indicates that telemental effectively addresses barriers such as distance
and transportation, thereby improving access to mental health services for
rural populations (Serafin, 2020). Additionally, telemental offers flexibility
around individuals' schedules, reduces travel and associated costs, and
diminishes the stigma linked to attending mental health facilities in person.

Effectiveness

Digital mental health has garnered significant attention for their potential
to address mental health challenges among adolescents. Research into
their effectiveness has yielded promising results, though outcomes can
vary based on several factors including type of illness or challenges (Lal &
Adair, 2014; Lattie et al., 2019).

In their 2010 review, Calear and Christensen (2010) evaluated four
Internet-based programs designed to prevent and treat anxiety and
depression in children and adolescents:

1. **BRAVE**: Tailored for children (8–12 years) and
   teenagers (13–17 years), this program addresses
   various anxiety disorders, including social phobia
   and generalized anxiety disorder.

2. **Project CATCH-IT**: A free online program
   integrating behavioral activation, cognitive
   behavioral therapy (CBT), and interpersonal
   psychotherapy (IP) to mitigate depression risk.

3. **MoodGYM**: An interactive platform aimed at
   preventing and reducing depressive symptoms
   among young individuals.

4. **Grip op je dip**: A cognitive behavioral therapy
   (CBT)-based initiative targeting individuals aged
   16–25 in Duch and provided for free.

The review found promising evidence that Internet-based
interventions can help reduce anxiety and depression symptoms.
However, the authors stressed that more in-depth research is needed to
understand the best ways to make these programs truly effective. They
also highlighted the importance of developing additional interventions to
address existing gaps and ensure that digital mental health tools provide
real, lasting support for those who need them.

In 2020, Newton et al. (2020) introduced "MindClimb," a smartphone
app designed to assist adolescents with anxiety. The app aims to reinforce
skills learned during cognitive behavioral therapy (CBT) sessions, such
as cognitive strategies, relaxation techniques, exposure practices, and
reward systems. By facilitating the planning and execution of exposure
activities between therapy sessions, MindClimb encourages consistent
practice. Initial evaluations indicated that both adolescents and therapists
found the app beneficial for reinforcing CBT techniques outside of
traditional sessions. However, the study emphasized the need for further
research with larger youth samples to optimize the app's integration into
clinical care and to assess its impact on treatment processes and patient
outcomes.

Further, another randomized clinical trial involving 59 young adults evaluated the efficacy of a self-guided mobile cognitive behavioral therapy (CBT) application. Participants who used the app experienced significant reductions in anxiety symptoms, suggesting that such digital interventions can be effective tools for managing anxiety in this population (Bress et al., 2024).

A recent randomized controlled trial by Bilan et al. (2025) investigated the effects of an "AI-driven digital cognitive program" on children aged 8 to 12 diagnosed with ADHD. Over a 12-week period, participants received either the "AI-based therapy or a placebo intervention." Assessments conducted before and after the intervention revealed that children undergoing the AI-driven therapy exhibited significant reductions in impulsivity and inattention. These behavioral improvements were associated with normalized brain activity patterns, particularly in regions linked to inhibitory control. The findings suggest that AI-driven digital cognitive therapy can enhance neurophysiological efficiency, offering a personalized, technology-based approach to ADHD treatment.

Research indicates that "Selfapy," "an Internet-based cognitive behavioral therapy (CBT) program" in German language, effectively reduces depressive symptoms in adults (Schefft et al., 2024).

Emerging research highlights the promising role of machine learning (ML) and artificial intelligence (AI) in psychological interventions and diagnostics, showing encouraging results in improving mental health assessments and treatment approaches (see Chen et al., 2023; Zhou et al., 2022). However, these technologies are still in their early stages, and their full potential remains to be explored and more clinical research and validation are essential to ensure their accuracy, reliability, and real-world effectiveness in diverse patient populations (Zhou et al., 2022). A recent systematic review by Park et al. (2024) analyzed studies from 2017 to 2022 to assess the effectiveness of artificial intelligence (AI) in managing depression. The findings highlighted that AI applications, particularly those utilizing biomarker-derived data, excelled in accuracy for monitoring and predicting depressive states (Park et al., 2024). Additionally, emerging research underscores AI's potential in diagnosing autism. A recent study has highlighted the potential of artificial intelligence (AI) in diagnosing autism and assessing the effectiveness of this approach. Researchers successfully distinguished between "typically developing children and those with autism spectrum disorder (ASD) with an accuracy rate of 76%," demonstrating AI's promise in enhancing early detection and diagnosis (Butera et al., 2025).

These advancements suggest that AI-driven tools could play a crucial role in enhancing mental health assessments and facilitating timely interventions.

Artificial intelligence (AI) and machine learning (ML) are increasingly being utilized to enhance the diagnosis of attention deficit hyperactivity disorder (ADHD). Recent studies have demonstrated the potential of these technologies to improve diagnostic accuracy and provide more objective assessments (Cao et al., 2023; Chen et al., 2023).

Personalization

Currently, with the integration of information and communication technology (ICT), mental health apps are evolving to meet the unique needs of teenagers by providing personalized support that aligns with their individual experiences. This level of personalization enhances user engagement and improves the effectiveness of interventions, ultimately making mental healthcare more accessible, relevant, and impactful for young individuals.

In today's fast-moving digital world, artificial intelligence (AI) is reshaping mental healthcare, making support more personalized, accessible, and responsive. AI-powered digital mental health services can adapt to an individual's unique needs, offering guidance and intervention exactly when and where they're needed.

By analyzing user behaviors, mood patterns, and preferences, these intelligent platforms provide customized mental health support, ensuring a more engaging and effective experience (Appleton, 2024; Valentine et al., 2024). This kind of instant, tailored assistance makes mental healthcare more responsive and user-friendly.

To make mental health support more engaging for teenagers, some apps integrate gamification to encourage regular use and habit formation (Cheng et al., 2019). One such example is Manifest (`www.manifestapp.xyz`), an AI-powered app designed to help Gen Z users build healthy mental wellness routines. Branded as the "Shazam for your feelings" on their website, Manifest's AI system listens to users, helping them identify their moods and understand the factors influencing them. Over time, the app tracks emotional patterns and provides personalized insights, guiding users toward healthier habits and improved self-awareness.

Cost-Effectiveness

Digital mental health services have emerged as a cost-effective solution to address the increasing demand for mental healthcare (Mohr et al., 2018). Traditional in-person services, involving professionals such as therapists, psychiatrists, and school nurses, often come with significant costs and face challenges due to a shortage of qualified providers (Patel et al., 2007; Federal Statistical Office, 2021; Singer, 2024). In contrast, digital platforms can offer reliable, evidence-based support around the clock, making mental health resources more accessible and affordable (see, e.g., Wies et al., 2021).

A meta-analysis by Gentili et al. (2022) explored the cost-effectiveness of digital health interventions, revealing promising yet complex findings. While evidence suggests that these tools can improve health outcomes while reducing costs, inconsistencies in study methodologies make it difficult to compare their true impact across different interventions.

The researchers stress the need for more standardized research to accurately assess the financial and health benefits of digital mental health solutions. As technology continues to evolve, it's essential to make sure that digital mental health solutions not only provide real, effective support but also remain affordable, ethical, and bias free.

Potential to Reduce Stigma

In many communities, discussing mental health challenges remains stigmatized, deterring individuals from seeking the help they need (Kaushik et al., 2016). Digital mental health platforms provide a private and accessible way for people to seek support, offering a safe space where they can connect anonymously without the pressure of face-to-face interactions (Wies et al., 2021). This discretion makes it easier for individuals, especially young people, to open up about their struggles and access help without fear of judgment. This anonymity can help reduce the fear of judgment and encourage more individuals, especially adolescents, to engage with mental health resources. A scoping research review highlighted that digital mental health technologies can increase accessibility and reduce stigma, making them a promising avenue for mental health support for adolescents (Wies et al., 2021).

Challenges and Limitations of Digital Mental Health for Adolescents

Like any new technology, digital mental health services come with challenges that must be carefully addressed to ensure they are effective, safe, and accessible for those who need them. While these solutions have the potential to transform mental healthcare, their success depends on continuous improvement and thoughtful implementation.

Some key challenges include low adaptability and engagement, as many digital tools struggle to keep users actively involved over time (Kaveladze et al., 2022; Boucher & Raiker, 2024). Many digital mental health applications lack rigorous evaluation, leading to concerns about their trustworthiness and effectiveness (see, e.g., Nicholas et al., 2016; Torous et al., 2018; Wies et al., 2021). A study by Larsen et al. (2019) on "Evaluation of mental health app store quality claims" found that a majority of mental health apps claim to diagnose conditions or improve symptoms, yet few provide scientific evidence to support these claims. This absence of validation makes it challenging for users and healthcare professionals to rely on these tools with confidence. The study emphasizes the need for comprehensive evaluations to ensure that these digital interventions are both safe and effective.

A comprehensive review by Weis et al. (2021) indicates that many mental health apps and Internet-based platforms "lack thorough professional evaluations or clinical validation studies," leading to uncertain impacts and outcomes.

Limited awareness and adoption of digital mental health resources among adolescents and their families can significantly reduce the effectiveness of these interventions. If teens and their caregivers are unaware of available digital tools, they cannot utilize them to support mental well-being (O'Connor et al., 2016; Zhao et al., 2025). Barriers such as age and disabilities further complicate access and engagement with these digital solutions (Torous et al., 2018; Sharma et al., 2020). For instance, younger adolescents may lack the digital literacy required to navigate these platforms effectively, while those with certain disabilities might find the interfaces challenging to use (see, e.g., Bauer et al., 2017). Addressing these issues is crucial to ensure that digital mental health interventions are inclusive and accessible to all adolescents.

Finally, privacy and data security remain significant concerns, as sensitive mental health information must be protected to maintain user trust (Martínez-Pérez et al., 2015; Bauer et al., 2017; Wies et al., 2021).

While digital mental health services face several challenges, addressing these obstacles can pave the way for more user-friendly, effective, and widely accepted solutions, ultimately enhancing mental well-being for adolescents and beyond. However, as we stand at the early stages of this transformative shift, it is crucial to invest in research and development to ensure these technologies are safe, reliable, and evidence based.

Examples of Digital Mental Health Solutions for Adolescents

The landscape of digital mental health solutions for adolescents is rapidly evolving, offering a variety of tools designed to support young people's mental well-being. These solutions range from app-based platforms to AI-driven chatbots, each utilizing different modalities to cater to diverse needs. Some are already available in the market, while others are in the conceptual or research and development phases.

These AI-driven solutions are designed to, for instance, streamline documentation, enhance assessment, track symptoms, personalize psychoeducation, deliver tailored therapy, and support crisis management and suicide prevention. Additionally, they provide personalized treatment recommendations, ensuring more effective and targeted mental healthcare.

Below, we explore some of the leading tools that are transforming mental health support through innovative AI technology.

Neolth (www.neolth.com) is an on-demand, personalized social and emotional learning platform designed specifically for middle and high school students in the United States. The app offers integrated community features, providing a supportive environment for adolescents to manage their mental health. According to data from Neolth, students experienced a 48% reduction in stress after 12 weeks of using the app (Neolth, 2025).

Headspace (www.headspace.com) is a widely recognized mindfulness and meditation app that helps users manage stress, improve focus, and build lasting mental wellness habits. Its intuitive design and diverse range of guided exercises make it especially appealing to adolescents looking for accessible ways to support their mental health. It is a subscription service and available worldwide in many languages (Headspace, 2025).

To offer a more personalized experience, Headspace has introduced Ebb, an AI-powered companion currently available in English for subscribers in the United States, Australia, and Canada. Ebb acts as a digital well-being guide, responding to users' thoughts, emotions, and concerns. Based on their input, it suggests tailored meditation sessions and content to help them navigate their feelings and develop healthier coping mechanisms (Headspace, 2025). Developed by mindfulness and mental health experts, Ebb encourages daily reflection and mindfulness as a way to support overall well-being. Headspace is not a substitute for professional therapy or medical treatment; it provides a convenient, everyday tool to help users maintain emotional balance and build healthier habits.

Earkick (www.earlick.com) is an AI-driven mental health platform designed to provide personalized, real-time support through biomarker analysis and an advanced large language model companion. Marketed as "Your Free Personal AI Therapist," Earkick is available 24/7 and can be downloaded from the Apple Store, offering in-app purchases for additional features (Earkick, 2025).

Key features of Earkick are given on their website as follows:

- **AI-powered tracking**: Earkick continuously monitors mental health trends using multiple data points to offer tailored support.

- **Real-time conversations**: Engages users in meaningful dialogue, responding to emotional cues and providing immediate guidance.

- **Guided self-care sessions**: Offers structured mental
  wellness exercises to help users develop healthier
  habits over time.

Earkick integrates various data sources to understand the user's
emotional state and overall well-being. It collects insights through audio,
video and text memos, typing behavior, sleep patterns, panic attack
insights, and menstrual cycle data.

By analyzing physiological markers and user interactions, Earkick
adapts over time to provide highly personalized support. Its AI-powered
approach makes it a convenient and insightful tool for anyone looking to
understand and improve their mental health in a more structured and
data-driven way.

Little Otter (www.littleotterhealth.com) is a digital mental health
platform dedicated to providing comprehensive care for children aged 0 to
18 and their families across 15 US states. The platform offers personalized,
evidence-based support through virtual therapy sessions, parent coaching,
and psychiatric consultations. By leveraging AI capabilities, Little Otter
enhances patient triage and tailors care to individual needs, making
mental health support more accessible to families nationwide (Little
Otter, 2025).

The key features given on their website are as follows: whole-family
approach, accessible virtual care, evidence-based treatments, and
family mental health checkup. Families using Little Otter have reported
significant improvements, with 80% of children showing clinically
significant progress after 12 sessions and 3 in 4 parents experiencing
reductions in anxiety and depression.

Talkspace (www.talkspace.com) is an online therapy platform in the
United States that connects individuals aged 13 and above with licensed
therapists through a user-friendly mobile app and website (Talkspace,
2025b). It offers various services, including individual therapy, couples
counseling, teen therapy, and psychiatric consultations and medication

management. Users can communicate with their therapists via text, audio, and video messages, as well as schedule live video sessions. This flexibility allows for personalized mental health support tailored to individual needs.

Talkspace has introduced several AI-driven features to enhance therapist efficiency and improve client care such as

- **Smart notes**: This feature generates concise summaries of client sessions, reducing therapists' documentation time by approximately ten minutes per session, thereby allowing more focus on client care (Talkspace, 2024).

- **Insights**: Launched in January 2025, "Insights" assists therapists in session preparation by synthesizing data from clients' therapeutic journeys, including symptom changes and key session details, to provide tailored presession briefings and postsession updates. This HIPAA-compliant technology ensures that user data remains confidential and secure with service provider (Talkspace, 2025a).

- **Suicide risk detection algorithm**: This proprietary AI analyzes real-time client messages to identify language patterns indicative of high-risk behaviors, enabling timely alerts to therapists for appropriate interventions (Talkspace, 2023). The AI tool has shown 83% accuracy in identifying individuals at risk of self-harm or suicide.

Clare&me (www.clareandme.com) is a Berlin-based digital mental health platform startup offering an AI-powered companion named Clare, designed to support individuals dealing with anxiety and mild depression. Clare provides "a safe, non-judgmental space for users to express their thoughts and emotions, promoting mental well-being" (Clare & me, 2025).

According to their website, it is a subscription-based service and accessible 24/7 via phone calls or messaging. "Clare" guides users through self-reflective exercises and grounding techniques. The platform emphasizes user privacy, ensuring that interactions remain anonymous, and data is handled with strict confidentiality. While Clare is not a replacement for traditional therapy, it serves as an immediate, accessible resource for those seeking support, especially in regions with limited access to mental health services.

Woebot (www.woebothealth.com) is an AI-powered mental health chatbot designed to provide accessible support through daily conversations (Woebot Health, 2025). Available on smartphones and tablets, it engages users in text-based dialogues to monitor mood, identify thought patterns, and develop effective coping strategies. Developed by experts, Woebot "incorporates elements from cognitive behavioral therapy (CBT), interpersonal psychotherapy (IPT), and dialectical behavioral therapy (DBT)" to assist users in managing symptoms of anxiety and depression. Features include mood tracking, progress reflection, gratitude journaling, and mindfulness practices, all aimed at promoting emotional regulation and self-awareness. While not a replacement for traditional therapy, Woebot offers a convenient and scalable solution for individuals seeking immediate mental health support.

Aiberry (www.aiberry.com) is an innovative mental health assessment platform that leverages artificial intelligence to enhance the accuracy and efficiency of mental health screenings. By analyzing user inputs, including text, audio, and video, during brief conversations using the solution, Aiberry provides real-time, quantified risk scores and health insights, facilitating early detection and informed clinical decision-making.

Limbic (www.limbic.ai) is a UK-based service provider, and it provides an innovative clinical AI-powered platform dedicated to enhancing mental healthcare by streamlining clinical assessments and providing continuous patient support (Limbic, 2025). According to their website, by automating administrative tasks, Limbic enables clinicians to focus more on therapy, ultimately improving patient outcomes.

Key outcomes include

- **Increased engagement and access**: A 15% rise in referrals, with significant increases among minority groups.

- **Shortened patient intake**: Patients experience a five-day reduction in waitlist times.

- **Streamlined clinical assessments**: Assessment times are cut by 50%, allowing clinicians to dedicate more time to therapy.

- **Improved treatment planning**: There's a 45% decrease in changes to treatment pathways, enhancing care consistency.

- **Enhanced treatment implementation**: A 30% reduction in no-shows and dropouts indicates better patient adherence.

- **High patient satisfaction**: Post-therapy satisfaction rates reach 94%, reflecting positive patient experiences.

The platform offers two primary products:

1. **Limbic Access**: This AI assistant manages patient intakes and assessments, reducing the time clinicians spend on these tasks by up to 50%. By efficiently collecting patient data and providing clinical decision support, Limbic Access enhances diagnostic accuracy and patient engagement.

2. **Limbic Care**: Serving as an AI companion, Limbic Care
   offers patients personalized support between therapy
   sessions. It integrates with individual treatment plans
   and provides on-demand conversational assistance
   through clinically validated AI, contributing to better
   clinical outcomes in less time.

AutMedAI is a cutting-edge machine learning model developed
by researchers at Karolinska Institutet in Sweden to support the
early detection of autism spectrum disorder (ASD) in young children
(Rajagopalan et al., 2024). Currently in the research phase, this model aims
to provide an efficient and accessible tool for early autism screening. By
analyzing 28 key developmental indicators, including milestones like the
age of first smile, first sentence, and eating patterns, AutMedAI can assess
autism risk with nearly 80% accuracy in children under the age of two.

BlinkLab (www.blinklab.org) is a mobile-based app designed to
make developmental diagnostics and care more accessible, especially
for children with autism and ADHD. Using AI and machine learning, the
app analyzes behavioral and neurological data to help detect early signs
of neurodevelopmental conditions quickly and accurately (BlinkLab,
2025). As a new player in the market, BlinkLab is on a mission to simplify
the diagnosis process for families and healthcare providers. The team is
currently working through the FDA approval process, aiming to bring this
innovative tool to more people who need it.

Conclusion

In this chapter, we gave an overview of the role of digital mental health
in bridging gaps in adolescent mental health services. They discuss how
adolescence is a crucial developmental phase during which mental health
challenges such as anxiety, depression, and behavioral disorders often
emerge. However, access to timely and effective mental health services

remains limited due to stigma, professional shortages, and financial barriers. Digital mental health solutions, including AI-powered tools, online therapy platforms, and mobile applications, have emerged as promising interventions to address these challenges.

The chapter explores the emergence of digital mental health solutions, highlighting how advancements in technology have facilitated online therapy, virtual interventions, and AI-driven diagnostics. We present present key benefits of digital mental health for adolescents, including improved accessibility, cost-effectiveness, personalization, and the potential to reduce stigma. Various studies and systematic reviews support the effectiveness of digital interventions in addressing mental health conditions, particularly through cognitive behavioral therapy (CBT)-based platforms and AI-driven applications.

Despite these benefits, the chapter also acknowledges the challenges and limitations of digital mental health, such as low user engagement, lack of clinical validation for many applications, privacy concerns, and accessibility issues for younger users or those with disabilities.

Finally, the chapter provides examples of existing digital mental health solutions tailored for adolescents, including AI-driven platforms like Woebot, Neolth, Headspace, Earkick, and BlinkLab. These tools offer various services, such as diagnosis, mental health tracking, therapy sessions, and community support, demonstrating the growing role of digital technology in adolescent mental healthcare.

References

Appleton, C. (2024, October 22). Revolutionizing Behavioral Health Through Technology and AI: The Promise of Personalized Care. Behavioral Health News. https://behavioralhealthnews.org/revolutionizing-behavioral-health-through-technology-and-ai-the-promise-of-personalized-care/

Bauer, M., Glenn, T., Monteith, S., Bauer, R., Whybrow, P. C., & Geddes, J. (2017). Ethical perspectives on recommending digital technology for patients with mental illness. International Journal of Bipolar Disorders, 5(1), 6. https://doi.org/10.1186/s40345-017-0073-9

Bilan, D. S., Chicchi Giglioli, I. A., Cuesta, P., Cañadas, E., de Ramón, I., Maestú, F., Alda, J., Ramos-Quiroga, J. A., Herrera, J. A., Amado, A., & Quintero, J. (2025). Decreased impulsiveness and MEG normalization after AI-digital therapy in ADHD children: A RCT. Npj Mental Health Research, 4(1), 1–14. https://doi.org/10.1038/s44184-024-00111-9

BlinkLab. (2025). BlinkLab | Revolutionizing Developmental Diagnostics and Care. https://www.blinklab.org/

Boucher, E. M., & Raiker, J. S. (2024). Engagement and retention in digital mental health interventions: A narrative review. BMC Digital Health, 2(1), 52. https://doi.org/10.1186/s44247-024-00105-9

Bress, J. N., Falk, A., Schier, M. M., Jaywant, A., Moroney, E., Dargis, M., Bennett, S. M., Scult, M. A., Volpp, K. G., Asch, D. A., Balachandran, M., Perlis, R. H., Lee, F. S., & Gunning, F. M. (2024). Efficacy of a Mobile App-Based Intervention for Young Adults With Anxiety Disorders: A Randomized Clinical Trial. JAMA Network Open, 7(8), e2428372. https://doi.org/10.1001/jamanetworkopen.2024.28372

Butera, C., Delafield-Butt, J., Lu, S.-C., Sobota, K., McGowan, T., Harrison, L., Kilroy, E., Jayashankar, A., & Aziz-Zadeh, L. (2025). Motor Signature Differences Between Autism Spectrum Disorder and Developmental Coordination Disorder, and Their Neural Mechanisms. Journal of Autism and Developmental Disorders, 55(1), 353–368. https://doi.org/10.1007/s10803-023-06171-8

Calear, A. L., & Christensen, H. (2010). Review of internet-based prevention and treatment programs for anxiety and depression in children and adolescents. Medical Journal of Australia, 192(S11), S12–S14. https://doi.org/10.5694/j.1326-5377.2010.tb03686.x

Cao, M., Martin, E., & Li, X. (2023). Machine learning in attention-deficit/hyperactivity disorder: New approaches toward understanding the neural mechanisms. Translational Psychiatry, 13(1), 1–12. https://doi.org/10.1038/s41398-023-02536-w

Chen, T., Tachmazidis, I., Batsakis, S., Adamou, M., Papadakis, E., & Antoniou, G. (2023). Diagnosing attention-deficit hyperactivity disorder (ADHD) using artificial intelligence: A clinical study in the UK. Frontiers in Psychiatry, 14. https://doi.org/10.3389/fpsyt.2023.1164433

Cheng, V. W. S., Davenport, T., Johnson, D., Vella, K., & Hickie, I. B. (2019). Gamification in Apps and Technologies for Improving Mental Health and Well-Being: Systematic Review. JMIR Mental Health, 6(6), e13717. https://doi.org/10.2196/13717

Clare & me. (2025). AI Self-Therapy | clare&me | Speak to an AI about your Mental Health over the Phone. https://www.clareandme.com/ai-for-mentalhealth-worries-and-overthinking

Earkick. (2025). Earkick—Your Free Personal AI Therapist Chat Bot. https://earkick.com/

Federal Statistical Office. (2021). Number of psychotherapists up 19% between 2015 and 2019. Federal Statistical Office. https://www.destatis.de/EN/Press/2021/03/PE21_N022_23.html

Gentili, A., Failla, G., Melnyk, A., Puleo, V., Tanna, G. L. D., Ricciardi, W., & Cascini, F. (2022). The cost-effectiveness of digital health interventions: A systematic review of the literature. Frontiers in Public Health, 10. https://doi.org/10.3389/fpubh.2022.787135

Gross, N. (2024). Mental health apps: Not a friend but a foe for public health - EPHA. Https://Epha.Org/. https://epha.org/mental-health-apps-not-a-friend-but-a-foe-for-public-health/

Headspace. (2025). Meet Ebb | AI Mental Health Companion. Headspace. https://www.headspace.com/ai-mental-health-companion

Kaushik, A., Kostaki, E., & Kyriakopoulos, M. (2016). The stigma of mental illness in children and adolescents: A systematic review. Psychiatry Research, 243, 469–494. https://doi.org/10.1016/j.psychres.2016.04.042

Kaveladze, B. T., Wasil, A. R., Bunyi, J. B., Ramirez, V., & Schueller, S. M. (2022). User Experience, Engagement, and Popularity in Mental Health Apps: Secondary Analysis of App Analytics and Expert App Reviews. JMIR Human Factors, 9(1), e30766. https://doi.org/10.2196/30766

Lal, S., & Adair, C. E. (2014). E-mental health: A rapid review of the literature. Psychiatric Services (Washington, D.C.), 65(1), 24–32. https://doi.org/10.1176/appi.ps.201300009

Larsen, M. E., Huckvale, K., Nicholas, J., Torous, J., Birrell, L., Li, E., & Reda, B. (2019). Using science to sell apps: Evaluation of mental health app store quality claims. Npj Digital Medicine, 2(1), 1–6. https://doi.org/10.1038/s41746-019-0093-1

Lattie, E. G., Adkins, E. C., Winquist, N., Stiles-Shields, C., Wafford, Q. E., & Graham, A. K. (2019). Digital Mental Health Interventions for Depression, Anxiety, and Enhancement of Psychological Well-Being Among College Students: Systematic Review. Journal of Medical Internet Research, 21(7), e12869. https://doi.org/10.2196/12869

Limbic. (2025). Limbic | Clinical AI for mental healthcare providers. https://www.limbic.ai/

Little Otter. (2025). Little Otter | Mental Health Services for the Whole Family. https://www.littleotterhealth.com/

Martínez-Pérez, B., de la Torre-Díez, I., & López-Coronado, M. (2015). Privacy and security in mobile health apps: A review and recommendations. Journal of Medical Systems, 39(1), 181. https://doi.org/10.1007/s10916-014-0181-3

Mohr, D. C., Riper, H., & Schueller, S. M. (2018). A Solution-Focused
Research Approach to Achieve an Implementable Revolution in
Digital Mental Health. JAMA Psychiatry, 75(2), 113–114. https://doi.
org/10.1001/jamapsychiatry.2017.3838

Naslund, J. A., Aschbrenner, K. A., Marsch, L. A., & Bartels, S. J. (2016).
The future of mental health care: Peer-to-peer support and social media.
Epidemiology and Psychiatric Sciences, 25(2), 113–122. https://doi.
org/10.1017/S2045796015001067

Neolth. (2025). Neolth-Science. Neolth. https://www.neolth.
com/science

Newton, A., Bagnell, A., Rosychuk, R., Duguay, J., Wozney, L., Huguet,
A., Henderson, J., & Curran, J. (2020). A Mobile Phone–Based App for
Use During Cognitive Behavioral Therapy for Adolescents With Anxiety
(MindClimb): User-Centered Design and Usability Study. JMIR mHealth
and uHealth, 8(12), e18439. https://doi.org/10.2196/18439

Nicholas, J., Boydell, K., & Christensen, H. (2016). mHealth in
psychiatry: Time for methodological change. BMJ Ment Health, 19(2),
33–34. https://doi.org/10.1136/eb-2015-102278

O'Connor, S., Hanlon, P., O'Donnell, C. A., Garcia, S., Glanville, J.,
& Mair, F. S. (2016). Understanding factors affecting patient and public
engagement and recruitment to digital health interventions: A systematic
review of qualitative studies. BMC Medical Informatics and Decision
Making, 16(1), 120. https://doi.org/10.1186/s12911-016-0359-3

Park, Y., Park, S., & Lee, M. (2024). Effectiveness of artificial intelligence
in detecting and managing depressive disorders: Systematic review.
Journal of Affective Disorders, 361, 445–456. https://doi.org/10.1016/j.
jad.2024.06.035

Patel, V., Flisher, A. J., Hetrick, S., & McGorry, P. (2007). Mental health
of young people: A global public-health challenge. The Lancet, 369(9569),
1302–1313. https://doi.org/10.1016/S0140-6736(07)60368-7

Rajagopalan, S. S., Zhang, Y., Yahia, A., & Tammimies, K. (2024). Machine Learning Prediction of Autism Spectrum Disorder From a Minimal Set of Medical and Background Information. JAMA Network Open, 7(8), e2429229. https://doi.org/10.1001/jamanetworkopen.2024.29229

Schefft, C., Krämer, R., Haaf, R., Jedeck, D., Schumacher, A., & Köhler, S. (2024). Evaluation of the internet-based intervention "Selfapy" in participants with unipolar depression and the impact on quality of life: A randomized, parallel group study. Quality of Life Research, 33(5), 1275–1286. https://doi.org/10.1007/s11136-024-03606-2

Serafin, M. (2020). Telemental Health Services for Youth in Rural Areas: Meeting Service Gaps and Best Practices. Wilder Research. https://www.wilder.org/sites/default/files/imports/DHS_SoC_TelementalHealthServices_LitReview_10-20.pdf?

Sharma, S., Avellan, T., Linna, J., Achary, K., Turunen, M., Hakulinen, J., & Varkey, B. (2020). Socio-Technical Aspirations for Children with Special Needs: A Study in Two Locations – India and Finland. ACM Trans. Access. Comput., 13(3), 13:1-13:27. https://doi.org/10.1145/3396076

Singer, S. (2024). Almost no Change in Waiting Times for Outpatient Psychotherapy After an Amendment to the Law (17.05.2024). Aerzteblatt. https://www.aerzteblatt.de/int/archive/article?id=239104

Sorkin, D. H., Janio, E. A., Eikey, E. V., Schneider, M., Davis, K., Schueller, S. M., Stadnick, N. A., Zheng, K., Neary, M., Safani, D., & Mukamel, D. B. (2021). Rise in Use of Digital Mental Health Tools and Technologies in the United States During the COVID-19 Pandemic: Survey Study. Journal of Medical Internet Research, 23(4), e26994. https://doi.org/10.2196/26994

Talkspace. (2023). Proprietary AI Algorithm Alerts Therapists to Suicide Risk in Patients Utilizing the Talkspace Platform—Talkspace, Inc. https://investors.talkspace.com/news-releases/news-release-details/proprietary-ai-algorithm-alerts-therapists-suicide-risk-patients/

Talkspace. (2024). Talkspace Launches Dedicated AI Innovation Group to Advance Provider Efficiency and Enhance Clinical Quality and Operations—Talkspace, Inc. https://investors.talkspace.com/news-releases/news-release-details/talkspace-launches-dedicated-ai-innovation-group-advance/

Talkspace. (2025a). Talkspace Launches AI-Powered Insights to Advance Provider Efficiency and Enhance Therapeutic Care. https://investors.talkspace.com/news-releases/news-release-details/talkspace-launches-ai-powered-insights-advance-provider/

Talkspace. (2025b). Talkspace—#1 Rated Online Therapy, 1 Million+ Users. https://www.talkspace.com/

Torous, J., Nicholas, J., Larsen, M. E., Firth, J., & Christensen, H. (2018). Clinical review of user engagement with mental health smartphone apps: Evidence, theory and improvements. BMJ Ment Health, 21(3), 116–119. https://doi.org/10.1136/eb-2018-102891

Valentine, L., Arnold, C., Nicholas, J., Castagnini, E., Malouf, J., Alvarez-Jimenez, M., & Bell, I. H. (2024). A Personalized, Transdiagnostic Smartphone App (Mello) Targeting Repetitive Negative Thinking for Depression and Anxiety: Qualitative Analysis of Young People's Experience. Journal of Medical Internet Research, 26(1), e63732. https://doi.org/10.2196/63732

Wetterlin, F. M., Mar, M. Y., Neilson, E. K., Werker, G. R., & Krausz, M. (2014). eMental Health Experiences and Expectations: A Survey of Youths' Web-Based Resource Preferences in Canada. Journal of Medical Internet Research, 16(12), e293. https://doi.org/10.2196/jmir.3526

Wies, B., Landers, C., & Ienca, M. (2021). Digital Mental Health for Young People: A Scoping Review of Ethical Promises and Challenges. Frontiers in Digital Health, 3. https://doi.org/10.3389/fdgth.2021.697072

Woebot Health. (2025). Woebot Health. Woebot Health. https://woebothealth.com/

Zhao, X., Schueller, S. M., Kim, J., Stadnick, N. A., Eikey, E., Schneider, M., Zheng, K., Mukamel, D. B., & Sorkin, D. H. (2025). Real-World Adoption of Mental Health Support Among Adolescents: Cross-Sectional Analysis of the California Health Interview Survey. Journal of Pediatric Psychology, 50(1), 20–29. https://doi.org/10.1093/jpepsy/jsad082

Zhou, S., Zhao, J., & Zhang, L. (2022). Application of Artificial Intelligence on Psychological Interventions and Diagnosis: An Overview. Frontiers in Psychiatry, 13. https://doi.org/10.3389/fpsyt.2022.811665

SECTION 2

Harnessing Data and AI Models: Reinventing Youth Mental Health Care Insights

CHAPTER 4

Empowering Mental Well-Being Through Data-Driven Insights

This chapter sheds light on how different sources of data can lead to transformation in mental well-being through meaningful insights, recommendations, therapies, and proactive alerts. This chapter focuses on different data collection mechanisms, including through mobile apps, gaming apps, and social media. In this context, the chapter explores how human-computer interaction has reinvented the field of adolescent behavior and mood detection to enable the development of a proactive and personalized approach to mental health. The chapter aims to empower individuals to take control of their well-being and make a positive impact on their mental health through the use of data. Furthermore, it advocates for a better parental and academic ecosystem by combining it with **cognitive behavioral therapy (CBT)** through step-by-step prescriptions and progress tracking based on users' activities.

In this chapter, these topics will be covered in the following sections:

- Aggregating data sources to assess mental health

- Addressing mental health risks through data collection using mobile apps

© Sharmistha Chatterjee, Azadeh Dindarian, Usha Rengaraju 2025
S. Chatterjee et al., *Revolutionizing Youth Mental Health with Ethical AI*,
https://doi.org/10.1007/979-8-8688-1186-9_4

Aggregating Data Sources to Assess Mental Health

Data plays a predominant role in assessing mental health conditions to flag a risk scenario and suggest remedial actions. The data sources can broadly be divided into four different categories:

- **First-party data**: This data can directly be sourced from the youth's family history, demographic details, age, location, and presence in social media.

- **Tracking apps**: This type of data can be collected from the vast majority of mental healthcare applications available in the market.

- **Device-based tracking**: This type of data can be tracked from sensors (from GPS, accelerometers, communication logs, Wi-Fi, and Bluetooth) plugged into an individual's body or mobile applications.

- **Third-party data sources**: This type of data can be tapped from social media or events, clubs, and other participation media in terms of social engagements or through feedback reports received from mental health diagnostic centers.

Figure 4-1 illustrates diverse sources of data that can be aggregated together to assess mental health in youth.

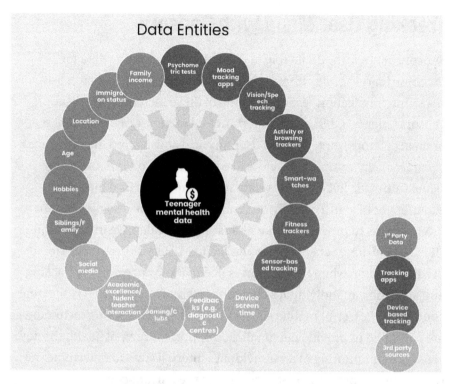

Figure 4-1. *Diverse data sources to assess mental health*

As first-party data directly involves data collection from individuals; it can be collected directly from individuals (youth and students) and their close family members through specific interview processes by asking them to fill in questions about symptoms, medical history, and physical examination. To enable data collection from third-party sources, psychological tests and assessment tools serve as a source to feed patients' history that can diagnose different mental health problems. However, this method cannot be used widely before obtaining consent from patients and without infringing the user's privacy.

Tracking User Mood with Sensors

Sensors play a formidable role in collecting and curating data for machine learning algorithms to predict mental health state, relapse risks, and medication response. Some of these include using location sensors (such as GPS) as an important tool to measure user mobility, or an accelerometer to track in user physical activity and the influence of external environmental factors like green space on mental health (Huckvale et al., 2020). Sensor data when integrated into the lived experiences of youth or individuals are capable of accessing different levels of risks to provide them with just-in-time adaptive intervention (JITAI). Real-time sensor data when fed to ML algorithms is able to predict the time when the support is needed to what extent by tracking and accessing an individual's changing internal and external contextual conditions. Powerful mobile sensors with advanced sensing technologies are equipped to provide personalized support for mental health through ecological monitoring of an individual's internal state to determine the right time and process to intervene given the current context.

One noteworthy example of JITAI is evident from a smartphone behavioral intervention FOCUS, which is equipped to support illness management in schizophrenia patients (Nahum-Shani et al., 2018). It operates in a push model where it prompts individuals through auditory signals and visual notifications. The individuals are prompted three times a day and given an opportunity either to ignore or engage with the prompt. The engaged individuals provide enough data points to access primarily five target domains: medication adherence, mood regulation, sleep, social functioning, and coping with hallucinations. The assessment report enables FOCUS to decide and recommend self-management techniques to improve their current state based on the type and level of difficulties recorded by the individuals (Strojny et al., 2024). Individuals reporting fatigue and interpersonal conflict are marked as high risk arising from a symptomatic relapse. In such scenarios of deteriorating illness, FOCUS determines the root cause—whether it is arising from too much fatigue, interpersonal conflict, insomnia (sleep difficulties), or skipped medication and offers the right kind of support to alleviate the circumstances and make the individual feel comfortable. FOCUS intervention decisions are driven by results of self-report of the random prompts which can be configured at prespecified time intervals daily or weekly. The design of JITAI methodologies equips FOCUS to monitor recovered as well as recovering patients every minute to assess changes in risks and take immediate actions.

Thus, we see interventions have a critical role to play in mitigating risks among patients. Along with CBT and other therapeutic treatments, transdiagnostic CBT (TCBT) is an efficient therapeutic treatment for different psychological disorders where we see interventions planned for one disorder have more probability of success for other emotional disorders as well and it addresses several common underlying root causes of mental health.

Figure 4-2 demonstrates how different JITAI components can be stitched together to produce both short-term and long-term health outcomes to patients. Tracking user mood with sensors (Nahum-Shani et al., 2018) helps in timely detection and interventions by effective monitoring.

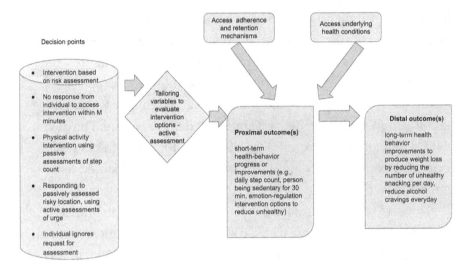

Figure 4-2. *JITAI components and operating methodology (Nahum-Shani et al., 2018)*

Wearable devices also play a prominent role in data collection that can be used for further analysis to predict mood changes. For example, electrocardiogram signals recorded from the patients are analyzed to discover heart rate features and predict outcomes related to bipolar disorder. Research results demonstrate classification algorithms like **support vector machines (SVM)** can achieve an accuracy of 69% in detecting bipolar disorder through mood changes. An SVM performs a classification task that selects the best hyperplane (a boundary line in a higher dimension) from complex patient records, to classify the patient into two classes, one class having mental disorder, anxiety, and depression and another class where mental well-being of patients is restored. The selected hyperplane maximizes the distance between the closest data points (support vectors) of each class.

Further, functional magnetic resonance imaging can help to detect bipolar disorder by recognizing the differences in the brain activity in patients and applying classification algorithms such as the Gaussian process (GP) to differentiate bipolar disorder from unipolar disorder. Experimental results demonstrate that up to 67% accuracy, 72% specificity, and sensitivity of 61% can be obtained through this. A GP is a statistical modeling process used for extracting complex relationships between various mental health conditions to identify intricate activities in the brain and assess mental health dynamics and probable impacts of different treatment effects.

Role of Data from Social Media

Social media data can also be considered to play a certain role in predicting mental health outcomes and assessing risk which can support regular clinical assessments. Digital data from social media platforms can result in a better understanding of behavioral psychology, by analyzing individual and interpersonal behaviors in public matters such as hate speech and political ideation. Researchers have been increasingly looking at early detection of mental health conditions, by using careful monitoring techniques from social media data to alleviate conditions that can range from both micro and macro scales, including crises on personal relationships as well as the global economy.

Collection of social media data daily can provide additional data points for clinical assessments and provide progress of medication on a daily/weekly/monthly level of the prescribed disease. More often, the data collection could be restricted to a clinical environment for commonly less serious mental health disorders such as schizophrenia or **post-traumatic stress disorder (PTSD),** where social media data can assist the data collection process by analyzing the sentiments of the posts.

By having a useful representation of the Twitter data as a time series, it becomes easier for ML models to identify optimal points for change-point detection that may be suitable for intervention in a high-risk mental state. In addition, the availability of ground-truth data for the same individual may enable the model to access the model performance and determine the addition/update/removal of features. Twitter attracts the most data scientists and researchers among popular social media platforms due to its public-facing interfaces. Twitter data can be crawled with Twitter Application Programming Interface (API). However, research results indicate that out of 119 datasets identified, only a small portion, 16 (13.4%), had access to ground-truth data (i.e., known characteristics) to access mental health disorders of users who had tweeted on social media through predictive modeling.

The challenge of not having high-quality social media data with supported ground-truth facts makes it difficult to design trustworthy AI algorithms that have wide-scale applications in research and clinical purposes. Moreover, it comes with the added disadvantage of how social data can collaborate across disciplines and contexts for better inference related to the prediction of mental health disorders. In addition, many times, when social media text is fed to ML models, we do not attach any weight to the intensity of the incoming tweets, fed into the system. This often contradicts the fact of how the predicted models generate their outcomes and what is the target of prediction, whether it is a single tweet or a set of tweets—multiple single tweets or sequence of responses from the same tweet provided by an individual.

Now, let us study a practical example of how an ML model can be built using social media data to detect stress.

Listing 4-1. This is an example of how an ML model can be used to detect stress from social media articles

In this part, we take a Reddit thread dataset data and perform human stress detection. The dataset is created using various articles from Reddit threads and each article is annotated with a label of "0" or "1," where "0" signifies a stress-negative article and "1" signifies a stress-positive article. There are more than 3000 articles in the dataset divided into title, body, and a combination of title and body.

This task can be viewed as a simple text classification task where we make use of the keras nlp library to easily complete this task. KerasNLP is a very easy-to-use and convenient library with many pretrained models that can be used as plug-and-play models per our requirements. So for this particular task, we will make use of the RoBERTa Classifier pipeline which we will just load and train it using our Reddit thread dataset to make our predictions.

1. In the first step, we install the necessary libraries.

    ```
    %pip install keras-nlp numpy sklearn
    ```

2. After installation, we import the necessary libraries.

    ```
    import pandas as pd
    import keras_nlp
    import numpy as np
    ```

3. Then, we load the Reddit thread dataset and view the column entries.

    ```
    df = pd.read_csv('Reddit_Combi.csv',sep=';')
    df.head()
    ```

4. After loading the data, our next step is to process it by assigning the labels (0 or 1 based on whether stress is negative or positive).

    ```
    text = df.Body_Title
    labels = df.label
    ```

5. Our next action is to train the data by creating a model from keras_nlp's RobertaClassifier.

```
classifier = keras_nlp.models.RobertaClassifier.
from_preset(
    "roberta_base_en",
    num_classes=2,
)
classifier.fit(x=text.values, y=labels,
batch_size=8, verbose=1)
```

6. Our last step is to make predictions and test the model by passing a passage of text data.

```
pred = classifier.predict([text.values[0]])
print(text.values[0])
np.argmax(pred[0],0)
```

This yields a score of 1 representing positive stress for the text shown in Figure 4-3, which demonstates the output of the prediction.

```
Envy to other is swallowing me Im from developingcountry, Indonesia , and for now i temporary work overseas for 3
years contract, it's a hard labor job, and stressful. Next year my contract is finish. But, during my stay here, b
ecause of job, and my social life, my depression got worse, and i envy this developed country. Why this country is
so good. I can afford anything i want here. Why can't we just have equality in currency exchange? I I need to work
15-20 years in big company in jakarta(our capital city) , just to get equal amount of saving money from what i got
from 3 years working here. Yes, that's right, it's saving money, not spending money. And yes, im going to be a ric
h person in such young age If I think about it, this society is sick, the gap of un equality beetwen developing vs
developed country, or the poor vs the rich is too big right now. Sorry if i look like an evil person , but because
of this,I almost wish for war to happen, or this world to end, and got reseted. So everyone can have another chanc
e to gain equality, the poor can finally have a chance to have a better life. Because if this world order stay thi
s way, it will eventually collapse on itself. Soon.
```

Figure 4-3. *Social media text representing stress using ML classifier*

For another example, with the below text, we again see a positive stress factor due to the presence of words envy, swallowing, and hard job.

Text: Envy to other is swallowing me Im from developingcountry, Indonesia , and for now i temporary work overseas for 3 years contract, it's a hard labor job, and stressful....

Prediction: 1

Location-Based Data As Behavioral Markers of Mental Health

Sociability (number and strength of social connections) or social connectedness and participation in social activities are strong influencers for mental fitness and well-being. Research data indicates when data from wearable sensors are mapped to a social network for a high school, adolescent girls with smaller social circles with a smaller number of network connections have higher depression and anxiety-related symptoms. In terms of social context, Bluetooth data also provides extremely beneficial indications to determine the human density, surrounding a participant from both previously known familiar Bluetooth-enabled devices and new Bluetooth-enabled devices that are coming in closer proximity to the individual.

We can safely conclude that both Bluetooth and location data gathered from mobile phones and sensors also serves as strong behavioral markers of mental health. Temporal and spatial data from Bluetooth demonstrates the intricacies and characteristics of proximal social networks, by distinguishing network connections from known friends vs. from unknown connections (Shin & Bae, 2023). In addition, larger details from GPS data can extract location-based features in minute details to portray a detailed image of visited places (location variance). It is often seen that when the youth population spend more time in fewer locations, with a lesser diversity of location change over a 24-hour window, remain prone to more severe depression symptoms. Therefore, parents and teachers become highly aware of the likelihood of mental health disorders when they examine decreased physical activity and higher degrees of sedentary conduct in children and adolescents. Sensors and wearable devices play a prominent role in extracting useful features like social context, location clusters, and circadian movement to infer behavioral markers for mental health. Even GPS data can provide four features: distance, entropy, irregularity, and location which helps in capturing the following details:

- Distance traveled between GPS coordinates.

- The entropy feature helps to compute the variations in time spent at various locations. It helps to capture allocated time to travel various locations and engage in numerous activities.

- The pattern of individual movements demonstrates different trends.

- GPS data also helps to compute the number and cluster of locations traversed at different times by an individual.

Along with location clusters, it becomes equally important to track human mobility patterns, influenced by demographic and socioeconomic backgrounds and geographical contexts, so that they act as additional inputs in determining the population density around participants. Overall, all these factors build a solid building block in evaluating risk related to mental health by synthesizing all metrics for location data.

Data Preprocessing and Feature Extraction

Social media has a lot of textual data, which needs to be preprocessed for better interpretation and feature extraction before feeding to the ML models. These may include

- Removal of noisy text data that do not aid in feature extraction or predictive tasks. This may include nonalphanumeric character elimination, stop words (common or filler words) removal, text (transforming words to their root) lemmatization, or tokenization (splitting sentences or documents into separate tokens delimited by spaces).

- Having sentiment dictionaries or stop word lists per language for non-English-based languages to fine-tune custom language-based preprocessing and feature selection.

- Accounting Valence Aware Dictionary references for sentiment reasoning for special characters and emoji in the text.

- Accounting Twitter's commonly used interaction characters such as hashtags and @-mentions.

- Focusing on special text like personal pronouns which turn out to be useful features for the prediction of depression.

Feature extraction helps in better identification of trends and patterns before feeding to the model to retrieve the predicted results. The most common ways of feature extraction include n-grams (groups of n words that appear sequentially) or term frequencies or word embeddings used through Word2Vec, GloVe, and Bidirectional Encoder Representations from Transformers.

Table 4-1 provides an overview of feature categories, the number of studies that used at least one feature from each category, and a description of the types of features they contain.

Table 4-1. *Feature extraction from Twitter data*

Feature type	Description
Text interpretation	Semantic dictionaries to interpret features from text
Demographics	Preknown or algorithmically inferred demographic information
Connectivity	User's social network, number of followers, or @-mentions
Sharing (when)	The interval between two tweets, tweet frequency
Sharing (what)	Type of content shared—retweets or URLs
Textual feature	Scoring with TFID, Bag of Words, LLMs
Keywords	Counts or distributions of keyword lists
Parts of speech	Labeled grammatical data referring to parts of speech
Images	Images, including profile pictures or shared images

To summarize, using Twitter data to design the right predictive model often comes with a choice of selection of

- Different preprocessing and modeling techniques used to include hyperparameter tuning and validation processes

- The process by which datasets were created by crawling Twitter data and how labeling techniques were used to label different mental health outcomes to arrive at the construct

However, data collection and training using social media hold challenges where there's a missing synergy between the label and the construct to be predicted. There have been instances where the model predicts instances of depression by referring to tweets that just contain any irregular state of mental health (e.g., a user tweeting "I have depression"). Even that could result from affiliations with accounts that mention different mental state disorders. Such affiliations might result when a user follows an account that mentions experiences of PTSD. Hence in addition to Twitter data, an external source of validation (e.g., reports supported by clinicians) would add to the veracity of the models and would increase the accuracy of predictions. Moreover, models that represent mental health outcomes as binary value are not able to access the different levels of risks involved based on the conditions related to comorbidity, anxiety, and depression.

The prime concern on social media data is respecting user privacy so that no data is collected and shared without the consent of users where models predict risks associated with user's mental health (Gordon, 2022). The well-known 2018 Cambridge Analytica Data Scandal infringed on millions of Facebook profiles of US voters, to analyze and infer the personal characteristics for political advertising. After witnessing such a serious data breach, where data and predictive models could influence choices at the ballot box, it becomes an ethical responsibility to design mental health framework in a manner that the system continuously learns from multiple data sources and aggregates and validates the data instead of relying on sparse data from social media that could result in data rights violations and biased predictions.

In addition, there have been concerns raised with respect to tracking technologies, where vulnerable groups feel threatened that their autonomy is claimed by social support networks or advocated by healthcare services. In such scenarios, self-reporting-based mobile apps pose the risk of exposing mental health disorder symptoms to unauthorized people, for example, information related to pregnant women's mental state getting leaked to social media. To combat such situations, design of suitable ethical systems should be enforced.

As a part of the design of ethical systems, it becomes increasingly important that mental health predictive systems coin the right word and make sufficient distinction when it refers to mental health and mental health disorders. Whenever a predictive model makes a prediction, that is, produces an algorithmic outcome, it should explicitly state whether the prediction results are catered to what type of mental health outcomes, mental health disorders, or well-being aspects such as those related to general well-being, happiness, life satisfaction, or self-esteem.

Leveraging Human-Computer Interaction (HCI) in Mental Healthcare

Recent research and technological advancements have improved the design and development of mental health solutions leading to the availability of efficient, user-friendly, cost-effective, and adaptable mental health applications in the market (Balcombe & De Leo, 2022). AI-driven mental health apps (5 Minute Journal app, Alan Mind Daily Journaling app, and others covered in Table 4-2), provided by the app providers, have played a key role to assist parents, teachers, children, and teenagers in prediction and identification of mental states. In addition, the digital platform supported in the backend of the apps has been equipped with critical messaging support, coordination, and treatment by mental healthcare to provide efficient risk assessment, self-help, and guided cognitive behavioral therapy (CBT) to prevent anxiety, depression, and life-threatening cases like suicide. Such practices may include the following:

- Reconditioning maladaptive associations, like controlling exposure to certain surroundings or activating certain kinds of behavior toward a certain environment.

- Increasing attention span on surroundings through the use of attention training, acceptance or tolerance training, and yoga or mindfulness exercises.

- Boosting metacognitive awareness and cognitive distancing, thereby facilitating decentering or defusion among individuals. Here, individuals are able to come outside of one's own mental state and be nonjudgmental toward themselves.

- Enabling cognitive reframing processes where individuals can challenge one's own ideas, thoughts, events, or surroundings or can even judge or review situations or surroundings.

- Improving overall behavioral, emotional, or cognitive self-management practices through rewards and motivation, thereby yielding an increase in self-efficacy.

Interactive AI has come up with innovative data collection, real-time screening, and notification techniques to reinvent prevalent manual, outdated, and time-consuming processes. Faster and more efficient amalgamation of human and machine interaction has led to the release of automated, highly effective mental health apps in the market with higher levels of efficacy, reliability, usability, and accessibility features. This space is rapidly transforming, and with AI regulations coming up, responsible AI has been a hot subject of debate. Mental healthcare providers are bringing in safe, secure, ethical, and sustainable apps to promote mental health-related education, training, and sociocultural adaptability. The purpose is to create interactive, real-time assistance in digital therapy and bring in the best of physical and digital collaboration to foster a hybrid approach. In the following subsections, we discuss more different machine learning algorithms, immersive technologies, and digital phenotyping, which has been used to enable personal sensing and capture the metadata to promote quick diagnostic actions to effectively manage, monitor, and notify mood changes in young people and alleviate various level of risks like suicidal actions and self-harm.

Assisting Mental Health Through Digital and Virtual Programs

Mental health assistance programs have seen a great boost through apps and virtual therapies. **Human-computer interaction (HCI)** can leverage **natural language processing (NLP)** techniques to interpret human sentiments from text and speech and suggest therapies by careful analysis of the text data. The mental health apps heavily rely on electronic health records, mood change scales, brain imaging data, and monitoring systems to cluster mental ill health categories.

Apps like Talkspace is well known for virtual psychiatric care visits, for adults and teens by offering therapy courses—both chat and video. This enables teens and their parents to join the courses virtually by means of individual workshops and get assisted with customized plans. It provides sufficient flexibility to send unlimited messages throughout the day to help youth through self-guided sessions, starting from 5 minutes a day.

This kind of virtual assistance is becoming increasingly popular, through mental-health-based apps and desktop devices to assist teens through Internet-based mental health screening, diagnosis, and aftercare to boost an individual's or youth's mental well-being. Traditional machine learning models, both supervised and unsupervised, have been able to detect mood swings to intervene in early symptoms of depression, suicidal feelings, or self-harm. User behavioral data from social media like Twitter can not only segment users based on mood change but also differentiate normal users from those at risk.

Figure 4-4 demonstrates how unsupervised clustering-based algorithms can be used to cluster different mood states.

Teenage Segmentation based on Mental Disorders

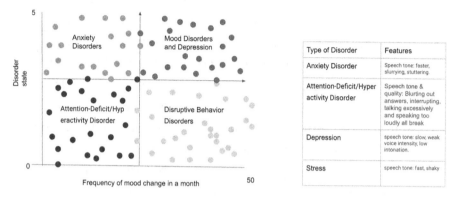

Figure 4-4. *Clustering algorithm to segment mental health disorders*

Along with a wide range of virtual assistance programs, the growing support from online peer-to-peer support platforms like Reddit has enabled youth and their parents to engage in more interaction and engagement to find solutions to their problems. Mental health subcommunities provided by Reddit and peer-supported mutual discourse provided by TalkLife have been able to create a safe space where parents and their children can come and talk about their problems and navigate the challenges, with like-minded people experiencing the same problems.

Since HCI plays a predominant role in data collection (clinical rating scores and self-rating questionnaires), the foundational principle of designing HCI systems relies on AI-powered user-centric design that encompasses responsible AI to respect human values and ethics. In this respect, let's try to understand how **digital phenotyping** operates to recognize personal sensing from captured metadata.

Digital Phenotyping

This method is used to infer the cognitive traits and affective states of a user to correctly deduce the mental state of the user and provide the right mental health diagnosis. The most commonly used techniques involve a psychometric assessment which works in combination with the monitoring apps (mental health) on smartphones, or sensor devices or wearables to track user swipes, taps, and keystroke events. These captured events through 24×7 monitoring apps, when aggregated with text, speech, and voice data, provide digital biomarkers for cognitive function to conclude the risk state, progress of the disease, recovery, or relapse.

Along with user-centric design comes security and privacy issues of data covered in Chapter 12, "Generative AI in Mental Well-Being: Balancing Pros and Cons." This can help in establishing trust, control, and collaboration to enable better use of the collected data and usage of the data in a safe and secure manner.

Immersive Technologies

Immersive technologies create an exciting experience for the users by combining the physical world with digital simulated reality, where video games, **augmented reality (AR)**, and **virtual reality (VR)** deserve special mention. One of the best ways to adjunct psychotherapy to teenage and adolescent students is to engage them on a video gaming platform to assess clinical intervention for mental conditions including schizophrenia, anxiety disorders, and **attention deficit hyperactivity disorders (ADHD)** (Bhoopathi & Sheoran, 2006). These platforms leverage increased HCI through emotion recognition by capturing speech and audio data. Such platforms come with recommendations to help youth with diagnosed eating disorders as well as prompting alerts to control overexcitement and movements that demonstrate absent-mindedness or hyperactiveness (Lukka & Palva, 2023). One example of the use of audio data can be cited by author C. R. Marmar, where speech-based markers have been used to detect post-traumatic stress disorders in US veterans. The sample audio recordings were collected from veterans exposed to warzones, and on being treated with feature engineering techniques were able to extract features like slower monotonous speech and less change in tonality. Random forest prediction model when applied to the extracted features was able to yield model accuracy of 89.1% with an AUC of 0.954, to detect PTSD (Chung & Teo, 2022).

A research study was conducted at Northwestern university researchers on patient-clinician interactions data for high-risk suicidal patients admitted to a hospital. The study leveraged Audio/Visual Emotion Challenge (AVEC) challenge German speech dataset, has varied attributes (including language, context, speakers, and recording conditions) that was trained to test the vocal parameters of speech to predict risk of suicide. This study, published in "Speech-based assessment of PTSD in a military population using diverse feature classes," demonstrated PTSD prediction can be done using frame-level, longer-range prosodic and lexical features and passing them to ML classifier models such as Gaussian, decision tree, or neural network classifiers (Mitra et al., 2015).

In addition to speech data, a support vector machine (SVM) can also aid in diagnosing PTSD when cortical and subcortical imaging data have been collected from war veterans to identify different kinds of early stress conditions in their lives. This model relies on the use of the posterior **cingulate cortex (PCC)**, one of the regions of the cortex, as one of the important features in modeling the data. PCC exhibits high levels of metabolic consumption, which play a determining role in processing human emotions and regulating behaviors.

Gaming Technologies, AR/VR

Interactive gaming techniques find a substantial space in improving mental health by shifting attention, lowering anxiety and depression, and providing users with a platform to concentrate on other things and forget unpleasant happenings (Lukka & Palva, 2023). Military veterans often use gaming to distract themselves, which acts as a sort of self-medication. In addition, online games give players an opportunity to socially connect and collaborate among multiple players—family members or peers, through chatting about game scores or other gaming styles. This kind of social interaction by means of digital interventions boosts social affinity among people and keeps high engagement when compared to other therapeutic interventions.

In the space of VR, gaming methodologies can successfully decrease anxiety levels and increase positive emotions, by engaging them in healthy breathing practices to cope with trauma-related chronic stress. For example, one of the VR gaming platforms provides teenagers with controllers to navigate the game, in a scene underwater, with a changing difficulty level, by assisting the gamers with alarms which prevent them from infringing into dangerous areas (ABC News, 2016). This works on the principle of biofeedback loop which echoes back breathing patterns, heart rate variability, or sweating patterns to the user to help them recognize psychophysiological processes and enable them to control bodily responses for different levels of stress or mental fluctuations. The role of the biofeedback (audio/visual) loop lies in providing diversity in input signals through **electroencephalography (EEG),** indoor positioning, and **electro-dermal activity (EDA)** among others to harness richness in outputs to help participants identify their body movements and patterns and control them to alleviate stress.

Similarly, the eye-tracking device in video games is an innovative HCI integration to capture psychological symptoms or disorders by recording different levels of visual attention for the user.

Some of the key design aspects of HCI are described below:

- The target audience includes the literate, illiterate, and those with reading difficulties to promote better engagement and ease of user experience.

- Measure different levels of attention for age-diverse teenage and youth populations by offering different types of genres and popularity levels of video games.

- Sufficient space and relaxation environment to engage an audience in virtual reality (VR)-based games.

- Enable individuals to react and communicate in a computer-generated environment and assist them with better and faster service.

- Enforcing quality, safety, and usability of the platform by adding human-in-the-loop in the environment of computer-simulated algorithms to adjust and tune the performance of predictive algorithms.

- Embedding guided digital interventions, phenotyping, and immersive technologies in an AI-driven secure, trusted multimodal digital platform.

The prime advantage of combining psychology and neuroscience through HCI is that it lets the users communicate and respond faster in a supported environment of digital technologies with interventions and feedback wherever needed. For example, in scenarios where CBT has failed to provide help to patients, compassion-focused therapy and AI-driven emotion regulation techniques can be introduced. More creative methods, ideation, and exploration in the HCI design space can be leveraged to reap its full potential.

AI-driven gaming practices/AR/VR come with its challenge when tracking technologies unfavorably impact the youth by providing negative emotional or behavioral patterns (Lukka & Palva, 2023). This exhibits a negative impact on mental ill health symptoms and makes them susceptible to more disappointment, guilt, stress, and embarrassment when they visualize or read about their personal data. To overcome the negative experiences and scenarios, HCI must be equipped with intervention methodologies and special care for impacted students, in a way that the negative data does not act as a tool to demotivate the students suffering from ill health; instead, it serves as a recommendation for remedial actions and behavior changes to drive them to positive sides of mental well-being.

Limitations of Gaming

The use of gaming technologies on mental health also demonstrates an effect on stress outcomes, more so for violent games. It is often found that players engaged in fighting in the gaming process are found to have higher cardiovascular stress response—where blood pressure increases and heart rate variations decrease. These findings demand more careful analysis to evaluate the risk of long-term health issues like chronic stress vs. the likelihood of displaying higher positive emotions among frequent players.

Addressing Mental Health Risks Through Data Collection Using Mobile Apps

One of the most common ways of data collection involves using mental health applications that aim variety of psychological disorders with their varied design style and architectural functionality. They serve a wide array of functions to enable ease of habit, usage, and reward-based motivation.

- Empowering youth for self-help and management through training, providing ample creative opportunities for mental health coping and social support

- Improving cognitive abilities by symptom tracking, alerts, and engaging users

- Provisioning clinical care support involving priority-based crisis intervention, diagnosis, first-hand treatment, and post-treatment options

Some examples include coloring, brainteasers, symptom tracking, deep breathing, journaling, and apps that involve daily schedule planning. In addition, there are coping tools that allow them to listen to music or audio clips to train them to control hallucinations related to auditory misalignments. Certain apps and computerized interventions also offer cognitive training practices to deal with remission memory and attention deficiency symptoms that may ultimately result in depression.

The primary purpose of the apps in this area remains concentrated on merging treatment for mental ill health by personalizing the app as much as possible to make it more flexible and targeted to patients at times when they need this the most. The full value of these apps can be realized, when more patient-centered customizations are embedded and feedback from patients is also received.

The distribution functionality for apps can be summarized as follows (Bakker et al., 2016):

- Psychoeducation (41%)

- Goal setting (38%)

- Mindfulness (38%)

- Surveys (45%)

- Dairy entries (34%)

- Microphones (21%)

- Tobacco usage (33%)

- Stress and anxiety management (28%)

- Mood disorders (20%)

These apps aim to handle challenges with respect to driving more engagement among users by zooming in on a single kind of mental health disorder in each app. This aims to drive in user participation through real-time interactions, reporting usage status reports, and gamified engagements. Additionally, it also emphasizes regulating data quality and privacy standards.

Reviews from mental health practitioners suggest certain best practices for designing apps that can increase active user participation by incorporating the following features to remediate and identify symptoms:

- Cognitive behavioral therapy (CBT) to control anxiety and mood disorders

- Increased accessibility options for usage by nonclinical segments of customers and care providers

- Equip automated tailored interventions to control risks

- Suggestions and activity set proposals concerning reports of thoughts, feelings, or behaviors

- Reports with regard to mental health states to increase mental health literacy

- Suggested activities linked to anxiety and mood disorders by identifying cause-and-effect relationship

- Logging app usage data

- App reminders or notifications to keep engagement and attention live

- Simple use-to-use app interfaces and interactions

- Provide help and support by embedding links to healthcare or support centers

- Experimental trial programs to boost efficiency

- Gamification involving reward-based programs and motivation to engage the audience by boosting positive reinforcement

- Promoting nontechnology-based activities to deal with problems related to mindfulness

However, there exist certain limitations when we see app makers driven toward profit-making become less interested to design apps for more concerning mental health problems like suicides or life-threatening mental health problems. Even so, some apps are designed to follow a specific therapeutic approach and fail to establish more confidence in the underlying principles of therapy. To address this issue, more evidence of the apps' underlying principle toward targeted therapies needs to be established to substantiate its relevance (Bakker et al., 2016). Further, multicomponent interventions in mental health app design also remain an active subject of research and experimentation.

Let's dive in to see some examples of how app designs might help different patients suffering from different problems.

Simple User Interface

Seamless user experiences built on advanced technology-driven simple intuitive UIs detect fast behavior changes for patients coping with depression or anxiety or situations where they feel their memory is impaired (Chandrashekar, 2018). A simple interactive UI can reduce cognitive load and can ease the mental activity stress on working memory. Hence, to boost learning capacities, such UIs should be designed with more embedded pictures than text, where sentence length is not only reduced but also has increased support of inclusive nonclinical languages.

Transdiagnostic Capabilities

Transdiagnostic abilities are better harnessed when mental health apps are coming up with methods to treat symptoms and psychological disorders that appear to be comorbid (Bakker et al., 2016). Such interventions are of higher order during delivery times where transdiagnostic apps can play a vital role in increasing patient engagement by providing an opportunity to interact with one app with better treatment plans for common symptoms. The objective remains to increase accessibility options and identify issues related to comorbidity and depression.

Self-monitoring Features

Apps providing features to self-monitor user behaviors, actions, and thinking patterns can substantially increase emotional **self-awareness (ESA)** and self-reflection by detecting symptoms related to abuse, depression, or anxiety (Cara et al., 2023). Reports from the app's logging and visualization dashboards can play an important role in reducing mental health problems and boosting coping power.

To understand how typically a journaling app works and can be used by teens or adults to record daily information from time to time, let us look at Figure 4-5.

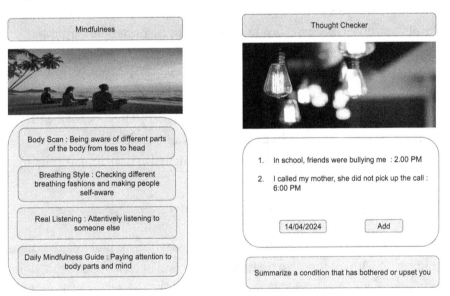

Figure 4-5. *Apps to boost mindfulness and record thoughts*

The above figure shows us how mental health apps can guide us with recording mindfulness and recording thoughts.

After having understood how a typical app journaling process looks like, let us look at some commonly available mental health apps in the market and their role in recording symptoms and providing suggestions to users.

Table 4-2. *Mental health app functionality (Bakker et al., 2016; Channel 3000 / News 3 Now, 2022; Chiauzzi & Newell, 2019; Hines, 2024)*

SI. no.	App name	Primary objective	Functionality
1	5 Minute Journal App	Daily reflection and mental wellness	Showcases gratitude to positively impact multiple areas of life, like personal growth, emotional intelligence, or meaningful relationships
			Track emotions, mood, and goals and provide export feature to export data to PDF, HTML, Dropbox, etc.
2	Alan Mind Daily Journaling App	Cognitive behavioral therapy	Promotes a mindful lifestyle during stressful periods by reducing specific anxieties. Even known for increasing focus and control. Records are automatically encrypted and provide flexibility to set their own encryption key

(continued)

Table 4-2. (*continued*)

Sl. no.	App name	Primary objective	Functionality
3	CBT Thought Diary App	Cognitive behavioral therapy	Equipped with an AI-powered chatbot to act as the user's interactive mental health assistant, available 24/7. It helps to deal with impostor syndrome, handle fears, and build personal resilience, through audio mindfulness meditations and breathwork
4	DailyBean App	Monthly calendar tracking to track and control mood	Provides a monthly calendar to track mood flow by a visual display that helps to represent the day with colorful icons, pictures, or lines of notes. Allows representation of moods through selected category blocks, which can represent the captured data on a weekly/monthly basis
5	Daylio Journal	User-friendly private journaling	Serves as a fitness goal pal or a mental health coach by providing a gratitude diary and mood tracker. Helps in minute health self-care through mental, emotional, and physical well-being. Further provides statistics on a weekly/monthly/yearly basis to boost productivity by understanding the habits set on the calendar

(*continued*)

Table 4-2. (*continued*)

Sl. no.	App name	Primary objective	Functionality
6	Day One Journal App	Private comprehensive journaling	Enables to form habits, display and stay relevant with streaks, calendar views, and programmable reminders. Equipped with daily journal prompts, tags, favorites, and search filters to facilitate easy search and revisit past memories
7	Jour	Therapeutic-based app	Improve mental well-being by boosting confidence, lowering anxiety and depression
8	Fabulous	Healthy habit-building and connecting people with shared goals	Provides a platform to record thoughts and introspect on progress. Network similar-minded people progressing toward similar types of goals
9	Gratitude Journal	Mindfulness and introspection	Encourage them to build positive outlook on life
10	MindDoc	Personalized mental health caretaker	Poll users three times a day, ask personalized questions, and come up with custom answers
11	Moodnotes	Mood and thought tracker	Provides a platform to record and reflect thoughts, helping users to become more self-aware

(*continued*)

Table 4-2. (*continued*)

Sl. no.	App name	Primary objective	Functionality
12	Reflectly	AI-powered journaling and introspecting app	AI to support the journaling process, understand their thoughts and behavior patterns
13	Tangerine App	Mood tracking, journaling, and recording app	Enables self-care, enhances clarity in thinking, and boosts memory and communication skills
14	VOS Mental Health Journal App	Holistic development platform for mental well-being	24/7 online therapy chat, personalized well-being plan to encourage positive mood. In addition, provide breathing exercises, meditation, inspiring quotes, and affirmations
15	Smiling Mind App	Promotes mental fitness skills for children, teens, young people, and adults	Provides support for meditation sessions in Indigenous Australian languages to reduce stress and anxiety and enhance relationships. Programs are targeted to provide sound sleep and mindful eating, help to develop emotional skills, and build resilience
16	SuperBetter App	Transformative gaming app to deal with real-life challenges and stress management	Encouragement to become happy, brave, and resilient by providing opportunities to challenge oneself. Further, the app enables users to create personal epic wins, adopt secret identities, complete quests, encounter bad people, activate power-ups, and check in with friends or allies

(*continued*)

Table 4-2. (*continued*)

Sl. no.	App name	Primary objective	Functionality
17	BetterHelp Therapy App	Anxiety and depression management with therapies	Messaging, one-to-one communication to scheduling sessions providing personalized support
18	Talkspace Therapy and Support App	Text, audio, and video-based therapy support	Licensed therapist support to guide individuals from depression, anxiety, stress, and PTSD
19	Together by Renee Mental Health App	Audio-based tracking for assistance	AI-based health assistance app that records mental vitals by analyzing voice, melody, vocal track, and vocal fold movement. Assimilates this information along with blood pressure (when the smartphone camera is pointed at the face) to help in early disease diagnosis and intervention
20	BeMe Mental Health App for Teens	Interaction-based app with facilities for coaching	Provides a safe space for teens and young adults to practice different activities and explore their emotions without the fear of judgment

After taking a glimpse through different sorts of mental health apps, now let us understand what makes a patient choose the right app, collect data, visualize the metrics, and improve overall mental well-being. Figure 4-6 illustrates such key selection factors for using an app daily (**Chiauzzi & Newell**, 2019).

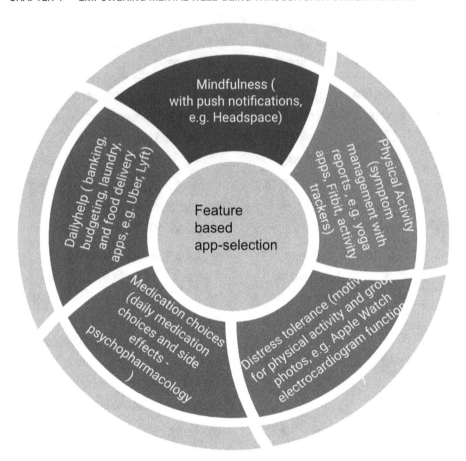

Figure 4-6. *Mental health app selection parameter*

Understanding Mental Health Conditions Through Causation Models

Diagnosis and causal factors leading to mental health problems serve as an important factor in offering treatment and support. CBT often relies on a formulation-based approach where predisposition, precipitation, perpetuation, and protection elements form a solid foundation in analyzing the causal model and doing a **root cause analysis (RCA)** in addressing the disorders. As detection of symptoms becomes inevitable in identifying the best treatment options, it also contributes in best understanding the bigger picture before deep diving into the problem. The main benefit of causation models is to understand the contribution of each of the underlying factors toward the risks of mental health and how alleviating one or more causal factors can improve mental health. For example, it may be seen that restricting screen time by 20% reduces mental health risks in youth by more than 10%.

We can see such an example in Figure 4-7 where different underlying factors as listed below remain responsible for resulting in different mental health diseases.

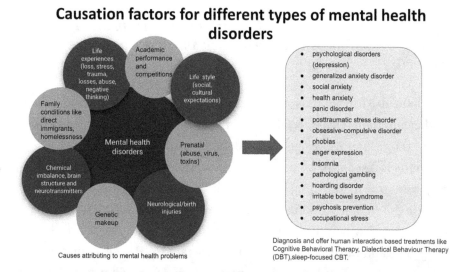

Figure 4-7. Causation factors for different types of mental health disorders

Causation for changes in brain morphology

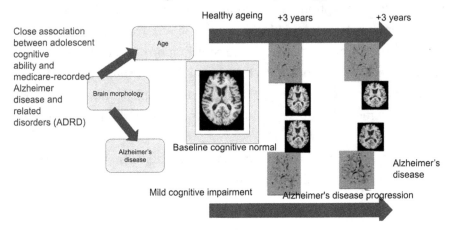

Figure 4-8. *Causation factors leading to a change in brain morphology for normal ageing vs. Alzheimer's disease (Sanchez et al., 2022)*

Figure 4-8 further illustrates the causation model in a specific medical domain where certain principal attributes yield change in brain morphology for normal ageing vs. Alzheimer's disease.

In the last section, we have seen how data aggregation plays an important role in detecting mental health causes and symptoms. To provide the right treatment and address the causal factors, it is important for us to know which segment of the population is suffering from which symptoms and the root cause behind those factors. Hence, to find which age group of the population segment is suffering from what sort of mental disease, let us understand with a practical working example to create two population segments, one segment of people having mental depression and another one not suffering from mental depression.

Studying Mental Health on Population Segments Through Clustering of Mental Health Survey Data

Listing 4-2. This is an example of clustering on mental health survey data

We have taken mental health survey data and perform clustering in an unsupervised manner to extract semantic relations from the dataset. The dataset is based on a 2014 survey that measures the occurrence of mental health disorders and awareness in the workplace.

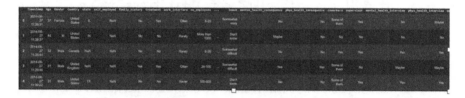

Figure 4-9. *Mental health survey data frame*

1. In the first step, we install the necessary libraries.

   ```
   %pip install pandas sklearn
   ```

2. After installation, we import the necessary libraries.

   ```
   ## Imports
   import pandas as pd
   from sklearn.preprocessing import LabelEncoder
   import seaborn as sns
   import matplotlib.pyplot as plt
   from sklearn.cluster import KMeans
   ```

119

3. In the next step, we load the dataset and preprocess
    the data using label encoding. Since there are a lot
    of categorical variables, we first convert the text
    classes into numerical classes which can be fed
    into the model using the label encoder of sklearn.
    This is very important since most of the machine
    learning models can't comprehend the text labels
    and they have to be mandatorily converted into
    integer labels.

```
df = pd.read_csv('survey.csv')
df = df.apply(LabelEncoder().fit_transform)
print(sns.heatmap(df.corr()))
```

This yields a correlation of different features which
can be seen as below. It shows treatment and family
history are closely correlated, whereas mental_
health_consequences are not correlated to family
history or treatment plans.

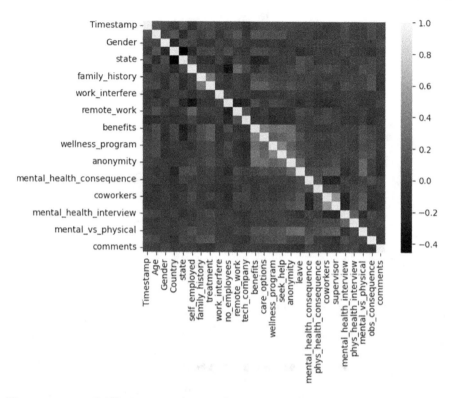

Figure 4-10. *Different attributes that could lead to mental health problems*

4. On the preprocessed dataset, we need to apply unsupervised clustering K-means as given below. K-means is one of the most popular and widely used clustering algorithms. We use K-means to sort the data into two clusters based on whether there is a mental health disorder or not.

```
kmeans = KMeans(n_clusters=2, random_state=0,
n_init="auto").fit(df)
df['cluster']=kmeans.labels_
```

5. Finally, we plot the results and visualize the output
    of people with age and self-employment into two
    different clusters.

```
plt.scatter(df['Age'],df['self_employed'],
c=df['cluster'])
plt.xlabel('Age')
plt.ylabel('self_employed')
plt.show()
```

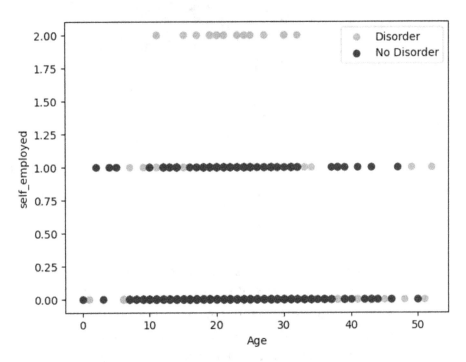

Figure 4-11. *Mental health survey data demonstrating two clusters—*
with or without mental health disorder

Studying Aggregated Data Sources Comprising of Various Linguistic, Psychological, and Behavioral Attributes Can Play a Role in Suicide Detection

Listing 4-3. This is an example of how an ML model can be used to predict suicide rates due to socioeconomic factors

In this part, we take a mental health dataset data and growing stress prediction. This dataset comprises various linguistic, psychological, and behavioral attributes of a very huge population. Nowadays, predictive modeling is very important, and using such extensive datasets, we can find similarities between attributes and perform predictions on various mental health issues including suicidal rates which are directly related to growing stress in individuals. So here, we use this dataset to predict the growing stress in individuals. The dataset comprises around 300k data points, and it contains attributes like gender, country, occupation, self-employment, family history, treatment availed, number of days spent indoors, growing stress, habit changes, mental health family history, mood swings, coping strategies, interest toward work, social weaknesses, mental health interview, and care options availed.

This task can be viewed as a tabular classification task where we can leverage tabular techniques like decision models. We use the labelencoder to convert the string labels into integer classes so that they can be fed into the models for training. Then, we move on to build our model for predictions. We make use of one of the most popular ensemble tree models, that is, random forest model which gives us very good results with very high accuracy.

1. In the first step, we install the necessary libraries.

    ```
    %pip install pandas sklearn
    ```

2. After installation, we import the necessary libraries.

```
import pandas as pd
from sklearn.preprocessing import OneHotEncoder,
LabelEncoder
from sklearn.ensemble import RandomForestClassifier
```

3. Then, we load the mentalhealth dataset.

```
df = pd.read_csv('/content/MentalHealthDataset.csv')
X= df.drop(columns=['Growing_Stress'],axis=1)
y= df['Growing_Stress']
```

4. After loading the data, our next step is to label
    encode the categorical variables.

```
X = X.apply(LabelEncoder().fit_transform)
le = LabelEncoder()
y = le.fit_transform(y)
```

5. Our next action is to train the data by creating a
    model from RandomForestClassifier.

```
clf = RandomForestClassifier()
clf.fit(X, y)
clf.score(X,y)
```

This yields predicted output as 2, as shown in Figure 4-12. It means
growing stress is present when we have contributing factors like mood
swings, mental health history, and social weakness, among others.

```
Timestamp                    39
Gender                        0
Country                      34
Occupation                    1
self_employed                 2
family_history                0
treatment                     1
Days_Indoors                  0
Changes_Habits                1
Mental_Health_History         2
Mood_Swings                   2
Coping_Struggles              0
Work_Interest                 1
Social_Weakness               2
mental_health_interview       1
care_options                  1
Name: 0, dtype: int64
#############
Prediction:   [2]
```

Figure 4-12. *Mental health stress predictions from socioeconomic factors*

Let us now summarize what we have learned in this chapter below.

Conclusion

In this chapter, we have learned about different data sources that can be aggregated to identify mental health risks. We have analyzed the most important data elements be it social media, sensor data, location data, or gaming data which can be collected from different mobile apps, to offer personalized treatment plans, just-in-time intervention options, CBT, and other therapeutic support to teenagers, youth, and young adults. We have gained a deeper understanding of the best styles and design techniques of mobile apps to gather user data through journaling and audio or video recordings. In addition, from this chapter, we get to know how sensor data from gaming apps can play a vital role in linking physical recordings of data in identifying mental health problems. On the darker side of it, we have seen the limitations of gaming/AR/VR technologies to capture data from individuals.

Now, let us delve into the next chapter to gain a deeper understanding of linking the captured data systems using modern digital technology to reshape youth mental health system design.

References

ABC News (Director). (2016, October 1). *Treating PTSD With Virtual Reality Therapy: A Way to Heal Trauma* [Video recording]. https://www.youtube.com/watch?v=QCCWH_CNjMO

Mental Health Smartphone Apps: Review and Evidence-Based Recommendations for Future Developments. *JMIR Mental Health, 3*(1). https://doi.org/10.2196/mental.4984

Balcombe, L., & De Leo, D. (2022). Human-Computer Interaction in Digital Mental Health. *Informatics, 9*(1), Article 1. https://doi.org/10.3390/informatics9010014

Bhoopathi, P. S., & Sheoran, R. (2006). Educational games for mental health professionals. *The Cochrane Database of Systematic Reviews, 2006*(2), CD001471. https://doi.org/10.1002/14651858.CD001471.pub2

Cara, N. H. D., Maggio, V., Davis, O. S. P., & Haworth, C. M. A. (2023). Methodologies for Monitoring Mental Health on Twitter: Systematic Review. *Journal of Medical Internet Research, 25*(1), e42734. https://doi.org/10.2196/42734

Chandrashekar, P. (2018). Do mental health mobile apps work: Evidence and recommendations for designing high-efficacy mental health mobile apps. *mHealth, 4*, 6. https://doi.org/10.21037/mhealth.2018.03.02

Channel 3000 / News 3 Now (Director). (2022, March 19). *VR game to help kids with mental health* [Video recording]. https://www.youtube.com/watch?v=NNfHYW20xnk

Chiauzzi, E., & Newell, A. (2019). Mental Health Apps in Psychiatric Treatment: A Patient Perspective on Real World Technology Usage. *JMIR Mental Health, 6*(4), e12292. https://doi.org/10.2196/12292

Chung, J., & Teo, J. (2022). Mental Health Prediction Using Machine Learning: Taxonomy, Applications, and Challenges. *Applied Computational Intelligence and Soft Computing, 2022*(1), 9970363. https://doi.org/10.1155/2022/9970363

Gordon, D. (2022). *Using A Mental Health App? New Study Says Your Data May Be Shared.* Forbes. https://www.forbes.com/sites/debgordon/2022/12/29/using-a-mental-health-app-new-study-says-your-data-may-be-shared/

Hines, K. (2024). *18 Best Journal Apps for Therapy and Mental Wellness.* https://kristihines.com/best-journal-apps-for-therapy/

Huckvale, K., Nicholas, J., Torous, J., & Larsen, M. E. (2020). Smartphone apps for the treatment of mental health conditions: Status and considerations. *Current Opinion in Psychology, 36*, 65–70. https://doi.org/10.1016/j.copsyc.2020.04.008

Lukka, L., & Palva, J. M. (2023). The Development of Game-Based Digital Mental Health Interventions: Bridging the Paradigms of Health Care and Entertainment. *JMIR Serious Games, 11*, e42173. https://doi.org/10.2196/42173

Mitra, V., Shriberg, E., Vergyri, D., Knoth, B., & Salomon, R. M. (2015). Cross-corpus depression prediction from speech. *2015 IEEE International Conference on Acoustics, Speech and Signal Processing (ICASSP)*, 4769–4773. https://doi.org/10.1109/ICASSP.2015.7178876

Nahum-Shani, I., Smith, S. N., Spring, B. J., Collins, L. M., Witkiewitz, K., Tewari, A., & Murphy, S. A. (2018). Just-in-time adaptive interventions (JITAIs) in mobile health: Key components and design principles for ongoing health behavior support. *Annals of Behavioral Medicine, 52*(6), 446–462. https://doi.org/10.1007/s12160-016-9830-8

Sanchez, P., Voisey, J. P., Xia, T., Watson, H. I., O'Neil, A. Q., & Tsaftaris, S. A. (2022). Causal machine learning for healthcare and precision medicine. *Royal Society Open Science*, 9(8), 220638. https://doi.org/10.1098/rsos.220638

Shin, J., & Bae, S. M. (2023). A Systematic Review of Location Data for Depression Prediction. *International Journal of Environmental Research and Public Health*, 20(11), 5984. https://doi.org/10.3390/ijerph20115984

Strojny, P., Żuber, M., & Strojny, A. (2024). The interplay between mental health and dosage for gaming disorder risk: A brief report. *Scientific Reports*, 14(1), 1257. https://doi.org/10.1038/s41598-024-51568-9

Capturing and Linking Data Systems Using Digital Technology to Redesign Youth Mental Health 360 Services

This chapter primarily discusses how to link different sources of data to better infer youth's or adolescent's behavioral and psychological patterns and support them with varying levels of service. This ranges from different stages of treatment, processing insurance claims to assisting individuals and their parents with assistive services through the use of knowledge graphs (KGs). This chapter lays a solid foundation of using knowledge graphs (like Neo4J) which play a prominent role in interlinking descriptions of concepts, entities, relationships, and events. This chapter helps us to dig a step deeper to decipher the relationship between different

© Sharmistha Chatterjee, Azadeh Dindarian, Usha Rengaraju 2025
S. Chatterjee et al., *Revolutionizing Youth Mental Health with Ethical AI*,
https://doi.org/10.1007/979-8-8688-1186-9_5

nutrients and medications, understand how **natural language processing (NLP)** plays a key role in extracting **named entity recognition (NER)**, and derive decisions for further medications and treatments. This chapter further emphasizes on identifying similar behavioral patterns in youth and offering them existing treatment plans and recommendations that work well in the community. With regard to designing a faster and more efficient youth care 360-degree service, this chapter familiarizes readers with knowledge graphs, ontologies, and their importance in the end-to-end pipeline right from prediction, to taking action to complete diagnosis, cure, and rehabilitation process. By equipping youth, parents, and doctors with a framework, this chapter provides a solid foundation where readers will get to understand how data and AI using **knowledge graph (KG)** can assist in early detection and intervention.

In this chapter, these topics will be covered in the following sections:

- Efficiency of knowledge graphs in the medical domain

- Customer 360 with knowledge graphs

- Designing conversational agents with knowledge graphs

- Designing explainable knowledge graphs

Efficiency of Knowledge Graphs in the Medical Domain

The COVID-19 pandemic has shifted our attention to addressing different mental health issues by effectively representing the aggregated data from different sources to help identify the root cause of the disease and provide treatment. In this context, KGs deserve special mention as they come with the huge capability of linking a semantic network of entities and concepts where it can carve out semantic relationships from mental healthcare data.

The representation languages like **Resource Description Framework (RDF)** and RDF Schema provide a framework to construct knowledge graphs from multiple knowledge-based applications in medicine. KGs rightly justify any decision derived from the nodes and edges by explaining well the relations with the available data. Popular examples of medical knowledge graphs include creating graphs in specialized domains such as a graph which includes all proteins (UniProt), known disease-gene associations (DisGeNet), and unified representation of drugs (DrugBank), synthesizing information to represent drug-drug interactions (Sider) (Chandak et al., 2023). The above stated massively integrated medical domain-specific knowledge graphs depict relationship between different entities like genes, diseases, and drugs built on the foundation of knowledge and data such as Bio2RDF1 and LinkedLifeData2, containing large volumes of semantic-ready biomedical datasets. Such massive knowledge sources help to expedite multidisciplinary exploration in life sciences and healthcare.

The potential of knowledge graphs being well understood by mental healthcare business communities has often been a proven choice in the medical community over vector databases using RAG applications. Users find it more compelling to understand KGs than using generative AI solutions, as KGs can stitch the story together between the different entities, and unlike simple LLM solutions, refrain from producing hallucinations, GenAI solutions built on knowledge graphs bring about unpredictability in the usage of generative AI models specifically LLMs. The power of transparency and enabling healthcare businesses to support multiple tasks like medical decision-making, literature retrieval, determining healthcare quality indicators, comorbidity analysis, and many others have proved the efficiency of KGs in medical health. In addition, KGs offer added flexibility by adjusting the data layer that grounds our LLM, which in turn increases the confidence and boosts the accuracy of our models.

Using Knowledge Graphs in Mental Health

Even so, in mental health, it becomes important to identify the stage-by-stage mappings of the following flow to connect patients, doctors, care providers, and the insurance system (Bickman, 2020).

- Identify the causal factors that lead to mental health problems like depression or anxiety from patient records

- Indicate mental health quality indicating factors, by analyzing comorbidity and other risk factors

- Providing the right course of treatment once the disease has been identified, with personalized medications

- Medical trials of psychobiotics or newly launched drugs and the efficiency of the trial programs, including options to support analysis of drugs and their interactions

- Availability of care provider and insurance services, how effective tie-ups can help patients in times of need

- Identifying similarities in patients' behavior and offering similar treatment plans

Challenges and Solutions Using Knowledge Graphs

Knowledge graphs have emerged as a powerful tool to represent specific diseases; however, the volume of data becomes a limitation in terms of efficiency and usability. Hence, there have been general recommendations to simplify its usage and extend its application:

- Design disease-centric subgraphs instead of referring to the original graph, for example, designing a smaller knowledge graph to represent anxiety or depression

- Use of disease-specific subgraphs to respond to clinical queries without having a hit in the recall

- Increasing the convenience for doctors (for mental health psychiatric doctors) to discover and understand the entity relationships among various knowledge resources for ease of highly accurate responses

- Integrating different knowledge and data sources such as clinical trials, antidepressants, medical publications, clinical guidelines, and treatment history to increase the knowledge base and authenticity of its usage

Another major challenge in using KGs is assimilating different knowledge resources produced by multiple creators as it leads to a lot of variability and heterogeneity in data formats. In addition, data from certain sources can be present in unstructured or semi-structured formats, with a lot of text which often leads to inconsistency and incorrectness. In order to integrate and synthesize the information, it is necessary to have a common understandable format (referred to as the common information exchange reference model), where we consider the interoperability among the source systems from where the knowledge resources have originated. This would enable us to have an efficient exchange of information between two or more sources. Further RDF and RDF Schema often find less application due to their limited expressiveness for medical knowledge representation.

Hence to overcome the existing challenges, we need to rely on NLP, a tool that is pretrained with medical terminologies and ontologies and is capable of extracting semantic relationships from textual data. Even the use of expressive representation of medical knowledge can help to represent and link medical information through KGs.

The first step before representing it through knowledge graphs is to integrate data sources to extract the entities and relationships. This can be achieved through (Huang et al., 2017)

- **Direct entity identification**: Offline procedure to identify entities with similar names from PubMed IDs and clinical trials.

- **First-hand concept identification**: Understand and evaluate the concept first hand by studying various knowledge resources offline. For example, locate a publication in PubMed and a clinical trial which is denoted with a MeSH term.

- **Semantic representation**: Leverage an NLP tool offline to semantically represent medical text involving identifying key concepts and extracting relationships. This can be done using Xerox's NLP tool XMedlan which facilitates extraction of the concept and relationship by using medical terminologies from SNOMED CT (organized collection of medical terms providing codes, terms, synonyms, and definitions referenced in clinical documentation).

- **Semantic queries**: Apply regular expressions to run semantic queries in online processes to integrate knowledge sources. This process runs for some time to determine linkages and connections among knowledge resources during run time.

Now having understood the sequences involved in knowledge extraction, let us now understand how we can create a DepressionKG to support a real-life use case in a clinical decisioning system. This DepressionKG makes use of **SPARQL (SPARQL Protocol and RDF Query Language)** (Huang et al., 2017) over the KG that can help psychiatric doctors in their day-to-day decision process. The extensive use of a standard query language for RDF (Resource Description Framework) database queries for mental health makes it simpler to discover and retrieve the underlying causal factors for depression.

To do away with manual intervention, it becomes important to design a software-aided mental health support system where queries can be designed using semantic web standards with an understanding of the structure of knowledge graphs. Such a system can be best used by psychiatric doctors to change the parameters in a preconfigured template to make their own customized queries.

Now let us look at a few case studies where we would get to see the application of SPARQL over KGs to find answers for psychiatric doctors

Navigating Case Studies to Demonstrate SPARQL Queries over KGs

In all the use cases demonstrated in the below subsection, we would refer to one patient who is the subject of our study and a psychiatric doctor who tries to find an appropriate treatment option for the patient over a WEBUI, software-aided system equipped with SPARQL (Bickman, 2020; Huang et al., 2017). The following examples demonstrate how a powerful query can ease a doctor's job by combining information from multiple sources, thus saving manual time and effort.

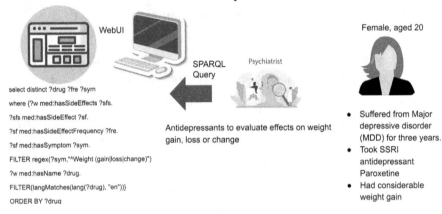

Figure 5-1. *Use case 1 with SPARQL query demonstrating usage of KGs*

In case study 1 as shown in Figure 5-1, we see that to treat a 20-year-old female patient suffering from depression and experiencing weight gain (on taking medications, paroxetine), the doctor executes a query to find the effect of antidepressants on weight gain or loss. By executing the query, the doctor retrieves the results, which suggest bupropion may lead to weight loss, whereas fluoxetine may lead to a modest weight loss.

SPARQL Query over DepressionKG to search for a clinical trial which investigates effect of Fluoxetine

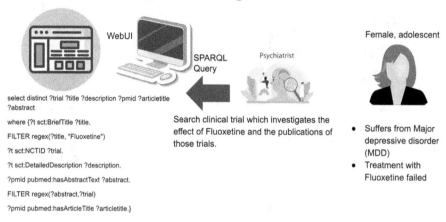

Figure 5-2. *Use case 2 with SPARQL query demonstrating usage of KGs*

In case study 2 as shown in Figure 5-2, we see a female adolescent with major depression was first treated with fluoxetine. Hence, the psychiatric doctor tries to evaluate the results of clinical trials that study the impact of fluoxetine and the publications of those trials. On executing the query, three relevant trial details were revealed with their reference publications.

In case study 3 as shown in Figure 5-3, we study a use case of an adult male who is suffering from a mood disorder and wants to take part in a clinical trial on mental depression. His psychiatric doctor searches for current trials that study the impact of drug intervention on neurotransmitter transporter activity, by referring to DrugBank and ClinicalTrial. On executing the query, trial the doctor can find 25 trials with specified conditions from 2016.

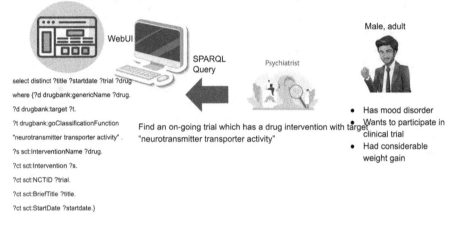

Figure 5-3. *Use case 3 with SPARQL query demonstrating usage of KGs*

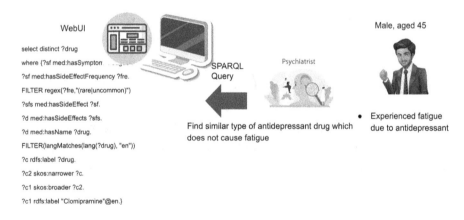

Figure 5-4. *Use case 4 with SPARQL query demonstrating usage of KGs*

In case study 4 as shown in Figure 5-4, we study a use case of an adult male, aged 45, who was under medication with the antidepressant clomipramine and had experienced a lot of fatigue. This has led the psychiatric doctor to evaluate similar kinds of antidepressants which do not cause the side effect of fatigue. The query below uses the predicate skos: narrower and skos: broader to search over the SNOMED CT (Systematized Nomenclature of Medicine–Clinical Terminology) consisting of a hierarchical based repository of medical terms used in healthcare. On executing the search, we can locate two sibling concepts (i.e., two antidepressants of the same class), one of which is "Dosulepin." rdfs: subClassOf is another predicate that can replace predicates skos: narrower and skos: broader to support built-in reasoning.

Prior research studies indicate medical knowledge graphs can be auto-generated and can be used for retrieving patient's records. As various knowledge resources can be integrated to construct depression KG, let us study a few examples of how we can design **Qualified Medical Knowledge Graph (QMKG)** (Goodwin & Harabagiu, 2013). QMKG is KG, created with vertices and edges where vertices act as triples of the form [lexical medical concept, medical concept type, assertion] (e.g., "atrial fibrillation," "EXISTING PROBLEM," "UNDER TREATMENT/MEDICATION") and weighted edges that functions as a bond between the connecting vertices weighted by their binding strength and dependency as specified in the EMR. Overall, the KG should be designed with semantic interoperability, which can be evaluated by psychiatric doctors in real-life situations and can have a few rounds of iterations for improvement.

Figure 5-5 demonstrates a query expansion procedure of a QMKG which has been generated from textual data of **electronic medical records (EMRs)**. As illustrated in the figure below, different orders or mental depression can be taken as a theme which is mapped to the parent node, to analyze the existing problems, problems which have been addressed with medication.

Query Expansion of a Qualified Medical Knowledge Graph

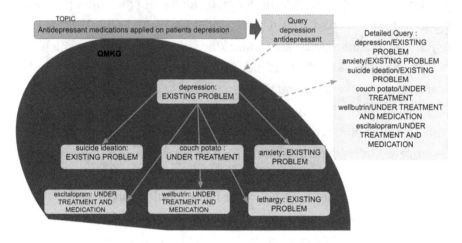

Figure 5-5. *Depression and antidepressant mapping in QMKG (Goodwin & Harabagiu, 2013)*

We have learned how queries can be expanded or executed on medical KGs. Now let us understand how to create medical KGs from scratch from varied knowledge resources.

Concepts Identification and Assertion Classification to create Medical Knowledge Graphs

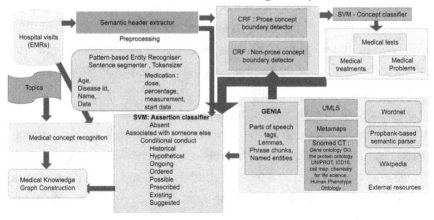

Figure 5-6. *Creation of qualified medical knowledge graphs (Goodwin & Harabagiu, 2013)*

140

Figure 5-6 illustrates the process of EMR identification and semantic header extraction to create medical KGs through a sequence of steps. The primary objective of the above figure is to retrieve data from patient cohorts specific to pertinent medical topics as mentioned in Text Retrieval Conference (TRECMed) and interpret the knowledge base from patients' hospital visits. As our study focuses on identifying and encoding different forms of medical knowledge, we need to rely on how efficiently natural language has been used to encode the different topics, for example, leveraging lexico-semantic medical knowledge encoding forms like Unified Medical Language System (UMLS). Often times, MetaMap and other open source software are able to parse the topics and EMRs to create specific items in UMLS through the process of defining unique identifiers. Other external sources like NLP, Genia annotations used for Biomedical text mining, Snomed CT, Wordnet, Wiki, and propbank-based semantic parser can also be leveraged and fed as input to classify the assertion classifiers. Lexico-semantic information, which helps in extracting the lemmas and multiword expressions, is mined from Wordnet.

The figure further illustrates the process of automatic assertion using an SVM classifier on retrieved medical terms either through means of automatic identification or by considering values which affect the course of disease identification, ongoing treatment options, conditional situations, improvements, or influence of past history of the patient. Medical concept recognition is one of the primal factors in constructing the graphs. This recognition helps to

- Identify medical problems such as irregular heartbeat due to mental depression

- Treatment options such as prescribed antidepressants

- Tests like psychometric tests to identify behavior and root cause of the disease

Medical concept recognition is associated with twofold process stated below:

- Evaluate the boundaries within the text that states a key medical concept

- Classify the medical concept to medical problems, treatments, or tests

To start with medical concepts recognized in EMRs either through narrative reports or those available within structured fields, are trained with two different classifiers, Conditional Random Fields (CRF) as illustrated in the figure. It includes the necessary extracts from the documents like TRECMed.

The feature selection follows a greedy selection approach where in each step, the feature that yields the highest score with an existing feature set is selected to produce concept boundary classifiers and concept-type classifiers.

QMKG has medical knowledge or medical concepts encoded within it that helps in classifying the assertion process. It is embedded with large volumes of data and finds better applications in large-scale deployments. To have an in-built automatic assertion, we may use ML classifier algorithms such as the SVM that relies on medical knowledge that helps to decide the assertions, in addition to the section header which has assertions embedded into it. Further to this, there is a large dependency on features provided by UMLS and features extracted from negated statements such as one derived through NegEx negation detection package.

In the overall QMKG generation process, medical concept recognition along with the assertion classifiers is leveraged both on the topics and the EMRs, which influences the edges and the strength of the edges between the vertices. KG becomes equipped with a strong knowledge base through its representation using nodes and edges. Here, nodes correspond to entities and edges map to entity relationships. With the QMKG generation procedure in place, the doctors find it greatly useful to retrieve the query results from the patient cohort.

We learned in this section the efficiency of KGs, the design of powerful queries, and their execution to design an effective retrieval system for doctors. Moreover, we also studied the step-by-step construction process of QMKG from different sources using machine learning algorithms. Now let us learn how KGs can be effectively used in the customer 360 system in the domain of mental health.

Customer 360 with Knowledge Graphs

In this section, let us study the applications of KGs in different spheres of customer's life centered around mental healthcare treatment and services, and their interaction follows a chain of processes. It is the goal of the service and care providers to ensure a smooth and seamless journey for the patients, such as in scenarios where patients avail a limited set of services instead of end to end, they do not experience any pain points in their journey. In the past, it has been difficult to gain a holistic view of patients and provide them with the best mental healthcare services, which has made it even more difficult to recommend patients with medication and behavioral counseling options.

To provide the best possible services, we need to collect the data for all events happening in the patient's diagnosis and treatment process (Stardog, 2023). This will help to build the KGs efficiently and train KGs using neural networks, or **graph neural networks (GNN)** to predict future events with high accuracy.

Figure 5-7 best illustrates different phases of KG creation through knowledge aggregation and how this gained knowledge has varied applications in mental health ecosystem, comprising of doctors, patients, healthcare providers, and the insurance system.

When we aggregate mental health illness information, such as schizophrenia, we tend to integrate data from text and speech where this high-dimensional data needs proper visualization by reducing the dimensions of data to a map. In such use cases, we make use of **self-organizing map (SOM)** networks to represent multidimensional data in much lower-dimensional space before forming the medical entity linkage or summarizing the medical knowledge. This helps us to better understand how the key contributing factors for mental disorders are organized. This is done by characterizing the multivariate mental health data through a semantic map where similar samples are assembled close together and dissimilar samples are placed apart. SOM is a neural network where a two-dimensional grid is needed to represent the featured space, where the complex, nonlinear statistical relationships between high-dimensional data are translated to relationships on a low-dimensional space.

Medical Knowledge Graph Formation and Applications

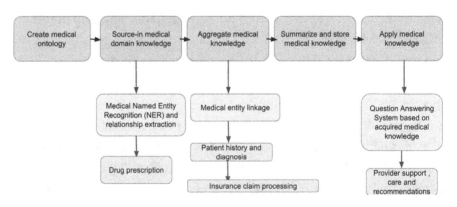

Figure 5-7. *Knowledge graph creation and its application in mental health—a holistic view*

Now let us look at a practical example of how a SOM can be used.

Listing 5-1. This is a comprehensive psychiatric differential diagnosis test. In this example, we take a psychiatric diagnosis dataset from the University of Science and Technology of China cdbd. This dataset consists of scores from 0 to 1 for all the study participants without disclosing their identity. It has more than 185k data points

	user_id	item_id	score
0	1615	12977	1
1	782	13124	0
2	1084	16475	0
3	593	8690	0
4	127	14225	1

Figure 5-8. *View of input dataset, for the below example*

SOM is an unsupervised technique, but like most of artificial neural networks, it operates in two modes, training and mapping.

1. In the first step, we install the necessary libraries.

```
%pip install pandas sklearn matplotlib numpy
```

2. After installation, we import the necessary libraries.

```
import pandas as pd
import numpy as np
from numpy.ma.core import ceil
from scipy.spatial import distance #distance
calculation
from sklearn.preprocessing import
MinMaxScaler  #normalisation
from sklearn.model_selection import train_test_split
```

145

```
from sklearn.metrics import accuracy_score #scoring
from sklearn.metrics import confusion_matrix
import matplotlib.pyplot as plt
from matplotlib import animation, colors
```

3. Then, we load the mental health dataset.

```
df = pd.read_csv('train.csv')
```

4. Our next step is to obtain the train and test datasets
    and apply normalization on the train dataset.

```
train_x, test_x, train_y, test_y = train_test_split
(df.iloc[:,:2].values, df.iloc[:,2].values, test_
size=0.2, random_state=42)
def minmax_scaler(data):
  scaler = MinMaxScaler()
  scaled = scaler.fit_transform(data)
  return scaled
train_x_norm = minmax_scaler(train_x)
```

5. We set the hyperparameters for tuning the model
    and evaluating it for the best set of parameters.

```
num_rows = 10
num_cols = 10
max_m_dsitance = 4
max_learning_rate = 0.5
max_steps = int(7.5*10e3)
```

6. With the hyperparameters set, our next step is to
    build the SOM model. We also define multiple
    helper functions that the model needs to access.

```python
# Euclidean distance
def e_distance(x,y):
  return distance.euclidean(x,y)
# Manhattan distance
def m_distance(x,y):
  return distance.cityblock(x,y)
# Best Matching Unit search
def winning_neuron(data, t, som, num_rows, num_cols):
  winner = [0,0]
  shortest_distance = np.sqrt(data.shape[1])
  # initialise with max distance
  input_data = data[t]
  for row in range(num_rows):
    for col in range(num_cols):
      distance = e_distance(som[row][col], data[t])
      if distance < shortest_distance:
        shortest_distance = distance
        winner = [row,col]
  return winner
# Learning rate and neighbourhood range calculation
def decay(step, max_steps,max_learning_rate,max_m_
dsitance):
  coefficient = 1.0 - (np.float64(step)/max_steps)
  learning_rate = coefficient*max_learning_rate
  neighbourhood_range = ceil(coefficient * max_m_
  dsitance)
  return learning_rate, neighbourhood_range
```

7. Following this, we train the SOM model as follows:

```
num_dims = train_x_norm.shape[1] # numnber of
dimensions in the input data
np.random.seed(40)
som = np.random.random_sample(size=(num_rows, num_cols,
num_dims)) # map construction
# start training iterations
for step in range(max_steps):
  if (step+1) % 1000 == 0:
    print("Iteration: ", step+1) # print out the
    current iteration for every 1k
  learning_rate, neighbourhood_range = decay(step,
  max_steps,max_learning_rate,max_m_dsitance)
  t = np.random.randint(0,high=train_x_norm.shape[0])
  # random index of traing data
  winner = winning_neuron(train_x_norm, t, som,
  num_rows, num_cols)
  for row in range(num_rows):
    for col in range(num_cols):
      if m_distance([row,col],winner) <=
      neighbourhood_range:
        som[row][col] += learning_rate*(train_x_
        norm[t]-som[row][col]) #update
        neighbour's weight
print("SOM training completed")
```

8. When the model is trained, we apply the model
    predictions.

```
label_data = train_y
map = np.empty(shape=(num_rows, num_cols),
dtype=object)
```

```
for row in range(num_rows):
  for col in range(num_cols):
    map[row][col] = [] # empty list to store the label
for t in range(train_x_norm.shape[0]):
  if (t+1) % 1000 == 0:
    print("sample data: ", t+1)
  winner = winning_neuron(train_x_norm, t, som,
  num_rows, num_cols)
  map[winner[0]][winner[1]].append(label_data[t])
  # label of winning neuron
```

9. Our last step is to visualize the predicted results.

```
label_map = np.zeros(shape=(num_rows, num_
cols),dtype=np.int64)
for row in range(num_rows):
  for col in range(num_cols):
    label_list = map[row][col]
    if len(label_list)==0:
      label = 2
    else:
      label = max(label_list, key=label_list.count)
    label_map[row][col] = label
title = ('Iteration ' + str(max_steps))
cmap = colors.ListedColormap(['tab:green', 'tab:red',
'tab:orange'])
plt.imshow(label_map, cmap=cmap)
plt.colorbar()
plt.title(title)
plt.show()
```

We get the following output, which provides a score between 0 and 1 for every data point.

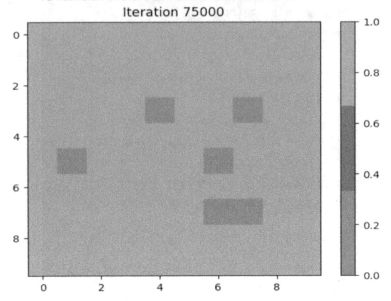

Figure 5-9. *Predicted outputs with probability representation*

As we get an understanding of the disparate information and its similarities, our next job is to extract them (neural network) and feed them to KGs for better evaluation of cause and effect impacts. The first use case we would like to look at is analyzing the role of KGs in understanding how drugs play a role in clinical trials.

First of all, let us gain an overall understanding of what it means to Customer 360 on the subject of mental health. Figure 5-9 explodes the different stages of a patient's journey starting from raising awareness, program consideration, providing access, service delivery, to ongoing care. These different phases of Customer 360 can be further decomposed in terms of three entities: in-patient hospital provider services, online or out-patient counseling or diagnostic services, and insurance services. As you can see in the figure below, each phase involves liking the data at different points, to connect the customer or the patient with different service providers to avail the services.

Patient Care Journey in Mental Health Treatment

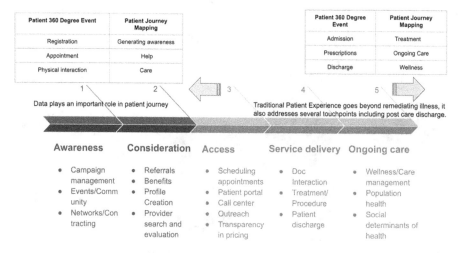

Figure 5-10. *Various stages of a patient's journey for mental health support and treatment (Qualtrics, 2025)*

If we further want to represent different types of treatment options available with varying provider services, we would like to link the information available with KGs as shown in Figure 5-11. As diagnosis is a repetitive step to monitor progress of treatment, we see a loop in the node of the KG. This gives patients a chance to evaluate the treatment options available among various providers and select the appropriate services relevant to their treatment.

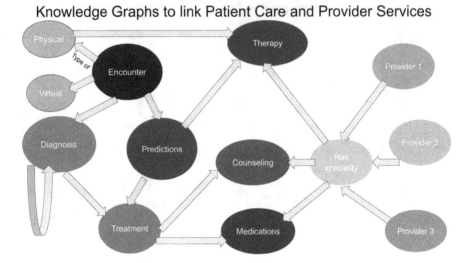

Figure 5-11. *KGs connecting patient diagnosis and treatment options with provider services (Stardog, 2023)*

Analyzing Clinical Trials and Impact of Drugs

An example of extracting relations from DrugBank and DrugBook is illustrated in Figure 5-12. The figure aims to showcase the relationships vividly from large volumes of textual data. In addition, it displays the process of searching and annotating different entities involved with a dashed arrow. Further, it also illustrates the impact and method of conducting clinical trials concerning different drugs on patients suffering from mental depression. This helps us to understand how semantic annotation can be used to extract relationships to infer mental health problems from psychometric tests. At the same time, the figure tries to link concepts after integrating knowledge sources by direct concept identification and connection with a direct arrow link.

Knowledge Graphs to demonstrate Clinical Trials and Impact of Drugs

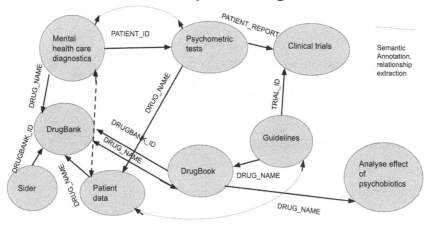

Figure 5-12. *Relationship extraction in KGs with direct concept recognition and semantic annotation*

Let us understand now how we can enable the reuse of mental health questionnaires through the use of KGs.

Designing Psychometrics Ontology Maps Through Questionnaires

To effectively evaluate the youth and teenage section of the population through a set of self-rated or parent-rated questionnaires, we need to design the Revised Children's Anxiety and Depression Scale (RCADS) properly (Santos et al., 2023; The Knowledge Graph Conference, 2023). This will enable the test takers to appropriately rate themselves on different subscales to clearly distinguish the mental health disorder, which can range from separation anxiety disorder, social phobia, generalized anxiety disorder, panic disorder, obsessive-compulsive disorder, and major depressive disorder. In addition, a typical questionnaire should also carry composite scales to infer the total anxiety scale or total internalizing scale through varying rating metrics like "Never," "Sometimes," "Often," and "Always."

Some example sentences in the questionnaire may include

1. I worry about exams always.

2. I feel lonely when I do badly at something as my parents will scold me.

3. I am not good at sports, hence feel my teammates ignore me.

4. I feel worried when I am staying away from my parents.

5. I feel puzzled when I meet a stranger and have to introduce myself.

6. Nothing is much fun anymore.

RCADS Questionnaire Items for Mood Assessment

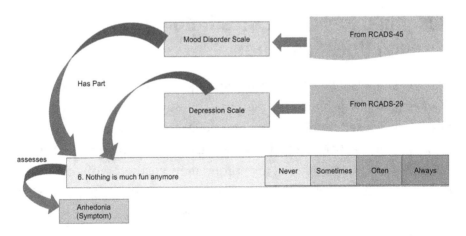

Figure 5-13. *Semantics detailing the questionnaire items on a 4-point rating scale (Santos et al., 2023; The Knowledge Graph Conference, 2023)*

Such a RCADS evaluator can be best described by KGs as illustrated in Figure 5-13. The figure evaluates sentence 6 to assess the symptoms of anhedonia. The rating factor can further be divided into CompositeScale and SyndromeScale to identify a symptom and its intensity.

Mental Functioning Ontology from Psychometric Tests

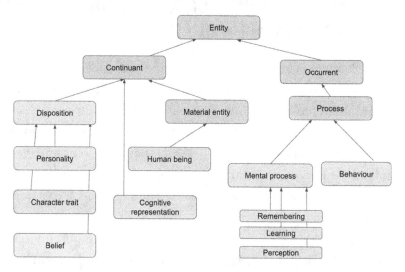

Figure 5-14. *Ontology map to demonstrate the functioning of mental health functions (Larsen & Hastings, 2018)*

This would also help doctors to detect the symptoms associated with and prescribe medications, leading to the development of efficient psychometrics ontology maps (Larsen & Hastings, 2018), that can be fine-tuned on receiving feedback from domain experts as explained in Figure 5-14.

This figure also highlights there are two major divisions of the ontology map—one being the core ontological types and the other covering domain ontology related to mental health dispositions, cognitive representations mental processes, and human behavior.

Now after having explored the effective use of KGs for mental health diagnostic tests, let us study with an example how we can understand the influence of daily dietary habits of patients suffering from mental health-related diseases.

Understanding the Role of Nutrition and Vitamins Through Mental Health KGs

We all know that nutrition plays a key role in overall physical growth and mental development, especially during our childhood and youth. With our intake of food and drinks, we intake different nutrients which leaves an effect on different chemical substances (like hormones and proteins) residing in our body (Dang et al., 2023). These nutrients and biochemicals not only impact our mental health conditions but also assist in nutrition-based therapies therapeutic assessment and treatment for psychological disorders and mental depression.

Here, we would see in depth how KGs play an important role in representing the relationships between nutrition and mental health when used with different entity types such as chemical, biochemical, and disease. It has a wide range of applications in mental well-being starting from dietary supplement knowledge graphs, diet-disease correlation analysis, and personalized dietary recommendations.

In Figure 5-15, we use a KG termed **GENA (Graph of Mental-health and Nutrition Association)**—to define the relationships relevant to mental health and nutrition. This figure concentrates on five different concepts including nutrition, biochemical, mental health, chemical, and disease obtained by extracting information from PubMed articles, using NLP. We can employ a model that runs a two-step process to derive the relationships between the entities. The first step is to identify the entities through **named entity recognition (NER)** and then followed by the second step which infers the relationships through **relation extraction (RE)**.

In Figure 5-15, the sentence "Vitamin C, Ascorbic acid and Vitamin E have beneficial effects on psychotic disorder and improve anxiety" has been interpreted by breaking down the elements and identifying "Vitamin C, Ascorbic acid and Vitamin E as entities and remaining of the sentence as Objects". We also have a relative clause modifier in the sentence—"which is a fat-soluble nutrient."

Further, we need to distinguish minor differences in the meaning of sentences by correct identification of the entities. Our purpose will be to extract the entity along the complete meaningful phrase that the entity defines. For example, if we have two sentences as stated below, then to identify two different entities as "Vitamin C" and "Vitamin C deficiency," we need to have an in-built postprocessing unit for NER.

- Vitamin C is an important vitamin for preventing mental stress.

- Vitamin C deficiency cases of stress-related diseases.

Information Interpretation and Subject Detection

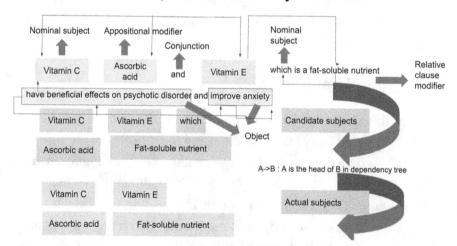

Figure 5-15. *Understand different nutrients and vitamins and their effect on human mental health functions (Dang et al., 2023)*

In addition, we need to analyze the results of syntactic **decision trees (DT)** to bring in added clarity and finer distinctions from model-predicted entities. A DT structure is capable of representing relationships among multiple words in a sentence. This is evident in the above figure where we show directed and labeled arcs from heads to dependents. To explore DT, relations we can leverage Python library spaCy **Dependency Parser (spaCy DP)** to connect two rules: the first rule does the entity discovery to discover the words that are nouns, and the second rule is used for discovering adjectives. We also need to use the DT to identify complete noun phrases (NPs) where the compound words modify the principal noun. In the above example, "*Vitamin C deficiency*," "*deficiency*" is the principal noun (its DEP tag is "*root*"), whereas "*vitamin*" and "*C*" are responsible for modifying the root word (Dang et al., 2023).

In addition, we may also look for verb-based DT patterns, by applying **PAS (predicate-argument structure)** patterns that can aid in extracting relations from biomedical text. It discovers relationship links between entity pairs in a sentence, by identifying grammatical patterns, one of them can be a linkage between the two entities through a predicate (verb). The characterization between the three units can then be stated as (Entity A, predicate, Entity B) (Dang et al., 2023).

Addressing Limitations

However, this method suffers from limitations when verb predicates have a predicate immediately following them and the meaning of the verb changes, based on the nature of the preposition. To differentiate between the two, we can say that "*A, be used with, B*" has a different meaning than "*A, be used for, B*." In addition, predicates are limited by single verbs in PAS patterns which fail to depict complex multiple actions and their relations. Even so, they fall short of drawing out meanings from complex relationships, having more than one **auxiliary noun phrase (aNP)**. To cast an example, we can see here that two relational phrases "*associated*

with behaviors related to" and "*associated*" have different interpretations.
To address the limitations and enable a wider impact in bringing forth
meaningful relations, the most recommended practice is to employ the
use of verb groups where more verbs can be linked with an optional
preposition, or linking verbs together with an adjective with an optional
preposition, following it. Further, we can also specify one or more aNPs,
where an (optional) preposition may follow the auxiliary noun phrase,
leading to multiple aNPs. Examples of that form can be cited respectively
as follows:

- "*Be mediated by combining with*"

- "*Be useful in improving*"

- "*Positive effects on cognition in individuals with*"

To handle the limitations, we need to design efficient DT parsers and
apply a complex multistep processing engine to derive the relationships
between the entities. To illustrate this further, let us see an example
of how DT-based patterns extract phrase dependency and meaning
containing multiple aNPs. The below example signifies the interpretations
of a sentence having complex relationships, between entities. In this
manner, both transitive and symmetric relations between entities can be
populated to KGs.

> *Ayurvedic medicines like Ashwagandha, Brahmi, and
> Jatamansi are considered beneficial for managing depression
> and anxiety, fostering a balance in mental health.*

This type of complex relation extractor can be broken down into
simple parts—("*Ayurvedic medicines- Ashwagandha, Brahmi, and
Jatamansi*", "*beneficial for managing*", "*depression and anxiety, fostering a
balance in mental health*").

The power of KGs is best enhanced with the power of knowledge
sources. We would want to keep the following points in consideration
while designing the KG pipeline, as illustrated in Figure 5-16.

Knowledge Graph representing relation between Nutrition and Mental Health

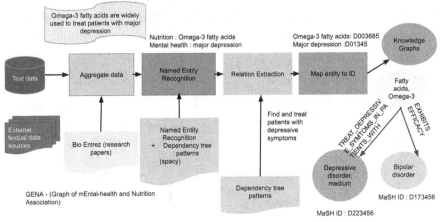

Figure 5-16. *Architectural components and processes to design knowledge graphs from knowledge sources*

1. Enhancing the dictionary matching techniques by integrating medical ontologies, including nutrition terms, food ontology, and biochemical terms. We can leverage primarily three ontologies, that is, APADISORDERS, ASDTTO, and MFOMD, to extract mental health terms.

2. Plugging in the dictionary matching scheme with an NLP pipeline of Python libraries like spaCY.

3. Employ a DT-based postprocessing engine to infer chemical and disease entities that link mental health nutrition.

4. Enrich new relations between nutrition and health entities from BC5CDR and other external corpus, which includes **chemical-induced disease (CID)** relationships and relations that mental health impacts on our bodily substances.

5. Leverage unsupervised ML models or DT pattern-based relationship extractors (verb and noun-based) capable of handling complicated sentences by identifying words that are grammatically related and can extract and explain the relations between them.

6. Normalizing entities (e.g., referring to serotonin as 5-hydroxytryptamine" or "5-HT") to keep the underlying meaning/concept intact while encoding the relationship information in our KGs.

7. Discover a match between entities encountered in reference text and their corresponding **MeSH (Medical Subject Headings)** concept by identifying the synonymous terms.

8. Apply cosine similarity matching for entities having no matching MeSH synonyms, by retrieving the embeddings of each term using the scispaCy model and then applying the similarity matching.

9. Populating the KG by normalizing and mapping each entity to a concept ID, where each edge specifies a unique relationship.

Through the above sequence of steps, we got an in-depth level of understanding how to embed different knowledge sources and DT parsers in a KG creation pipeline. We further got to learn a few new concepts, like normalization, entity matching, and cosine similarity, and their role in designing a comprehensive KG as shown in Figure 5-16's output.

After studying the principal components of the processing engine, we are now at a stage to carve out the multirelational entities involved in disease diagnosis and prescribing medication. Now, with the complex

161

dependency representation in mind, let us look at studying another use case of how multiple contributing factors can be represented with KGs to demonstrate major depression and depression-related symptoms.

Knowledge Graph to offer Medication on Mental Health

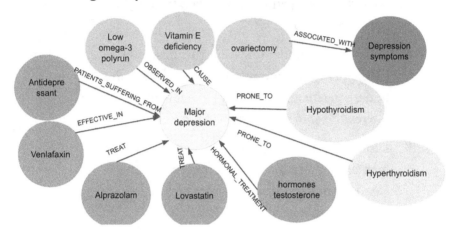

Figure 5-17. *Analyzing the impact of medications and their effectiveness on major depression-related symptoms*

Figure 5-17 further illustrates different antidepressants that can be used for treatments for major depression. It also highlights other causes like hypothyroidism/hyperthyroidism along with vitamin and omega-3 deficiencies that often lead to major depression. We also see that ovariectomy is also associated with depressive symptoms.

We can also bring in causal analysis to study the impact of medications on mental health. Under this scenario, psycho doctors would like to better understand whether the medication is causing any side effects or if there's some other unknown third cause that is leading to the presumed side effects.

Here in Figure 5-18, we represent what type of medications help handle excessive sleep disorders. However, this can be further decomposed to explain better. Fluoxetine was able to lower depression for children and adolescent patients, but administering fluoxetine has shown decreased weight gain. Hence to influence weight gain and have a positive influence on depression, the dose needs to be monitored, intervened on medical needs, and explained through AI causal reasoning and KGs. Similarly, on administering modafinil, any report of dizziness or drowsiness involves revisiting the dosage and intervening to see if it's modafinil that is the main cause behind it.

Causal Impact of Mental Health Medications

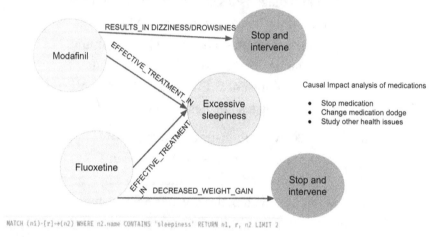

```
MATCH (n1)-[r]->(n2) WHERE n2.name CONTAINS 'sleepiness' RETURN n1, r, n2 LIMIT 2
```

Figure 5-18. *Studying the causal impact of medications for excessive sleepiness*

Improving Accessibility Options for Mental Health

In this subsection, we will learn the importance of KG for "Patient 360" services from the angle of data entities involved like social media sentiment, therapy sessions, and written entries (Ashwin Kumar & Veena Kumar, 2022). By connecting factors responsible for mental health disturbance with actions, we would be able to link patients with suitable care providers. In addition, we can leverage similarity algorithms to design recommender systems that refer patients exhibiting similar behavior to similar care provider services.

Through Figures 5-19 and 5-20, we explain two different use cases of mood analysis and treatment recommendations using KGs.

1. Figure 5-19 captures a variety of data sources for an individual patient, based on the patient's subscription to social media, gaming sites visits to mental health diagnostics centers for therapy. This helps us to extract insights into the individual's mood changes through social media posts, gaming scores, mental health diagnoses, and treatments and efficiently represent a "Customer 360" view of the patient and suggest the best course of treatment.

2. In Figure 5-20, we demonstrate how similarity algorithms of KGs can be used to recommend patients' personalized best treatment plans by comparing them with patients experiencing similar symptoms.

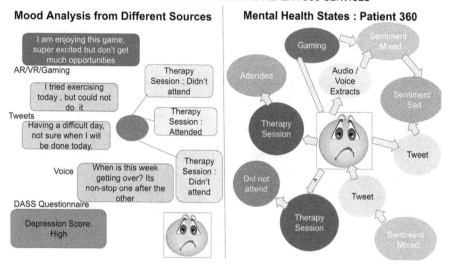

Figure 5-19. *Patient 360 mental health state representation obtained
from different data sources (Ashwin Kumar & Veena Kumar, 2022)*

Figure 5-19 demonstrates a specific use case where data is collected
from multiple avenues of life like social media sentiment, presence in
therapy sessions, and written comments, diaries, or notes. It further
explains how negative reactions/moods from social media, audio extracts,
gaming, and psychological questionnaires can be effective in inferring the
sentiments of users. This kind of KG offers the flexibility of multiple data
points/vertices originating from the same patient and merging to the same
target vertex. This reinforces the fact that a patient having three similar
gaming scores and expressions points to a specific sentiment with a larger
weight than pointing to three different sentiments.

Figure 5-20 goes one step further to compare two similar patients by
analyzing their tweets, gaming reactions, audio extracts, and attendance
in therapy sessions to conclude their mood states. Employing graph
embedding (Fast RP embedding) or Jaccard similarity measures is useful
for comparing a patient's 360 holistic view with other patients. Such
inferences especially for new patients or patients having limited records

in the system can help to discover the mood change transitions, or to evaluate the type of psychological behavior sessions efficiently with KGs and recommend similar treatment plans.

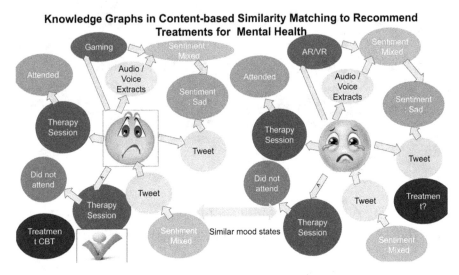

Figure 5-20. Patient 360, similar patient treatment recommendation (Ashwin Kumar & Veena Kumar, 2022)

The similarity problem can be extended to bring in multiple stakeholders from the mental health domain to evaluate similar problems and provide solutions. The stakeholders can include

- Doctors who access psychometric tests, do diagnoses, and provide treatment

- Therapists who provide customized care to patients

- Insurance companies to sell insurance plans and offer claim reimbursement facilities

- Public health programs and awareness campaigns sponsored by the government catered to patients

Redesigning Patient 360 Insurance Claim with KGs

There has been a long debate over including mental health problems in insurance (Tong et al., 2023). Owing to long-drawn discussions debating the importance of it, insurance programs assisting mental health problems have been introduced in many countries. One such example is the Mental Healthcare Act of 2017 in India, which has advocated support for mental well-being and insurance claims. The introduction of insurance support has been able to help in the following areas:

- Assistance to patients during hospitalization covering treatment, diagnostic cost, medication reimbursement, hospital room rent, ambulance charge, and pre- and posthospitalization care for X number of days.

- Insurance claims related to mental disorders like mental depression or anxiety, mood or psychotic disorder, schizophrenia, and other problems related to mental health affecting thinking or memory power, decisioning, and cognitive abilities.

- Outpatient patient counseling (OPD) for mental health, which includes counseling sessions, psychometric testing, and even reimbursement for rehabilitation.

- Insurance claims exclude preexisting diseases before onboarding the patients in the platform which may include mental retardation, recurring mental problems, and issues with mental health due to alcohol or drug.

Knowledge Graphs to demonstrate Insurance Claims related to Mental Health

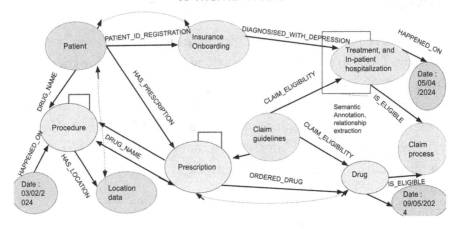

Figure 5-21. *Patient 360, insurance claim processing with eligibility check (Tong et al., 2023)*

In Figure 5-21, we examine, after a patient is onboarded to a mental health insurance network, how the claim is processed based on diagnosis dates, diagnosed disease, prescribed treatments and drugs, hospital admittance, and the patient's eligibility conditions. The KG can further be expanded to capture a sequence of events to reflect a patient's journey with the insurance claims process, to and fro hospital visits, tests availed, and medications purchased within a certain period of hospitalization.

After having the background on the use of KGs in applying medications, let us now study how KGs can be used effectively to design conversational agents.

Designing Conversational Agents with KGs

Mental health responses from conversational agents can be challenging as they lack empathy and care while providing the answers to the audience (Schneider et al., 2023). To design an authentic responsible chatbot, we should take into consideration the intelligence of the system, as well as sufficient knowledge enrichment so that the responses are reliable and aptly suitable for mental health support. Let us learn the sequence of steps involved in designing a knowledge graph, comprising the eventualities from **activities, states, events, and their relations (ASER)** more so the contextual relations to generate responses that do not cause any harm to its audiences. Figure 5-22 here demonstrates how leveraging the ground-truth information from its neighbors along with proper knowledge enrichment in the input sequence is able to improve the KG performance, in terms of output responses (Tong et al., 2023).

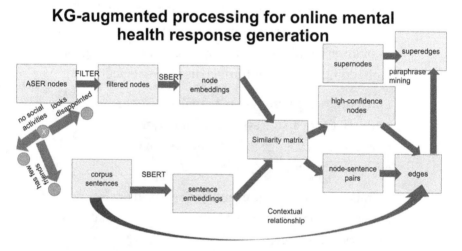

Figure 5-22. *ASER nodes, with NLP to generate KGs for mental health support responses*

Some of the key design principles we should follow are listed below:

- Design the online mental health response generation system by plugging in NLP libraries, embedding processing engines like **SBERT (Sentence Bidirectional Encoder Representations from Transformer)**, and filtering engine to filter out

eventualities with psychological-related keywords
and eliminate stop words and words unrelated to
mental health.

- Evaluate the correct number of relevant ASER nodes
  that can be plugged into the system to enrich our
  knowledge base.

- Incorporating unique dialogues with multiple loops,
  along with social media posts containing diverse topics
  and rich in knowledge sources.

- Implementing efficient similarity scoring evaluation
  pipelines to differentiate high-confidence nodes
  from lower-confidence ones, by carefully evaluating
  embedding similarities through use of SentenceBERT.

- Efficiently design the system to make the right
  trade-off between the system efficiency, the
  amount of knowledge fed into the system vs. the
  computation time.

- Extract and analyze the eventualities based on
  contextual relationships extracted from the textual data
  to connect the edge.

- Use of subgraphs possessing ground-truth neighbors to
  improve the relevance and coherence of the response.
  Incorporate evaluators for adjacent sentence identification
  by comparing their presence in similar documents and by
  validating the number of sentences in between.

- Merge the synonymous nodes through a paraphrase
  mining technique and then applying the same
  process for every pair of nodes, to yield the resultant
  knowledge graph.

- Build and extend the linkage between contextual
  eventualities on a continuous basis to augment the
  knowledge base. Iterate for continuous improvement
  by ensuring the maximum enriched input sequences
  with ground-truth neighbors.

Event-based Knowledge Graphs in Support Response Generation

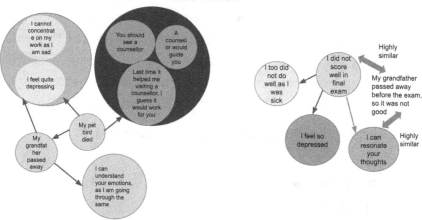

Figure 5-23. *Similar node identification and categorization to generate improved responses (Tong et al., 2023)*

We go one step ahead to visualize, through Figure 5-23, how similar nodes generating similar kind of responses can be grouped together.

In the above figure, we have marked them with the same color orange and green on the left side of the figure. Further, the figure on the right side explains how different negative events of life can lead to depression.

It is now time for us to understand how to architect knowledge graph systems to design a mental health chatbot.

Architecting Knowledge Graph-Based Question Answering Systems

KGs with their knowledge base act as a prominent tool to provide answers to dialogue systems and NLP tasks (Schneider et al., 2023). The power of KGs lies in improving the performance of various dialogue systems through accurate responses, by linking similar/dissimilar semantically rich topics and their contexts. On submission of complex queries, the graph structure follows knowledge discovery on a topic or domain by allowing retrieval of the most relevant and contextual answer. This makes exploratory search, knowledge-based reasoning, and answering systems fast and convenient to take in input queries, along with enhanced response time. The four-step sequences of an exploratory search would involve

- **Understanding**: Speech recognition and semantic matching

- **Dialogue management**: Dialogue tracking and query formulation

- **Entity matching**: Entity matching and query execution

- **Response completion**: Response generation and speech synthesis

During the design of system architecture, we should carefully ensure query interpretation is correct with the right semantic matching algorithms in place. This can kick off the dialogue management system to formulate the queries. Once the queries are in place, we can trigger the entity-matching pipelines to start executing the queries for the final response to the end users.

One of the prime factors for consideration in designing the chatbot is to have an intelligent NLP engine which can classify the type of question being asked, extract and expand useful keywords to derive the entity relationships, and further proceed with semantic analysis. We also need to consider how we feed in unstructured, structured, and semi-structured data from varying data sources, aggregate them, maintain proper lineage, and apply rules before proceeding with knowledge engineering. Once knowledge engineering is completed, we would like to store the intermediate graphs and subgraphs in Neo4j or other graph databases that can be used for online/offline training purposes. In addition, question answering chatbots should be capable of automatically generating visual knowledge graphs. After the KG construction, the chatbot can reply to the question based on the content of knowledge used to build the graph and play an important role in knowledge retrieval and intelligent recommendation.

Further, the graphs stored in the system can be trained using **graph neural networks (GNN)** that helps to boost the performance of the chatbots, by capturing the relationships between different parts of the text.

Figure 5-24 explains the different components involved in architecting an end-to-end question answering system where input questions can be fed through a graphical interface, and the rest of the processing tasks can be passed on to back-end processing units and graph databases. The design of an end-to-end system offers many advantages:

- Providing patients universal assistance with mental health support and services and better diagnosis facilities.

- A well-defined monitoring system with visualization charts helps to monitor the risk associated with the disease and progress in due course of the treatment.

- Early intervention leads to reduced costs on patient expenses, thereby leading to a reduction in the severity of the disease.

- Continuous updates and real-time displays of patient disease state can be used to provide inputs to mental health research and the development of enhanced medical decisioning systems.

- Better introspection by leveraging massive knowledge that comes through KGs to offer the best treatment options and better step-by-step guidance to patients.

Knowledge Graph Question Answering System Flow

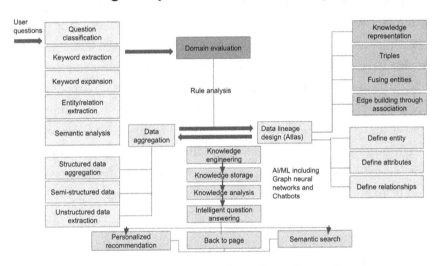

Figure 5-24. *Question answering-based chatbot using knowledge graphs*

After gaining thorough insight into KG creation and applying it in different Customer 360 applications, it is time for us to understand how we can make the KGs more explainable.

Designing Explainable Knowledge Graphs

Designing explainable, transparent KGs becomes highly relevant in medical code prediction, which includes ICD-10 medical billing codes, labels, and words. To induce explainability in the predicted codes, the best way is to visualize the predictions on word-to-word and word-to-code levels via a knowledge graph. It generally makes human cognition to interpret the predictions from the level links and paves the way for the development of a reliable and trusted medical coding application. It is designed using a hierarchical label-wise attention network (H-LAN) deep learning model that is capable of predicting multiple labels with attention to particular words and sentences per label, in a way that is self-explanatory for medical coders to read and understand from the billing codes. Further, it can be further annotated using pretrained ClinicalBERT, based on the mental health problem domain.

One of the popular ways to enhance the knowledge base is to apply a technique termed "G-coder" where a multilayer CNN was employed with an attention mechanism, to yield a knowledge graph that can map ICD-9 description with Freebase ontology data, enabling the coding description and medical terminologies interpretable with limited explainability. The solution had a limitation where terms could be explained only based on the type of attention networks getting used, but at the same time failed to provide explainability at the level of the prediction. Hence, efforts to incorporate domain knowledge to boost model performance were confined to the representation of the enriched knowledge in the output in the attention layer.

Improving Design to Bring in More Explainability

With research and the evolution of building in KGs came **Computer-Assisted Coding (CAC)**, where the translation of clinical notes to medical codes is made easier using ML and deep learning algorithms. This has made the job of the medical coders much easier. These models feed in unstructured textual data and use multilabel text classification predictive algorithms to predict medical billing codes, thus saving in manual effort of medical coders. Figure 5-25 illustrates the following sequence of steps for medical ICD-10 code prediction automatically with KGs (Khalid et al., 2025).

1. Text data ingestion like ingesting data contains patient records and discharge summaries.

2. ICD-10 code representation with a black box model.

3. Bringing in explainability by word and sentence level attention.

4. Knowledge enhancement with semantic enrichment by feeding in medical ontology repository and ICD10-cm library, to feed in the model output's predicted code descriptions. This phase involves the removal of stop words, duplicate words, stemming, and lemmatization followed by the determination of the meaning of unknown and attention words that help to link the connections with each other and with the predicted labels.

5. Semantic knowledge consolidation by aggregating in sentence-level, word-level, and ICD-10 description attention semantics by node segregation based on the nodes types, for example, mental healthcare terms, synonyms, definitions, patient summary, medical code, medical code words, and model attention words.

6. Derive the word-to-word and word-to-code level
    connections (weak or strong) by applying semantic
    matching algorithms to determine the relevance of
    the medical concepts.

7. Resultant explainable KG creation, where the
    relationships between the graph nodes emphasize
    the context and semantics and the "connected
    links" emphasizes the interpretation of the attention
    methodology and model prediction.

Knowledge Graphs to explain ICD10-codes

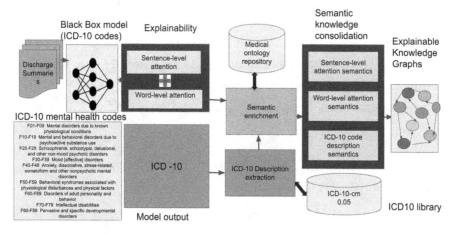

Figure 5-25. *Bringing in explainability through knowledge graphs
(Khalid et al., 2025)*

In a nutshell, using "Symbolic AI" such as "Knowledge Graphs" comes
with a better ability to explain the results of a deep-learning model for
CAC. At the same time, it boosts the accuracy and precision of the models;
some of the best accuracies noted are 64% word-to-word and 53% word-
to-code levels.

We have seen above how bringing in explainability in CAC can not only predict applicable medical codes in inpatient discharge summaries but also allow flexibility in configurations to view and understand the relationships between concepts based on the relevance of the context. This can serve as a powerful tool to review the logic behind the predictions, giving room for medical coders to accept or reject with confidence due to the transparency and explainability of the predictions. With transparency comes trust and reliability, which can further be improved by enriching the graph with more detailed domain knowledge, for example, by adding in long-term prognosis records, hospital admission, and processing of mental health insurance claims. In the below section part of the code, we illustrate an example of using graph database FalkorDB. Here, the example demonstrates llamaindex within FalkorDB as it is bears a unique advantage of streamlined search-and-retrieval, in comparison with LangChain which is a versatile, modular platform useful for building multiple use cases. Moreover, we have selected FalkorDB over Neo4j to do comparative analysis for easier understanding of readers with a simple use case.

Listing 5-2. This is an example of building a mental health QA chatbot using FalkorDB knowledge graph and LlamaIndex (Nayak, 2024)

1. The first step to build the chatbot application is to start FalkorDB as a graph database by leverage a falkordb docker image, which has an in-built browser to visualize graphs.

```
docker run -p 6379:6379 -it --rm falkordb/falkordb:edge
```

2. In the next step, we install the necessary libraries.

```
%pip install llama-index-llms-openai
%pip install llama-index-graph-stores-falkordb
%pip install pyvis
```

3. After installation is complete, we need to have the
    required imports of the libraries and set the OpenAI
    API key by having account registration at https://
    platform.openai.com/api-keys.

```
import os
from llama_index.core import Settings
from llama_index.llms.openai import OpenAI
from llama_index.core import StorageContext
from IPython.display import Markdown, display
from llama_index.core import SimpleDirectoryReader,
KnowledgeGraphIndex
os.environ["OPENAI_API_KEY"] = "<your OpenAI API
key here>".
```

4. Now we need to instantiate the knowledge graph,
    using FalkorDBGraphStore, which internally
    setups redis-server at port 6379. We also set up the
    necessary logger to log the requests.

```
from llama_index.graph_stores.falkordb import
FalkorDBGraphStore
import logging
import sys
logging.basicConfig(stream=sys.stdout,
level=logging.INFO)
graph_store = FalkorDBGraphStore(
    "redis://localhost:6379", decode_responses=True
)
```

179

5. Our next step is to load the mental health-related documents and convert the loaded documents into Triplets and Store FalkorDB. Here, we have used LLM model as gpt-3.5-turbo.

```
documents = SimpleDirectoryReader("./data").load_data()
llm = OpenAI(temperature=0, model="gpt-3.5-turbo")
Settings.llm = llm
Settings.chunk_size = 512
storage_context = StorageContext.from_defaults(graph_
store=graph_store)
index = KnowledgeGraphIndex.from_documents(
    documents,
    max_triplets_per_chunk=20,
    storage_context=storage_context,
)
```

6. We then query the knowledge graph using llama_index.

```
query_engine = index.as_query_engine(
    include_text=False, response_mode="tree_summarize"
)
response = query_engine.query(
    "Tell me more about Interleaf",
)
```

7. We obtain the response to the query as follows:

```
Mental health refers to a person's
emotional, psychological, and social well-
being. It involves how individuals think,
feel, and act, and also helps determine
```

180

how they handle stress, relate to others,
and make choices. Good mental health is
essential for overall well-being and can
impact various aspects of a person's life.

8. Now we have to visualize the KG as follows:

```
from pyvis.network import Network
g = index.get_networkx_graph()
net = Network(notebook=True, cdn_resources="in_line",
directed=True)
net.from_nx(g)
net.show("falkordbgraph_draw.html")
```

9. We can visualize the query results on the web server
by running the URL http://localhost:3000.

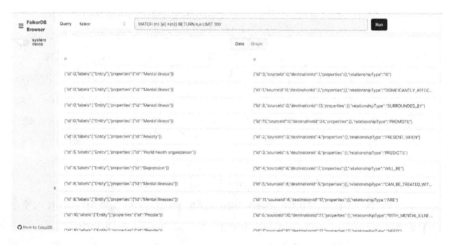

Figure 5-26. *Query responses from FalkorDBGraph*

We have seen how, with an example, KG can be used to retrieve
information catering to the mental health domain. Now let us summarize
what we have learnt from this chapter.

Conclusion

We have learned how mental health data can be used to build KGs and the different applications of the KGs in the medical mental health industry. The chapter lays a solid foundation for building a semantic network of entities and concepts to derive and represent semantic relations. The graph networks built can be used in patient diagnosis, treatment, clinical trials of newly launched drugs, creation of psychometrics ontology maps, connecting patients to suitable care provider services, or helping patients to reimburse claims related to the treatment of mental health.

We have also got insights into how SPARQL queries can be executed over KGs, to retrieve information related to disease, medications, or doctors. This chapter also provides a detailed understanding of the role of nutrition and vitamins on mental health through mental health KGs by applying named entity recognition and relationship extraction and highlights the importance of disease-centric subgraphs. The chapter brought to the forefront concepts related to eventualities which can be tracked from activities, states, events, and their relations to generate responses for chatbots to guide patients in different awareness programs and ongoing virtual/physical support services available. Further, the chapter introduces the concept of recommender systems by identifying similar patterns in patients' behavior to offer them similar treatment options. In the end, the readers get equipped with an understanding of building explainable KGs for the prediction of medical codes.

Now let us delve into the next chapter to gain a deeper understanding of how we can preserve data privacy in mental health to ensure confidentiality and security of our systems.

References

Ashwin Kumar & Veena Kumar. (2022). *Mental Health Hero*. Devpost.
https://devpost.com/software/mental-health-hero

Bickman, L. (2020). Improving Mental Health Services: A 50-Year
Journey from Randomized Experiments to Artificial Intelligence and
Precision Mental Health. *Administration and Policy in Mental Health*,
47(5), 795–843. https://doi.org/10.1007/s10488-020-01065-8

Chandak, P., Huang, K., & Zitnik, M. (2023). Building a knowledge
graph to enable precision medicine. *Scientific Data*, *10*(1), 67. https://
doi.org/10.1038/s41597-023-01960-3

Dang, L. D., Phan, U. T. P., & Nguyen, N. T. H. (2023). GENA: A
knowledge graph for nutrition and mental health. *Journal of Biomedical
Informatics*, *145*, 104460. https://doi.org/10.1016/j.jbi.2023.104460

Goodwin, T., & Harabagiu, S. M. (2013). Automatic Generation of a
Qualified Medical Knowledge Graph and Its Usage for Retrieving Patient
Cohorts from Electronic Medical Records. *2013 IEEE Seventh International
Conference on Semantic Computing*, 363–370. https://doi.org/10.1109/
ICSC.2013.68

Huang, Z., Yang, J., van Harmelen, F., & Hu, Q. (2017). Constructing
Knowledge Graphs of Depression. In S. Siuly, Z. Huang, U. Aickelin,
R. Zhou, H. Wang, Y. Zhang, & S. Klimenko (Eds.), *Health Information
Science* (pp. 149–161). Springer International Publishing. https://doi.
org/10.1007/978-3-319-69182-4_16

Khalid, M., Abbas, A., Sajjad, H., Khattak, H., Hameed, T., &
Bukhari, S. (2025). *Knowledge Graph Based Trustworthy Medical Code
Recommendations*. 627–637. https://www.scitepress.org/Link.aspx?d
oi=10.5220/0011925700003414

Larsen, R. R., & Hastings, J. (2018). From Affective Science to
Psychiatric Disorder: Ontology as a Semantic Bridge. *Frontiers in
Psychiatry*, *9*, 487. https://doi.org/10.3389/fpsyt.2018.00487

Nayak, P. (2024, March 26). Building a Mental Health QA Chatbot Using FalkorDB Knowledge Graph and LlamaIndex. *The AI Forum.* https://medium.com/the-ai-forum/building-a-mental-health-qa-chatbot-using-falkordb-knowledge-graph-and-llamaindex-d78be8d50f97

Qualtrics. (2025). *Your complete guide to patient journey mapping.* Qualtrics. https://www.qualtrics.com/en-au/experience-management/industry/patient-journey-mapping/

Santos, Henrique, Rook, Kelsey, Pinheiro, Paulo, Gruen, Daniel M., Chorpita, Bruce F., & McGuinness, Deborah L. (2023). *Facilitating Reuse of Mental Health Questionnaires via Knowledge Graphs.* https://dspace.rpi.edu/items/ca73de16-d821-4e4d-b07c-d98790c3b4ba

Schneider, P., Rehtanz, N., Jokinen, K., & Matthes, F. (2023). From Data to Dialogue: Leveraging the Structure of Knowledge Graphs for Conversational Exploratory Search. In C.-R. Huang, Y. Harada, J.-B. Kim, S. Chen, Y.-Y. Hsu, E. Chersoni, P. A, W. H. Zeng, B. Peng, Y. Li, & J. Li (Eds.), *Proceedings of the 37th Pacific Asia Conference on Language, Information and Computation* (pp. 609–619). Association for Computational Linguistics. https://aclanthology.org/2023.paclic-1.61/

Stardog (Director). (2023). *Demo Day Patient 360, Connecting Data from Electronic Health Records* [Video recording]. https://www.youtube.com/watch?v=QsYZ5QC_Yqs

The Knowledge Graph Conference (Director). (2023, September 6). *AI Psychology—HCLS Platforms Reuse of Mental Health Questionnaires via Knowledge Graphs* [Video recording]. https://www.youtube.com/watch?v=TDntQRoOOks

Tong, L., Liu, Q., Yu, W., Yu, M., & Jiang, M. (2023). *Improving Mental Health Support Response Generation with Event-based Knowledge Graph.* Knowledge Augmented Methods for Natural Language Processing, in conjunction with AAAI. https://knowledge-nlp.github.io/aaai2023/papers/006-MHKG-oral.pdf

Preserving Data Assessment, Privacy in Mental Healthcare: Ensuring Authenticity, Confidentiality, and Security in Data Integration from Diverse Source

This chapter highlights the criticality of data privacy in mental health as confidentiality and integrity of data are of paramount importance both for patients and doctors. With data privacy laws and regulations in place across the globe, specific to different countries, this chapter brings about

© Sharmistha Chatterjee, Azadeh Dindarian, Usha Rengaraju 2025
S. Chatterjee et al., *Revolutionizing Youth Mental Health with Ethical AI*,
https://doi.org/10.1007/979-8-8688-1186-9_6

clarity in incorporating privacy into advanced data and AI techniques while building mental health predictive algorithms. During the chapter, readers get an overview of the flaws in the measurement processes of data collected by the Internet of medical devices, along with potential risks of systems and algorithms designed for patients, care providers, and hospitals that could lead to data loss. In this context, this chapter also raises awareness of the significant implications of data leaks and security vulnerabilities when such leakage occurs and necessitating the urgency of safeguarding mental health information in a secured information, by and large for mental health apps (monitoring day-to-day activities of teenagers and youth) as well as for algorithms and services deployed at the cloud. The readers are further enlightened of the best practices of designing an algorithm with a code sample, along with a general understanding of the federated learning (FL) environment using aggregated mental health data. Towards end of the chapter, readers get a deeper understanding of how to safeguard psychological health.

In this chapter, these topics will be covered in the following sections:

- Privacy, security, and authenticity of data collected from mental health apps

- Privacy policies, regulations, and laws

- Building attack-resistant systems

- Privacy-enabled federated learning for mental health

At first, let us glance through the potential challenges to privacy, security, and authenticity of data recorded through mental health apps.

Privacy, Security, and Authenticity of Data Collected from Mental Health Apps

Mental health app discovery processes occur through social media, web searches, online reviews, or through recommendations available from networks, friends, and family. In many cases, it has been observed that the reviews are not authentic and app developers try to gamify the process by paying money to the users to provide better reviews. Such cases had been observed where mental health apps had been rated highly with inaccurate blood pressure reportings, which had led to the withdrawal of the app from the market. The most popular 25 iPhone apps on anxiety and worry lacked authentic evidence-based treatments. Often, apps had used scientific languages without providing strong support on the effectiveness of the treatments.

In addition, people having chronic mental health problems shy away from reporting their mental states through apps and wearables using self-reporting mechanisms due to internal stress, depression, and forgetfulness. Sometimes, it becomes a natural tendency of patients to prefer apps that enable relaxation and time management, instead of recording the severity of their disorders. More so, people do feel concerned about the privacy risks associated with the apps that threaten their safety and mental well-being. In addition, passive monitoring mechanisms to collect data from patients without requiring direct patient input pose a question in their minds and often create reluctance. In such cases, patients remain concerned about private sensor-based data, which collects their activities, mobility and social interaction patterns, physiological metrics, and speech conditions and interprets the results to evaluate the risk elements.

Now let us understand common security threats and data breaches for apps or wearables.

Security Threats and Data Breaches for Apps or Wearables

Cyberattacks are extremely popular in the healthcare industry as criminals infringe on the systems to monetize health information which leads to disruption of services. Irresponsible data handling by mental health groups had put them in front of lawsuits, with a penalty of several million dollars. More than 500 data breaches have resulted in the disclosure of 519,935,970 healthcare records between 2009 and 2023 (Alder, 2025). Further, the volume of data breaches has doubled in the United States from 2018 to 2023 leading to 364,5771 healthcare records getting exposed from 1.99 healthcare data breaches.

Crisis Text Line, a global nonprofit organization providing free and confidential text-based mental health support, was found to leak sensitive data from people's online conversations in 2022. The organization used these chat conversations to train its customer service chatbots. Federal Trade Commission fined BetterHelp $7.8 million, a mental health platform known for its direct online counseling and therapy services (over the web or phone) when it was found to leak people's health data to Facebook, Snapchat, Pinterest, and Criteo, an advertising company (Dorner et al., 2023). Even though BetterHelp was formed to better assist people with mental health and navigate the existing challenges of receiving expensive and crucial data, they failed to obtain consent from users by informing them about the data privacy policies.

In 2023, a California-based mental health startup called Cerebral exposed the personal data of more than 3 million people, online (Lyngaas, 2023). The firm was connecting anxiety and depression patients with mental health professionals via video calls. The leaked data primarily contained participants' responses from self-assessments where they demonstrated their experiences with panic attacks, alcohol abuse, or a personality disorder.

A recent cyberattack on NHS Dumfries and Galloway in Scotland in May 2024 has led to the publication of children's mental health data which is being investigated by the National Crime Agency and National Cyber Security Centre (Hunter, 2024).

In May 2024, the Los Angeles Department of Mental Health notified that on March 20, 2024, when an employee had clicked a phishing email, it compromised some of the accounts' privacy, leading to access to personal details like date of birth, address, telephone number, Social Security number, and medical record number.

Let us now understand that the absence of regulations imposed on apps and wearables along with lack of standardization practices leads to inconsistencies in readings.

Absence of Regulation of Apps and Wearables

In the absence of proper regulation of medical and mental health apps, there has been an increasing concern in the community worldwide, owing to the safety of patient data. The major concerns remain around the following:

- FDA review is limited to a few medical apps and wearables, including for mental health.

- The regulation applies only to software which will enable the platform to serve as a regulated mobile device.

- The software quality and commitment are governed by the maturity level of the organization, which helps the FDA Digital Health Software Pre-certification Program to certify only the manufacturers instead of the software.

- Controversial consumer medical products are released
  in the market such as the Apple Watch to diagnose
  atrial fibrillation, raising concerns among physicians.

Authenticity of Measurements of Apps and Wearables

Sensors in smartphones and wearables have often reported inaccurate readings, which has led to underfitting and overfitting ML models yielding high misclassification errors. In the absence of rigorous testing of embedded sensors, calibration, and certification processes, the data collected from sensors and wearables would be biased, owing to the type, make, and model of the device. The inconsistency in the collected data is highly influenced by the hardware based on manufacturer design along with the features of the smartphone that include hardware power consumption, speed of the processor, smartphone model, make, size, the operating system of the device, the position of the sensor in the device, and sensor's reaction to internal and external environmental conditions.

In the following code example, we see how data from wearables merged with other user attributes can be used to predict anxiety level and depression among students.

Listing 6-1. This is an example of prediction of student stress factors: a comprehensive analysis

In this part, we take a student stress factors dataset data and try to predict the stress levels of students based on various aspects of their daily lives. This dataset consists of 20 high-impact features of around 1000 students which fall into the following major categories:

Psychological factors: 'anxiety_level', 'self_esteem', 'mental_health_history', 'depression'.

Physiological factors: 'headache', 'blood_pressure', 'sleep_quality', 'breathing_problem'. This can be collected from sensors and wearables plugged into their body.

Environmental factors: 'noise_level', 'living_ conditions', 'safety', 'basic_needs'.

Academic factors: 'academic_performance', 'study_ load', 'teacher_student_relationship', 'future_career_ concerns'.

Social factors: 'social_support', 'peer_pressure', 'extracurricular_activities', 'bullying'.

This task can be viewed as a tabular classification task where we can leverage tabular techniques like decision models to predict the stress level of students based on the psychological, physiological, social, environmental, and academic factors. We use one of the most popular and best models available, the XGBoost model.

1. In the first step, we install the necessary libraries.

```
%pip install pandas
%pip install xgboost sklearn
```

2. After installation, we import the necessary libraries and model XGBoostClassifier.

```
import pandas as pd
from sklearn.model_selection import train_test_split
from xgboost XGBoostClassifier
```

3. In the next step, we load the dataset and drop
    the stress_level column which is used as the
    target column.

```
df = pd.read_csv('/content/StressLevelDataset.csv')
X= df.drop(columns=['stress_level'],axis=1)
y= df['stress_level']
```

4. After loading the dataset, we need to preprocess the
    dataset, to obtain the train and test data subsets.

```
X_train, X_test, y_train, y_test = train_test_split(X,
y, test_size=0.33, random_state=42)
```

5. Then, we fit the data to XGBoost model to predict
    new data points.

```
bst = XGBClassifier()
bst.fit(X_train, y_train)
bst.score(X_test,y_test)
```

This yields the following prediction on being queried with

```
bst.score(X_test,y_test)
```

```
anxiety_level                      21
self_esteem                        15
mental_health_history               1
depression                         26
headache                            3
blood_pressure                      3
sleep_quality                       1
breathing_problem                   5
noise_level                         3
living_conditions                   1
safety                              2
basic_needs                         1
academic_performance                2
study_load                          5
teacher_student_relationship        1
future_career_concerns              5
social_support                      1
peer_pressure                       5
extracurricular_activities          5
bullying                            5
Name: 328, dtype: int64

Prediction:   [2]
```

This code listing illustrates a simple use case of evaluating stress factors. An important point to consider is that if the dataset is skewed toward any of the above factors that lead to depression, then the results can be biased or inaccurate leading to misdiagnosis. In future chapters, we will study how fair datasets play an important role in fair predictions.

Now let us understand how at the common threat to mental health patients in this digital era.

Threat to Patients with the Rise of the Digital Economy

Commercialization and the boom of the digital economy have given us a new source of revenue where personal data is collected, aggregated, and mapped to other data and abstracted as data products for markets to consume. Selling personal data from medical apps and wearables, facial recognition, as well as biometrics used in authentication schemes poses a direct threat to patients' privacy. Human interpretation leads to a wide range of biases and prejudices, infringing user rights and transparency of the processes. While the rise of the market value of personal data has been exponential over the years, the absence of appropriate controls would increase the threat of personal mental health data getting shared and exposed, resulting in the breakage of confidentiality, integrity, and trustworthiness of AI-driven mental health digitization processes.

A survey released by Facebook in the United States in 2018 revealed the statistics related to the ignorance of people and their discomfort in sharing personal data (Bauer et al., 2020). Seventy-four percent were ignorant, 51% remained uncomfortable with personal data sharing, and 27% stated the inaccuracy of the classifications. Further US survey results revealed that 60% of the people had an assumption that algorithms were unfair and the predictive algorithms reflected human bias (Bauer et al., 2020). In addition to the United States, users from Germany refrained from sharing mental health data to help organizations or clinicians develop personalized health recommender systems. This is due to the fact that patients are not only concerned about the privacy of their data being available to researchers and AI algorithm experts, but they also remain worried about data leakage to the external world, due to commercialization processes of personal data.

Along with malicious attacks and hacks and the nonconsented commercialization of personal data, people get anxious when certain surveillance programs are introduced by the government that aim to combine facial recognition technology with DNA fingerprinting. This raises questions in people's minds on the safety of being free and expressive in public environments.

Increasing concerns emerged among patients when mental health data comprising of clinical and counseling sessions is collected, digitally written and stored in systems. Most patients fear that EMR mental health records may lead to the exposure of their personal information, as data gets collected from patient portals, apps, or wearables. Hence, policies need to be enforced to store EMR by being fully compliant with **Health Insurance Portability and Accountability Act (HIPAA)** (U.S. Department of Health and Human Services, 2017). HIPAA comes with well-equipped federal standards to protect sensitive health information from disclosure without patient's consent.

Health Insurance Portability and Accountability Act (HIPAA) protection laws have been applied to some of the traditional healthcare and clinical environments in the United States which prevents mental data leakage from healthcare providers, insurers, and business associates. The majority of adults show concern that sensitive data from mental health notes exhibits a threat of getting leaked to other healthcare providers beyond their chosen ones. More than 30% of the participants in Canada and the UK show a tendency to refrain from availing of mental healthcare services due to fear of loss of personal data.

Research studies on ubiquitous data collection aim to deploy sensors where certain signals (like **electrocardiograms (ECG), electroencephalograms (EEG), temperature (TEMP), electrooculography (EOG), electromyograms (EMG), blood volume pressure (BVP), respiration (RESP), arterial blood pressure (ABP), electroglottography (EGG), mechanomyograms (MMG))** from the sensors could be integrated with smart buildings or smart cities (Alwakeel et al., 2023). Bringing in inputs from multiple sensors helps to run multimodal feature fusion to measure and analyze multiple signal types in real time. This helps to detect and predict mental issues such as anxiety, mood state recognition, suicidal tendencies, and substance abuse. Though this initiative would help a hospital or a military unit to use the signals and identify people at risk, the question remains how safely the collected data and predicted outcomes of the model are private and ethical.

In spite of having federal standards like HIPAA (U.S. Department of Health and Human Services, 2017), it fails to prevent patient-generated data from the apps, wearables, or data transferred over the Internet. So the risk of data getting lost from firms remains high, who collect, store, and analyze the data before selling to third parties. Further, HIPAA does not also protect the diverse range of nonmedical digital data, scheduled for collection in the medical predictive algorithms. This limits to concluding someone's mental health, such data include social media posts, patterns of phone usage, or selfies. However, tech giants like Meta have applied smart AI techniques to flag, and review and then delete, posts about self-harm and suicide. Oftentimes, this leads organizations to identify medical patterns from large, diverse datasets without taking individual consent.

With privacy, security, and authenticity of the data becoming a major concern, let us explore some of the best practices for designing privacy-enabled mental health apps.

Need for Enhanced Privacy on Mobile Apps and Wearables

We have seen in previous chapters the role of digital mobile apps and wearables (or fitness trackers) that can log user steps, heart rate, sleep patterns, and social media usage, to draw a comprehensive image of the mental health condition of a patient. We have a definite set of mHealth apps which includes wellness and fitness apps (e.g., calorie counters, exercise trackers), personal health apps (e.g., diabetes monitors, symptom checkers), and medical resource apps (e.g., drugs catalogs, medical terminology libraries), which are involved in a lot of data transmission through unsecured Internet communication. Many of the apps do not follow the laws and guidelines for sharing protected information and pose serious threats due to the grant of unnecessary app permissions and employment of insecure cryptography techniques along with hard-coding and logging of sensitive information (Bauer et al., 2020).

With the growing popularity of mHealth apps, we observe that mental health information appears to have captured far more granular details than that are recorded through a single conversation recorded by a therapist, in a weekly once-off session. The most important areas of consideration in in-person counseling sessions are listed below, which demonstrate increased time commitment and cost through physical sessions. Mental health counseling sessions may not have insurance coverage.

- Each physical session may cost a minimum of $100.

- The frequency of receiving in-person appointments remains restricted to a small proportion of psychiatrists, which can increase the average wait times for in-person appointments for more than 2 months (recorded by a research study in 2022 that only 18.5% of psychiatrists had appointments from new patients).

Mental health apps continue to innovate themselves by harnessing their hardware, like sensors, and front-facing or rear-facing cameras to trigger the button to capture selfies. Such automated selfies are capable of tracking user mental states when users answer questions related to their mental well-being. Even apps deployed with AI models are capable of correlating the responses, by linking common human emotions (e.g., sadness and hopelessness) with the facial gestures visible in the images, along with image color, and background objects, to infer any signs of depression. Any kind of depressive symptoms further received assurance on the user's depressive state when apps are powered to take in photos, when the device is unlocked, and the photos are then compared with sleep hours, text messages, and social media posts. The aggregated data can then be shown to a therapist to receive a correct diagnosis; however, if the sensitive data has been collected without a user's consent, then it infringes the user's rights and results in violations. The only way to collect the data with informed consent is to monitor users who are at greater risk of threat or self-harm.

Let us now take a look at how the ecosystem of mental health apps is designed and what kind of data they collect and share with different entities to further assess risk and remediations.

Data Transfer Ecosystem in Mental Health Apps

Research studies show phones often pose a threat to many mental health problems, but as we have seen in Chapter 4, "Empowering Mental Well-Being Through Data-Driven Insights," many mental health apps have been designed as a part of the solution to resolve the existing crisis that aims to ameliorate the risk of unintended harm. With the rapid growth of mental health apps, data privacy risks are also on the rise, due to the threat of security breaches and adversarial attacks. The different entities involved include the following whose data processing, storage, and transmission need to be accessed end to end for privacy considerations. Figure 6-1 given below illustrates privacy threats involved in data exchange process with mobile apps in an enterprise business model.

- Users (i.e., data subjects), which provide information to the mHealth or wearable apps, for logging and sharing.

- Different companies and service providers act as data controllers along with other service providers (i.e., data processors) which receive the data.

As personal PII information flows from an app to an organization's infrastructure, both the app developers and system architects/cloud engineers need to have a better understanding of the data privacy standards, authentication techniques, and practices for the data at rest, in transit and in use. Along with client- and server-side data privacy and security infrastructure, developers need to have a full understanding of the nature of the personal data, necessary consents obtained, and usage of the data in creating ML models and serving patients and healthcare service providers.

In addition to developers and architects, testers should be introduced to the data flows in the trusted ecosystem so that they can independently act as hackers and adversaries to simulate attacks and breaches to test the system resiliency and degree of privacy maintained by the system. In the trusted ecosystem of mental health apps, it is imperative that we also consider apps' network traffic and data transmissions which gives us an opportunity to closely inspect data transmitted to third parties such as marketing bodies and advertising services. By using logs, stored data, and reverse-engineered code, we can reveal possible instances of data leaks and weak access control and provide fixes for insecure coding.

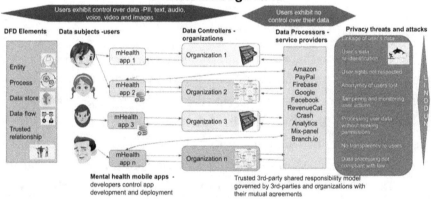

Figure 6-1. *Privacy threats in a data exchange platform with different entities*

One of the key challenges to respecting user privacy breaks when tech companies and organizations, in the absence of strong regulations, use private data for their business purposes leading to misuse of highly sensitive data. This may be the most confidential data received from people at their most vulnerable moments.

In addition, personal data may be flowing to partner third-party service providers who collect and process the user's data to generate insights. Often, the data processing is offloaded to public cloud infrastructures including Amazon AWS, Google Cloud, or Azure, where the data is stored in servers and databases and the insights are derived from several APIs from the data provided by the apps. Such analytics tools include CrashAnalytics, RevenueCat, PayPal, and Firebase, where the control and processing responsibilities are shadowed by the developers. Developers need to work closely with service providers to ensure trust and confidentiality of their data are fully maintained. The key role for mental health app developers is to earn the trust of their users, so that users feel comfortable downloading the apps and provide information that can track their mental states. This may enforce strict privacy and security audits with the providers at a regular cadence to fully operate in a shared responsibility model.

To further exacerbate the risk, app advertising plays a predominant role. Advertising remains a major concern when ads are displayed on mental health apps which adds fuel to the privacy threats. The app's business model relies on advertising as a major source of revenue, with this objective in mind when the user's private data are sold or distributed to the app developer(s), or third-party sites used for functionality reasons, it poses a direct threat of private information getting leaked without receiving the owner's consent. Hence, it becomes important for users and health professionals to educate themselves on the privacy loopholes and operating model of the business to carefully evaluate the conditions and provide their consent to control data leakage. Otherwise, users completely lose control of their private data which becomes available to other third-party developers, mobile advertising platforms, and data brokers. With the growing trends of personal advertising and the introduction of modern techniques like **cross-device tracking (CDT)**, users are getting monitored on multiple devices on their browsing activities receiving better retargeted ads. In order to effectively deliver ads based on screen size and resolution, a user downloading a mHealth app on a mobile device misses out on the same ad, while browsing social media networks using a PC, which violates users' online privacy and minimizes their anonymity. This generates a high risk of exposing user data to people with serious mental health conditions by supporting online advertising businesses, without having a proper regulation to manage and reuse the data gathered from the users.

With multiple risks associated with today's rapidly changing political and business environment, it is largely felt to be more proactive in building systems, meeting the regulatory requirements for data protection and respecting individual privacy. Hence, recommendations have been pouring in from system architects and designers to follow "privacy by design" from day one to define, evaluate, and foresee the digital vulnerabilities beforehand to resign operations, management methods, and processes that can be ideally suited for the organization's privacy standards as well as meet regulatory requirements. The goal is to enable proactive, preventive practices ingrained in the system's lifecycle by default without compromising on privacy and security standards. This facilitates and prioritizes user-centric privacy workflows with more visibility and transparency brought into business practices.

As privacy policies become important in the governance of mental health data, let us look into some of the data protection standards and guidelines that can enable us to rigorously test mental health processing data systems and frameworks and certify them with confidence.

Data Protection Privacy Policies, Standards, and Regulations

Mental Health Hub has mentioned high standards of data protection and security practices to govern the confidentiality and security of personal data. It operates in accordance with the Data Protection Act 2018 and the General Data Protection Regulation (GDPR) guidelines (Ienca & Malgieri, 2022), by respecting user privacy and lawful and transparent processing of personal data entrusted to them. We have listed below certain best practices for data usage in mental health processing systems.

Collection and Use of Personal Data

Personal data collection becomes an important component of mental health processing systems which include personal information such as name, email, ID numbers, address, physical, physiological, mental, economic, genetic, social, medical, and cultural factors regarding the past, present, or future of the patient.

However, such data is collected based on the relevancy and necessity of the systems based on end-user consent without indulging in unlawful usage of the data or driving services through automated decision-making or profiling. Any of the PII data collected should have a definite data retention policy, which would enforce a guideline to purge the data beyond the period for which it was collected (Martínez-Pérez et al., 2014). In addition, the data controller and processor entity hold high accountability in case of a data breach where user PII data has been compromised. In such a scenario, the system should allow to raise an alarm as a breach notification obligation without any delay, as this would have seriously impacted a user.

Apart from the popular PII data, within the purview of mental health, certain data cannot be revealed, including abstract and physiological conditions of the brain: for example, nonpathological emotional information or information related to thoughts, preferences, or memories. Though emotions are not regarded as sensitive data, the presence of emotion detection tools based on biometric methods such as facial recognition tools or **consumer brain-computer interface (BCI)** is capable of determining the mental health conditions of the user. Even they can reveal details about religious beliefs, political opinions, and sex life or sexual orientation.

Data Security Measures

Respecting the security and integrity of end-user data becomes highly important and demands robust technical guidelines closely tied up with organizational goals and metrics. These can be governed by (Martínez-Pérez et al., 2014)

- **Secure infrastructure**: Platform and system design based on industry-standard secure servers with proper network infrastructure, highly compliant with GDPR or country-specific privacy laws, to prevent unauthorized access, data breaches, and cyber threats.

- **Encryption**: Apply the right data anonymization and encryption technologies to protect your data during transmission and storage. This ensures incoming data from mobile, wearables, and external third parties do not get leaked during transmission and maintain full confidentiality, remaining inaccessible to unauthorized individuals.

- **Access control**: Enable a zero-trust policy to restrict authorized personnel's access to personal data. By and large, allowing role-based access and providing necessary training to employees helps to maintain data protection through strict confidentiality agreements.

- **Data minimization**: Following and maintaining a data retention policy guideline to retain personal data based on system needs and the purpose of the collected data. When the access to the data gets too infrequent or void, then the data can be archived for a certain period, before it is finally disposed of.

Third-Party Data Processing

Data protection standards set by the system hold the same significance in front of other trusted third parties who are service providers and are responsible for the efficient delivery of mental health services. They work in tandem with the overall system and are educated with the necessary instructions to protect the privacy and confidentiality of data. Hence after a careful selection, of the third parties, time-to-time audit is essential where they interact and process data components present in the system.

User or Owner Rights

Individual rights govern and respect the rights of end users whose data is processed by mental health systems. It grants specific rights to users' personal information, which may include

- **Control to access**: Users can choose to request access to the personal data that the system has and are entitled to obtain a copy of the personal data.

- **Right to rectification**: Users can validate the personal data being held by the system and possess the necessary rights to request its correction, in case of inaccuracy or incomplete information.

- **Control to data erasure**: Users hold the right to remove their personal information by requesting the erasure of their personal data from the systems.

- **Control to restriction of data fine-tuning**: Users possess the right to request the restriction of data fine-tuning, filtering or further processing of their personal data in case users believe the data is inaccurate or unlawfully handled.

- **Control to data portability**: Users can request a copy of their personal data in a common data exchange standard or machine-readable format or transfer the data to another controller, as and where it is suitable and technically feasible.

- **Control to object**: Users have the discretion to stop the usage of their personal data for business and marketing purposes.

- **Control to lodge a complaint**: Users are also entrusted with the right to complain to the data regulatory bodies or the relevant supervisory authorities, in case they feel their data is misused.

Let us now take a look at the standards set by HIPAA, when it comes to sharing or disclosing private mental healthcare data.

HIPAA Privacy Rule Related to Mental Health

HIPAA is one of the foremost privacy rules (U.S. Department of Health and Human Services, 2017) known in the healthcare industry that is equipped to provide consumers how their private health information is used by doctors and care providers and instances where it is disclosed for any health plans. The benefit of HIPPA is to empower individuals with privacy rights and protections to question the usage of their confidential information. By enablement of this control, a greater trust is established between data collectors and individuals. It adds more safety to patients where data including mental health data may be shared or disclosed for the benefit of offering better treatment to individuals. Further, the privacy policy provides governance for a healthcare provider to share the protected health information of a patient receiving treatment for a mental health condition by adding clarity concerning the following areas:

- Right to share patient's mental health information with their family members, friends, or others connected with patient care and treatment

- Ability to transfer information of an adult patient to their family members

- Ability to transfer information of a minor patient to their parents/guardians

- Take the required patient's concerns and agreement before sharing their confidential mental health information

- Ability to discuss and communicate with a patient's family members about do's and don'ts and scenarios when the patient fails to follow the medication, therapies, and other procedures.

- Keep in touch and listen to the patient's family members, keeping them informed on the progress of the mental health treatment being provided.

- Frequent communication and status updates to family members, law enforcement, and others who are involved when the patient's mental health degrades or it becomes serious and threatful to self-harm or others

- Keeping informed law enforcement bodies for patients receiving treatment in emergency psychiatric hold, when the patient is getting released.

HIPAA does not differentiate between different types of protected health information and uniformly applies the law about the sharing of any health-related information. However, certain exceptions exist, especially for psychotherapy sessions when notes are recorded by healthcare providers, which has additional protection guidelines. It

applies to the notes taken and analyzed during the counseling session
as they are sensitive and private. They are not considered for further
treatment processes as they do not contain medication, prescription,
and other diagnostic summaries. Hence, HIPAA governs the disclosure of
psychotherapy notes with the patient's authorization.

HIPAA allows to help in collaborative treatment and progress by
providing the flexibility to share a patient's mental health information,
with family members that would act as immediate help in case of
emergency, for example, a patient develops warning signs due to a specific
drug. However, sharing this sensitive information with immediate family
members cannot be extended freely when the patient objects. Similarly,
HIPAA allows the state laws to govern children to allow the selection of
their representatives, more so in scenarios when an adolescent obtains
permission to receive mental health treatment without parental consent
and provides their due consent to receive the treatment. State privacy
laws override HIPAA to influence the age of majority and parental rights to
decide a child's mental healthcare decisions. Based on the state's allowable
permits, a healthcare provider may choose to disclose protected mental
health information of a child receiving treatment to parents or may choose
to refrain from sharing the information.

Designing Compliant, Protected Mental Telemedicine Services

As we move into the era of digitization and adopt faster modes of treatment
in mental health, we rely on telemedicine services where treatment can be
provided to patients at remote locations. The common ways of offering this
type of treatment include

- **Interactive medicine:** Technology tools like video
  conferencing and phone consultations in real time,
  offering treatment and counseling sessions to patients.

- **Distributed diagnosis**: The psychiatrist receives information of a patient from another location to offer treatments to patients who are traveling to geographically distant locations.

- **Remote patient monitoring and diagnosis**: Aggregating patient information from patient portals and IoT/medical devices such as EKG, ultrasound, dermatoscopy, and pulse oximeters.

All of these methods involve a lot of information exchange related to patient mental health conditions for diagnosis of the patient, offering treatment and prevention of any diseases or injuries. The data containing a lot of PII data should be HIPAA compliant, and the process should be facilitated from a well-architected system which does all kinds of data sanity checks and enforces privacy guardrails.

In a nutshell, the system architecture should allow different eligibility conditions to be fulfilled from different governing laws and policies (including state confidentiality statutes) applicable before sharing the information. Our goal remains to design the individual system components and API which considers professional ethics and underlying risk factors before disclosing the information. Hence, when we encounter a scenario where a patient is experiencing temporary psychosis or is intoxicated with drugs or alcohol and goes beyond normal conditions to give consent to disclose the treatment plans and medications, then HIPAA allows a healthcare provider to take discretionary actions in the best interest of the patient to disclose the private information. Even conditions of serious mental illness with a high risk and probability of patients committing suicide can be dealt with utmost urgency by the care providers. Within HIPAA's purview, the primary goal for care providers is to prevent or lessen the threat of harm to the health or safety of the patient.

To design such a compliant system, certain major factors for consideration include the following:

- Maintain a checklist of items to validate authorization and consent received from the patient.

- Provide built-in support for validating eligibility criteria for sharing information based on HIPPA and state privacy laws and guidelines.

- Evaluate patients' risk conditions whether they have stopped taking medications or are under psychosis or the effect of drugs or alcohol.

- Assess the seriousness of the risk using predictive AI models, and notify care providers to share information.

- Have support for APIs that can send alert messages or notifications to police, a parent or other family member, school administrators, or campus police.

- Incorporate all in-built cloud security features for data ingestion and data storage during model prediction to avert any loss of confidential information other than those associated closely with patient care, treatment, and payment.

- Automate the manual process of summarization for counseling sessions using speech-to-text AI models with patient consent, to drive system efficiency and reduce manual analysis and intervention.

Keeping in mind the regulations and privacy laws in place, we need to define a structured privacy policy which can identify the security loopholes that exist in today's mental health apps and make users vulnerable to different attacks and threats. The primary ways of incorporating privacy analysis in our mental health ecosystem will be to include a series of penetration tests, with static and dynamic analysis, that thoroughly scan and inspect all apps' permissions, network traffic, and identified servers.

It is equally important to scan reverse-engineered code, databases, and generated data to rule out any chances of vulnerability. In addition, the privacy analysis frameworks can also demand companies and software developers to retrieve **Privacy Impact Assessments (PIAs)** of the apps which is a mandatory requirement for GDPR.

Other Information Security and Privacy Compliance Strategies

Other high-priority user data security and privacy compliance include ISO 27001, 27701, and GDPR standards which all emphasize encryption and explicit user consent for data processing. In the context of mental health data, GDPR brings about indications that can help us to differentiate numerous variables that impact legal decisions and can be summarized as follows (Ienca & Malgieri, 2022):

- Present or future circumstances (commercial or medical) under which the data is collected for processing

- Medical or business justifications including diagnostic, observational, or targeting why the data needs processing

- Explaining the clear intent and objective involved in processing the data which may include

- Public interests in diagnoses or data analyses

- Private interests in enhancing mental functioning or improving one's well-being

- Commercial interests in exploiting the cognitive biases of consumers, etc.

GDPR also adds clarity with respect to the assessment and mitigation of data processing impact in Article 35 (the "Data Protection Impact Assessment"). It emphasizes that any personal data getting processed and exhibiting higher risk for fundamental rights and freedoms of data subjects must have a regular impact assessment with detailed risks laid down, where risks are assessed and mitigated.

While GDPR formulated by the European Union regulation on information privacy is being used to protect mental health data in Europe, countries like India have shown keen interest in digital mental health technologies. Data protection guidelines being outlined in the GDPR, DPIA, and IEEE Global Initiative on Ethics of Autonomous and Intelligent Systems have generated enough attention on data privacy risks for individuals having mental health issues, to create a framework within India, governing highly sensitive information to combat threats arising out of Aadhaar linkage.

GDPR rules are standardized for any system and would thereby apply to mental health system architecture, and AI model predictions, including GenAI predictive services. All new data processing has to comply with infosec and privacy policies to measure the confidentiality of user data in the following ways:

- **Data encryption and user consent**: Ensuring all data encryption, preventing misuse of PII data through deletion, and obtaining user consent in case data is sent to an external server like OpenAI.

- **Algorithmic guardrails**: Incorporate all GenAI and non-GenAI guardrails to proactively prevent malignant intents like prompt injections.

- **Compliance reviews**: Setting up external compliance to ISO 27001 (information security) and 27701 (user privacy) (Mukherjee, 2023), as well as GDPR, so as to ensure any GenAI release goes through the listed checkpoints and adheres to appropriate auditable documents. This adds robustness and compliance checkmark to new data flows and processing units.

In addition, GenAI agents and therapeutic chatbots should follow clinical safety rules to prohibit use in high-risk mental health scenarios (e.g., sensitive conversations related to trauma, self-harm, and suicidality). All GenAI responses to external users should be subject to rigorous and regular testing with different datasets to ensure that the responses generated comply with regulations. This also applies to human clinician oversight, following NHS's DCB0129 protocol for clinical safety, where human clinician review and evidence become a mandated practice. Regular audits would enable safeguard patients seeking help from unhealthy coping mechanisms (any form of abuse or alcohol), on being prompted by therapeutic chatbots.

Let us now explore the primary strategies for how we build attack-resistant systems by being compliant and at the same by introducing comprehensive guidelines and checkpoints through different threat modeling frameworks.

Building Attack-Resistant Systems

Being aware of user rights to data is the foremost step, to prevent damage or misuse of data. With mental illness treatment plans being widely adopted by private care providers, it becomes immensely important that all the providers respect the privacy laws concerning sharing confidential information of patients, whether it is a general medical healthcare privacy rule such as the **Health Insurance Portability and Accountability Act**

(HIPAA) or it is a country or state specific privacy rule. As we are largely dependent on mHealth apps for data collection, we should also audit for GDPR compliance for the apps and the infrastructure they assess, which can validate the type of collected data, and apps' permissions, data portability, and data deletion features. In addition, the consent management and governance process helps to identify any unawareness and noncompliance issues which can raise the risk of privacy threats.

Optimizing Privacy Awareness with the Service Agreement

With more and more innovative ways coming up and mental health apps getting smarter through research, it has become easier for researchers and developers to upload the apps to app stores. With the nuances of privacy threats, regulators need to be more conscious of educating the developer pool of regulatory actions in case the app fails to meet the privacy design guidelines. It becomes immensely important to have interventions to certify the usage of data collected from the apps in a way that is efficient in striking a balance to keep the user data safe and at the same time does not deprive them of social media platforms. The key essence will be in proactive risk and threat management arising from serious mental health conditions, in a way that does not isolate someone from valuable support networks, but at the same time provides support at the most crucial time, using the data collected with due consent of the individual.

Even organizations claim that protected health information (PHI) obtained from users is treated with utmost confidentiality, but there's no governing (like federal) law protecting its usage. Such data may flow from healthcare service events including insurance or medical records or hospital or pharmacy bills. Mental health apps providing conversational platforms (e.g., Woebot and BetterHelp) can make users feel safe and secure about their data if they opt into HIPAA compliance. In the absence of legal classification of user data as PHI, judicious use of personal data becomes even more difficult to control. One solution that organizations provide is lengthy service agreements that people do not read. One of the mental health apps Woebot has been known to provide reader-friendly terms of service, but its excessively lengthy text of 5,625 words often keeps readers less engaged and prevents them from reading it and signing it after understanding the clauses. Though major (46% of dementia apps and 19% of diabetes apps) top-user base mHealth apps had a privacy policy, they lacked transparency, consent, and intervenability. They fell short in their privacy policy readability scores (Gunning Fog, Flesch-Kincaid Grade Level, and SMOG formulas), which raised several questions about their transparency and consent. Even they presented unfair clauses, like "contract by using" and "unilateral change," which is contradictory to EU GDPR that emphasizes explicit informed consent as few or many people who reveal their secret data may belong to a vulnerable and fragile state.

The key challenge remains to certify the apps' privacy policies, using readability scores and AI-assisted privacy tools. This helps to identify noncompliance threats and have the right governance in place for incorrect or insufficient privacy policies. This in turn increases the developers' responsibility to understand the privacy policies of the OS (Android's permission system) and use it correctly. For example, this implies that app developers should start slow and incrementally seek permissions "in context" as and when a user is found to interact with features that demand the necessary permissions. In addition, app developers should get more security-related training including security testing to be more effective in managing threats.

As it becomes imperative that smartphone apps and wearables should have adequate consent pop-ups, forms and disclosures, and privacy policies for users, app designers have come forward to include these policies. It is well understood that the role of apps in establishing human connection is incredible and it plays a central role in collecting the right data to assist people to overcome mental health struggles.

However, the policies vary internationally and by app store. Some of the key design aspects in designing privacy policies include

- Making the privacy policies easy to comprehend, as it often remains unclear to people how to understand the postsecondary level technical information, along with legal terms

- Obtain a user's consent to data usage in a manner which is freely obtained, reversible, informed, enthusiastic, and specific, with a clear mention of the data subject's intent to use the data in systems, the data collected, and all the processing involved

- Designing the privacy policies for apps and wearables in a manner that prompts for authorization of users during any attempt to sell, transfer, analyze, and disclose consumer data to third parties

- Employing user agency to give autonomy to users and adding transparency in the end-to-end processes, so that users can agree which monitoring services they are comfortable with and are aware of the potential side effects in case of any interventions

- Balancing privacy and user experience where inclusion of privacy policy does not disturb end-user experience

- Authorization of the sale of personal data in the case of a merger, acquisition, or bankruptcy

- Securing data transfer process from patient apps to EMR, by keeping informed the users, answering any issues or concerns, and involving a thorough review of the system

Let us now investigate how we can follow privacy engineering principles to leverage technology for introducing and developing privacy considerations into product design. Having "privacy by design and by default" helps to narrow the boundary between technology and regulations to merge the requirements and integrate them into policies, processes, and products.

Defense Strategies Using Privacy Threat Modeling Frameworks

To understand the importance of privacy engineering frameworks, we must understand and study one of the most popular privacy engineering frameworks like LINDDUN (included in the NIST Privacy Framework), which sets the guidelines in front of organizations to integrate privacy engineering into their products, processes, and policies (Iwaya et al., 2022). The process emphasizes assessing and controlling the privacy risks and meeting regulatory requirements.

The working methodology of LINDDUN privacy threat analysis comprises three main steps (Iwaya et al., 2022):

- System modeling is aimed at defining all the data components and their interactions using **data flow diagrams (DFDs).**

- Evaluate system components to identify and extract possible privacy threats, over each DFD element, and describe it with proper terms within the context.

- Determine appropriate tools and solutions to manage and control threats.

LINDDUN commonly handles the following seven threat categories
which enables organizations to analyze privacy threats and act proactively
to protect the users or termed as data subjects with regards to threat
analysis (Iwaya et al., 2022):

- **Linkability**: Enables an adversary to link two **items of
  interest (IOI)**, without having prior knowledge of the
  identity of the data subject(s) involved (e.g., attackers
  can link data coming from different apps based on
  some common fields or PII), which may include both
  transactional and contextual data.

- **Identifiability**: Prompts an adversary can identify a
  data theme from a set of data subjects through an IOI
  (e.g., attackers can map the entire dataset of a user by
  reidentifying a user based on leaked data, metadata,
  and unique IDs).

- **Nonrepudiation**: Sets the standard guidelines where
  the data controllers or processors cannot refute a claim,
  where it has performed an action or sent a request, by
  which the data gets deleted and the user's actions are
  exposed.

- **Detectability**: Helps an adversary to have sufficient
  clues to distinguish whether an item of interest on a
  data theme is present, with or without interpreting the
  content. This leads to pinpointing devices which are
  subscribed to or having communication with mental
  health services.

- **Information disclosure**: Enables an adversary to read
  raw text about any data theme and disclose/leak it to
  the outside world.

- **Unawareness:** Process of data collection, processing, storage, or sharing where the subject matter (or the user) remains completely unaware of the key purpose of data collection, due to nontransparency of organization policies.

- **Noncompliance:** Process by which the organization accomplishes data processing, storage, or transmission of personal data remains noncompliant with legislation, regulation, and policy, as a failure on the part of the organization to conduct **Privacy Impact Assessments (PIA).**

With the general understanding of the LINDDUN framework, let us explore the different analysis processes and the tools involved using Figure 6-2.

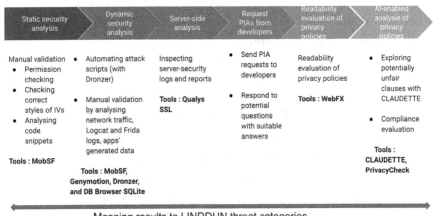

Privacy Analysis Processes

Figure 6-2. *LINDDUN threat analysis process with tools (Iwaya et al., 2022)*

As listed above in the principal privacy analysis process, let us summarize the sequences of how the analysis has to be actioned (Iwaya et al., 2022):

1. First, select the mental health apps on which privacy analysis is to be actioned.

2. Initiate static and dynamic security analysis to expose the security vulnerabilities in the selected apps. For example, deploying Qualys SSL for evaluation of servers, recognized in the previous dynamic analysis step.

3. Focus on any threats related to linkability, identifiability, nonrepudiation, detectability, and disclosure of information.

4. Scan through permissions recorded by every app and manifest files to discover unawareness and noncompliance threats.

5. Retrieve PIA reports of apps under investigation by requesting developers, or companies.

6. Conduct readability analysis using AI-powered tools on the privacy policies that the app has made a statement/declaration.

7. Determine unfair clauses and noncompliance areas that the app has promised.

8. Prepare a report ranking threats in orders or priority and share it with the app developers and companies to request review and correction.

As demonstrated in Figure 6-3, the threat disclosure process happens in a proper template report with enough evidence supported by the attached report, server log files, and web traffic to facilitate further discussion and progress with improvements on successive iterations. The process allows companies' feedback, which gives them an opportunity to report all the issues found for different apps to their respective stakeholders. With the feedback shared, it becomes easier to hear viewpoints from software developers, lead engineers, and privacy officers and then track the follow-up actions (e.g., forwarding the reports to the technical and legal teams) and address the fixes to yield a secure and privacy-preserving app.

Responsible Disclosure Process

Figure 6-3. *LINDDUN threat disclosure and iterative improvement process (Iwaya et al., 2022)*

Continuous feedback and improvement strive to maintain privacy standards. The standards set by information privacy regulations, like GDPR and the Australian Privacy Act, along with threat modeling frameworks, are used to stop data leakage to the external world. We as community developers should also encourage the publication of PIA reports to demonstrate to stakeholders the app has been a subject of critical privacy analysis, and concerns are being addressed to reduce issues on app privacy. In the first place, developers can check the list of available

apps from the Play Store and request details of the public reports of their
PIAs. Now let us discuss some readily available tools in the market which
can help us with static and dynamic analysis of our systems and how they
efficiently assist in preventing different types of threats.

Static Security Analysis

The most commonly used for performing the static analysis is MobSF
which can unfold a wide range of security issues on Android OS (Iwaya
et al., 2022). It is an open source tool and provides flexibility and simplicity
to use a fully automated framework for pen-testing on each APK file
for each app. The report generated by running automated testing using
MobSF includes the apps' average Common Vulnerability Scoring System
(CVSS) score, trackers, certificates, Android permissions, hard-
coded secrets, and URLs. Common false positives of vulnerabilities are
further analyzed manually to flag them as dangerous or safe based on the
protection level of the permission and the purpose that the application
provides. MobSF's static analysis method serves as an effective tool to
detect information disclosure threats where data remains unencrypted
and has a high risk of getting leaked through insecure channels of
communication. Moreover, this tool helps in classifying information
disclosure risks as critical, high, medium, and low and signifies how
authentically developers are adhering to best practices of secure coding
(Iwaya et al., 2022).

The Research Centre on Privacy, Harm Reduction and Adversarial
Influence Online (REPHRAIN) also suggests having a testbed for
privacy assessment, which would provide a drag-and-drop facility to
help developers drop their app files into a user interface. The testbed
is designed to run multiple static and dynamic privacy tests against the
dropped-in app and generate a report in CSV format which would help
developers to run their analysis.

Researching mental health apps' reverse-engineered code often
revealed detailed statistics of the apps using insecure pseudorandom
number generator (PRNG) and insecure cyphers (i.e., MD5 and SHA1,
ECB). Employing static analysis tools on multiple iterations helps to test
the correct usage of initializing vectors (IVs) and validate code snippets
that could have used insecure random number generators and cyphers.
Static analysis and evaluation reports help developers to prohibit using
unsecured IVs (like a hard-coded string of fixed value that generates weak
ciphertexts) and instead force them to use an authentic cryptographic
pseudorandom number generator that can securely generate an
initial state.

The code sample below gives an example of generating IVs using
insecure random (i.e., java.util.Random)

```
public IvParameterSpec getIv () {
byte[] bArr = new byte[this.writer.getBlockSize()];
 System.arraycopy ("fldsjfodasjifudslfjdsaofshaufihadsf"
 getBytes(), 0, bArr, 0, this.writer.getBlockSize());
return new IvParameterSpec (bArr);
}
```

The below code snippet gives a demonstration of generating an IV
as a hard-coded string of fixed value, which poses threats on resulting
ciphertexts when used for a fixed key, many times.

```
if (this.encrypt){
byte[] bArr = new byte[16];
this.random.nextBytes (bArr);
dataOutputStream2.write (bArr);
 this.cipher.init (1, this.secretKeySpec, new IvParameterSpec
 (bArr));
}
```

Dynamic Security Analysis

One of the prominent dynamic analysis tools is Genymotion Android emulator which scans the apps to perform black box security testing on the app by triggering malicious operations on it. Both MobSF and Genymotion provide an interface to emulate the apps, schedule security checks on the app, and record data which includes, logs, network traffic, and databases. Another dynamic security analyzer is Drozer which simulates an application by running an attack script against a target app. The attack scripts were sufficient for checking for attack surfaces, SQL injection, and directory traversal vulnerabilities.

In our threat modeling framework, we have an overview of linkability threats, and these threats are further likely to be disclosed during the dynamic analysis of an app. Dynamic analysis in contrast to manual inspection of network traffic can reveal IP addresses, device IDs, session IDs, or details specific to location, frequency, and browser settings. Dynamic threat analysis can prevent user behavior leakage from web traffic logs. In instances where apps try to pseudo-anonymize users through anonymous IDs or hashed advertisement IDs, dynamic analysis can safeguard against linking these IDs with data from third parties.

The dynamic analysis also helps us in assessing identifiability threats where attackers can sufficiently recognize a subject within a set of broader themes, examples of which include recognizing a web page reader or an email sender. This helps an attacker to reidentify the subject by combining the data attributes with the known identities retrieved from network traffic, logged or stored data. Leveraging Logcat dumps and DB Browser SQLite can provide greater assistance in determining threats related to the identifiability of login used and contextual data. Careful dynamic scanning can offer a great benefit in preventing the reidentification of leaked pseudo-identifiers, such as usernames and email addresses. On top of dynamic analysis, strong access control needs to be maintained to control identifiability threats.

Static and dynamic analysis becomes an inevitable part of our privacy ecosystem; let us glance through some commonly available AI-based tools.

Analysis of Privacy Policies with AI-Based Tools

Two of the most commonly used tools, having in-built support of AI, include CLAUDETTE which identifies unfair clauses with a 78% accuracy for identifying unfair clauses and an accuracy of 74%–95% reported for distinguishing between unfair clause categories. We also use PrivacyCheck to compute GDPR compliance scores and access an individual's control. This tool evaluates an assessment privacy score and reports an accuracy of 60%, by carefully analyzing ten user control questions and the ten GDPR questions.

Leaking Device Data

The example below demonstrates a linkability threat in how device data and the user's usage and activities in the app or device are transmitted to third-party vendors, like AppsFlyer, a mobile marketing analytics and attribution platform. Even though fields such as app ID, AppsFlyer key, and UID have been masked (XXXXXX), other sensitive information marked in bold is transmitted to the third party.

POST https://t.appsflyer.com/api/v5/androidevent?buildnumber=4.
9.17&app_id=**XXXXXX**
Content-Length: 1586
Content-Type: application/json
HTTP/1.1
User-Agent: Dalvik/2.1.0 (Linux; U; Android 9: Samsung
b'{"country":"uk","af_timestamp":"1599497460531",
"appsflyerkey":XXXXXX" "af_events_api":"1","isFirstCal":"
true","registeredUninstall":false,**"operator":"T-Mobi le"**
,"network": "WIFI", "timepassedsince lastlaunch":"-1",

"af_v2": "bb46f3ca34e18e212b2a47d9f7afb500a97bc4b3","uid":
XXXXXX" "isGaidWithGps":"true", "lang_code ":"en","install
Date":"2020-09-07_164952+0000", "app_version_code":"112",
"firstLaunch Date":"2020-10-07_165102+0000", "model":"Samsung
A10", "currency":"POUNDS", "lang":"English", "brand":"Android",
"b atteryLevel":"100.0", "deviceType":"userdebug",
"product":"vb ox86p","af_sdks":"0000000000", "deviceData":
{"sensors":[{"sT ":1,"sVS": [0,9.776219,0.813417], "sN":
"Genymotion Accelerometer", "SV":"Genymobile", "sVE":
[0,9.776219,0.813417 1}],"cpu_abi": "x86", "build_display_
id":"vbox86p-userdebug 9 PI 451

Leaking User Activity Data

Here, we cite an example of a linkability threat, demonstrating how a third
party receives a user's activity information.

POST https://api.iterable.com/api/users/update HTTP/1.1 Accept:
application/json
Content-Type: application/json
API-Key: XXXXX
 b'{"email": "XXXXXX" {"Amplitude.device_
id":"XXXXXX","dataFields" "device_preferred _languages":"en",
"device_default_language":"en", "bedtime_re minder_
enabled": false, "mindfulness_reminder_enabled": false,
"timeZone":"America\\/New York"}}'

In the below example, we demonstrate a nonrepudiation threat where
a third-party logging service can first record an app's activity and then send
the data over HTTP.

Leaking User Activity Information Through Third-Party Logging Service

POST

http://logs-01.loggly.com/inputs/**XXXXXX**

Host: logs-01.loggly.com

HTTP/1.1

Proxy-Connection: keep-alive

Content-Length: 189

Origin: http://localhost

User-Agent: Mozilla/5.0 (Linux; Android 9; Samsung A10

b'{*"message":"A/B Journaling Reflection"*,"level": "verbose", "appVersion": "8.03.002", "plat formVersion":9,"platform": "android", "online": true, "sessionId":"3d282505-8eca-415d-a2a2-cf6674ee1b21"}"

Leakage of User Personal Data

In the below example, we showcase an example of an identifiability threat where a user's email ID is exposed to third-party services.

POST https://sdk.iad-02.braze.com/api/v3/data HTTP/1.1 Accept-Encoding: gzip, deflate

Content-Type: application/json

X-Braze-Api-Key:"XXXXX"

b'{**"device_id":"XXXXX"**,"time":1599059514, "api_key":" ","sdk_version":"3.7.0","app_version":"8.12.0","app_ver"}],"events":[{"name" sion code":"81200007.0.0.0", "attributes":[**{"email":" @gmail.com", "user id":"XXXX}],**
events[{ "name" : "ce", "data":{"n":"LOGIN"},"time":1 .599059514722E9, "user_id":"l","session_id":"36e7203e-cb58-2091-12c9-1ec36c9

Let us now see, with an example code listing, common practices of encoding data making it vulnerable for attackers to infringe. Later on in the chapter, we would see how to make the ML models private.

Listing 6-2. This is example where we study an example of statistical research on the effects of mental health on student's CGPA

In this part, we take a student mental health dataset data and try to predict the effects of students' mental health on their CGPA. A university generated this dataset through a Google Form survey to examine the current academic situation and mental health of their students. This dataset contains various physiological and mental health attributes and more than 200 data points.

This task can be viewed as a tabular classification task where we can leverage tabular techniques like decision models to predict students' CGPA based on their physiological and mental health conditions. We use the label encoder to convert the string labels into integer classes so that they can be fed into the models for training. Then, we move on to build our model for predictions. We make use of one of the best models available, that is, XGBoost model.

1. In the first step, we install the necessary libraries.

    ```
    %pip install pandas
    %pip install xgboost sklearn
    ```

2. After installation, we import the necessary libraries
    and model XGBoostClassifier.

    ```
    import pandas as pd
    from sklearn.model_selection import train_test_split
    from xgboost XGBoostClassifier
    from sklearn.preprocessing import OneHotEncoder,
    LabelEncoder
    ```

3.   In the next step, we load the dataset and drop
the stress_level column which is used as the
target column.

```
df = pd.read_csv('/content/Student Mental health.csv')
X= df.drop(columns=['What is your
CGPA?','Timestamp'],axis=1)
y= df['What is your CGPA?']
```

4.   After loading the dataset, we need to preprocess the
dataset, to obtain the train and test data subsets.

```
X = X.apply(LabelEncoder().fit_transform)
le = LabelEncoder()
y = le.fit_transform(y)
```

5.   Then, we fit the data to XGBoost model to validate
existing data points.

```
bst = XGBClassifier()
bst.fit(X, y)
bst.score(X, y)
```

This yields the following prediction on being
queried with

```
bst.score(X_test,y_test)
```

```
Choose your gender                                    0
Age                                                   0
What is your course?                                 17
Your current year of Study                            3
Marital status                                        0
Do you have Depression?                               1
Do you have Anxiety?                                  0
Do you have Panic attack?                             1
Did you seek any specialist for a treatment?          0
Name: 0, dtype: int64

Prediction:   ['3.00 - 3.49']
```

Figure 6-4. *Interview questionnaire posed to patients*

We can see the model perfectly predicts the CGPA range. This kind of textual/categorial data encoding without adding noise dangerous can be discovered by static analysis tools. In this example, attackers can directly infringer CGPA, in the same way income details and other sensitive behavioral traits can be exposed to attackers in naive label encodings. Later in this chapter, we would see how to add noise to all attributes of the dataset to anonymize our ML models.

Dynamic analysis plays a significant role in identifying nonrepudiation threats initiated by third-party systems such as marketing and advertising, cloud service provisioning (e.g., Amplitude, Facebook), and payments services (e.g., Stripe and PayPal) when mental health apps enter into communication with these third-party services using insecure HTTP protocol. When the attacker takes up the role to prove the user has known or has uttered something involving mental health apps and services leading to the exposure of decrypted network connection logs. Such scenarios arise in situations when persons waiting for deniability are prevented from modifying database entries. Nonrepudiation threat detection is key to preventing unknowing ways of recording user actions by third-party logging services through any mental health app.

Further, detectability threats can also be detected by dynamic analysis of network traffic or scanning existing weak access controls that threaten to expose data from file systems or databases. To protect against the detection of additional data of user data from using inferential methods (where attackers knowing specific user profiles for a mental health service can further infer about psychological support they must have availed or will be availing soon), dynamic analysis provides a layer of protection of sensitive user information that would have leaked otherwise on communication through insecure communication channels.

A lot of dependency exists on a system's provisions to educate users in making privacy-aware decisions when data is shared with third parties. This kind of unawareness threat is better handled by dynamic and static analysis by issuing requests of PIAs and entering into a free discussion with developers. MobSF static analysis does a thorough investigation, to audit the restricted actions (READ_EXTERNAL_STORAGE and WRITE_EXTERNAL_STORAGE which grant indiscriminate access to the device's external storage) and limit the grant of unwanted permissions that pose potential threats.

Dynamic threat analysis, in spite of providing a useful tool to detect different types of attacks, poses limitations. There is a risk in running this Android emulator with MobSF due to incompatibility, as the app is not capable of running on a rooted device. Further, developers also tend to block the app, to stop it in emulation environments. Nevertheless, using the tools helps us run automatic operations instead of manually navigating the app, page by page, entering text, and recording entries with the apps. MobSF is capable of generating Logcat log, Dumpsys log, Frida API monitor log, and HTTP/S traffic log for the entire period of interaction with the app and is equipped with a download button to allow the data generated by each app and stored in its device's storage. Logs are additionally analyzed to check if a user's personal sensitive information or secret tokens of the app are disclosed, along with the app's functionality, usage, and actions pursued on the app. This analysis procedure also enforces to classification of the data (as encrypted and not encrypted) by leveraging DB Browser SQLite to search data in any storage locations on the device including apps' folders, files, and databases.

Beyond static and dynamic analysis, we need to look at the following resources to determine any chances of vulnerabilities.

- Analyze the security of web servers deployed in place, by assessing the HTTPS data transmission and security aspects governing the transmission. Data transmitted to and from the apps can be scanned using tools like Qualys SSL Labs, which can perform remote security testing on the web server by validating against known vulnerabilities.

- Evaluate the easiness in which the app privacy policies are readable when used by psychologically and cognitively challenged users; some metrics of importance are the number of words, sentences, number of complex words, etc.

- Privacy policy analysis using AI-enabled tools like CLAUDETTE and PrivacyCheck to detect biased, discriminatory, illegal clauses in apps' privacy policies, for example, jurisdiction disputes, choice of law, unilateral termination, or change.

After getting an overview of threat modeling frameworks and privacy analysis techniques, we should also have a fair amount of understanding of how our AI-based mental health frameworks are being designed to be compliant with the laws and we should choose to deploy large-scale systems to the cloud (such as AWS, Azure, or Google Cloud) where the data is protected and stored as per the standards and guidelines of healthcare and state/country privacy rule.

Our next step as system designers should be to respect user privacy, whether the solutions are deployed in the cloud or on-prem. Now let us deep dive into some primary considerations while storing and deploying mental health solutions on the cloud.

Design Considerations for Storing Mental Health Data on the Cloud

With most of the mental apps and wearables using cloud services, the foremost question that developers and cloud architects need to be worried about is how we ensure data security and privacy of the confidential mental health information getting stored in the cloud. The greatest challenge remains in the data transmission processes; when data is transferred through the cloud, it possesses the maximum risk of tampering, hacking, and intrusion. Let us understand some of the key architectural considerations when we store personal data and mental health-related information in the cloud.

- Leverage end-to-end encryption techniques by enabling a strong encryption mechanism which can make data visible to sending and receiving parties only.

- Enable multifactor authentication to ensure data confidentiality by prompting to ask username and password, followed by a question related to the subject through the OTP (one-time pad).

- The right selection of cloud service, choosing a private or hybrid cloud deployment, works best for storing sensitive clinical application's data.

- Employ **infrastructure as a service (IaaS)**, a cloud architecture model, for mental healthcare sectors to offer scalability, security, data protection, and backups.

- Establish the best-secured communication mechanisms to enable and prevent man-in-the-middle attacks from browsers and encrypt data to ensure safe and secured transfer over the network and through the cloud.

However, the selection of the right encryption algorithm enables it to save computing resources and time. Encryption can be performed using a fresh, randomly generated 256-bit symmetric key chosen by the client-side application. AES256 ensures semantic security by generating a fresh randomly chosen 128-bit initialization vector. "Lazy re-encryption" is a powerful technique to safeguard encryption keys for files and directories changing rapidly, where group membership also changes, due to the addition/removal of members from the user group.

In terms of employing cloud data security, the additional measures which we should be concerned about, apart from "Data Security Measures" discussed in the previous subsection, are service-level agreement, security audits performed by third-party auditors, **Virtual Private Networks (VPC)**, an efficient intrusion detection system, and a zero-trust-based cloud dependability model. A popular security mechanism is to employ homomorphic encryption that provides confidentiality, authorization, and availability at the cloud data center. It operates on a key split model residing in the cloud with separate key delegations and healthcare requirements. Any access request is validated by the private key jointly with the delegated key for successful authorization.

For mental healthcare deployed on AWS, **Identity Access Management (IAM)** is one of the popular techniques for controlling access to AWS resources through role-based permissions. The best way is to maintain a zero-trust environment by granting the least privilege to users and incrementally allowing access, based on their roles. This helps to control data breaches. It also supports multifactor authentication and **single sign-on (SSO)** to further control and secure centralized user access.

233

Taking AWS as an example cloud provider, let us look at different security tools available through Figure 6-5 to safeguard different services.

AWS Account Security Tools

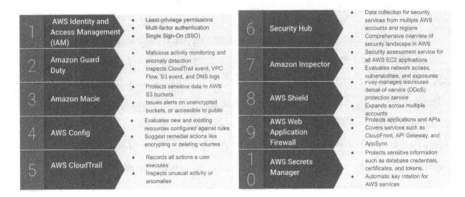

Figure 6-5. *AWS security tools to protect sensitive mental health data*

With the above design considerations in mind, let us now study how we would design an encryption-based mental healthcare system, using a cloud provider by leveraging a one-time pad-based multifactor authentication process.

End-to-End Encryption of Mental Healthcare Data Transaction

Any data transactions containing secure mental healthcare data have to be done under strict security protocols to eliminate adversarial attacks. With an illustrative example in Figure 6-6, we show how the sequence of steps involved in a transaction process to transfer and retrieve the data securely at the client's and doctor's end using a cloud environment (Settu, 2017; Sivan & Zukarnain, 2021).

End-to-end Encryption-based Secured Mental Healthcare Transaction Process

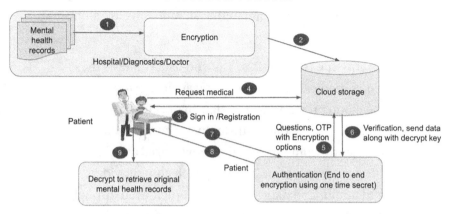

Figure 6-6. *Patient and hospital system-secured data transmission (Settu, 2017)*

The encryption is a two-entity exchange process occurring at the hospital and the client end that helps to authenticate the client and establish trust between both parties.

- **Hospital- or doctor-side transaction**: These transactions operate on the Noise Protocol Framework, where both parties exchange handshake messages, using the Diffie-Hellman key agreement process. The handshake-based exchange process functions with the help of a single message composed of static key patterns formed by aggregated tokens. The interactive protocol supports public key exchange with the data getting stored in the cloud. The resultant message from the Diffie-Hellman algorithm is hashed with the shared secret key before initiating the data transfer process. The fixed message size offers benefits like testing, error reduction, or controlling memory integer overflow. In addition, private mental health data can also be encrypted using cypher and hash functions in addition to DH functions.

- **Client-side transaction:** The end user requests the cloud medical data by initiating an access request to the cloud provider using a sign-in or sign-up process. This allows safe data transfer in a secure manner protecting from intermediate adversarial attacks (that can come from Internet or telecom providers) and authorized data access. This is achieved through an end-to-end encryption process and a one-time challenge generator algorithm that fully supports authenticated data exchange between the cloud provider and requester. Initially, the authenticated parties are identified using a one-time pad (OTP), followed by message exchange leveraging the Diffie-Hellman key exchange process. After the user sign-in, the OTP verification process helps to validate the user by matching it with the provider's secret key. It then helps to establish a private channel of communication between the user and the provider. To provide the data to the user, the corresponding decryption key is also exchanged during the key exchange process, which helps in decrypting the data. The Noise Protocol Framework along with its handshake message and hash function adds extra layers of security in the decryption process.

The two-entity exchange process can be further summarized as a sequence of steps as follows:

1. Initiate the sign-up process from the client end to the cloud provider.

2. Trigger the user authentication process with the secret key sent on the phone, by the provider.

3. Secret key entry by the client, using the provider OTP.

4. Start the key exchange between the client and the cloud provider.

5. Process exchange of DH key values to provide access to the decryption key.

6. Utilize the corresponding secret key and handshake message to decrypt the data.

Other than encryption-based methods, federated learning (FL) also plays a crucial role in collating protected mental health information by collaborating with other devices to learn and update the shared model, facilitating mental disorders' detection, diagnosis, and treatment. Their unique approach of keeping sensitive data localized reduces the risk of data breaches helping them to stay compliant with privacy regulations. Now let us explore in the below section how privacy can be embedded in FL to safeguard sensitive information related to mental disorders.

Privacy-Enabled Federated Learning (FL) for Mental Health

As we have seen in previous sections, different predictive ML models can aid in the diagnosis and treatment of mental health problems like depression; the principal challenge remains when traditional centralized

ML models aggregate patient data, violating user privacy. Though this poses a limitation in front of us, to properly leverage ML algorithms for clinical applications, researchers and data scientists have made an extensive study to come up with a system architecture that is built to support medical data privacy for mental health (like depression). In addition, researchers have also come up with an application level privacy to train models, termed **Differential Privacy (DP)** where a change in one data record between two datasets does not substantially change the outcome from the predicted result, which cannot help the attacker to guess the input dataset or determine any record which can be present in the input (Wang et al., 2024). We would see that in an example with a code sample. Privacy is of paramount importance. We will also see in Figure 6-8 how DP can be infused in the FL environment.

The system architecture for FL, as illustrated in Figure 6-7, showcases a multiview federated learning framework setup, that uses multisource data, to build both traditional and deep learning models by aggregating data across different institutions or parties. The variety associated with federated learning data can yield an accuracy of 84.32%, with an improvement rate that is about 9% higher than local centralized training.

Federated Learning Sequences in Mental Health

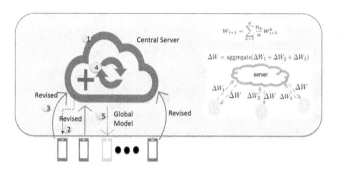

1. The coordinator sends the initial model W to all the participants
2. Each participant trains locally using W and obtains an update ΔWi
3. All the participants send their updates to the coordinator
4. The coordinator aggregates the updates and use them to update W
5. Repeat the Step 1-4 loop until converge.

Figure 6-7. *Federated learning environment to protect sensitive mental health data*

Let us now study how specialized FL operates for mental health predictions.

Specialized FL for Mental Health Monitoring

FL with its ability to train global models by aggregation of local models trained on data collected from devices (utilizing sensor and PPG signal data from smartphones and wearables including location, accelerometer, and call logs) can help in real-time psychological state assessment and categorize individuals as stressed and nonstressed using readings like heart rate (HR) and heart rate variability (HRV) measures. However, plain vanilla FL can also suffer from a range of cyber threats, notably backdoor attacks, model inversion attacks, and membership inference attacks where very sensitive information of local mobile devices can be compromised. As stated above, in these situations, we adopt DP through the introduction of noise into the combined updates before transmitting back to the central server. The added noise is calibrated to the sensitivity of the model's gradients and the amount of privacy safeguards provided by the model is denoted by privacy budget ϵ (epsilon).

In the federated averaging (FedAvg) algorithm of DP (Xu et al., 2022), we can add Laplacian noise to prevent loss of user data confidentiality. FedAvg operates by averaging model updates (weights) from multiple clients before the local models are aggregated to update the global model. FL with DP can further be extended to clustered FL where the group of clients can be clustered to form minor subgroups. This clustered version of FL helps to have better data balance, by the inclusion of minority members in every cluster. In addition, we can also add homomorphic encryption during model weights transmission followed by secure parameters aggregation during the update of the global model to add a layer of security at the transport layer. Figure 6-8 illustrates an enhanced version of FL with clustering, DP, and encryption in place for mental health monitoring.

Encryption and Differentially Private enabled Clustered Federated Learning

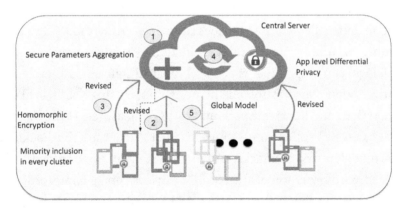

Figure 6-8. *Enabling encryption, DP, and clustering in FL for mental health*

Data Distributions in FL

The data distributions in FL can be independent and identically
distributed (IID) or nonindependent and identically distributed (Xu et al.,
2022). When we train models in FL setup, data about non-IID category
experience different challenges due to the following attributes (Khalil
et al., 2024):

1. **Skewed quantity**: Data quantity skewness happens
   due to an imbalance in data class distribution
   among different clients, leading to bias in training
   data and consequently in predicted results. This
   is more observed when one type of client collects
   more varieties or classes of data than the other.

2. **Label skew**: Label skew is observed when within the
   same data class, data size ranges vary among clients,
   leading to uneven data class sizes. This is visible
   in big hospitals or mental diagnostic clinics when
   they collect more data related to depression when
   compared to smaller medical centers.

3. **Feature skew**: Feature skewness is more serious
   as data collected by clients may not be uniform
   concerning the number of features. This difference
   arises due to the nature of mental health diseases,
   due to differences in machines getting used like the
   usage of MRI scanners, which may not come from
   the same manufacturer.

In contrast to the non-IID dataset, the IID dataset is balanced for all
the attributes. Non-IID datasets can overcome the challenges of data
skewness using data augmentation and feature skewness using data
imputation methods.

Privacy-preserving federated learning can be used for depression detection by aggregating the data from multiple mobile applications. In the IID version of FL, greater number of mobile clients participating in the learning collaboration the volume of data increases which leads to increased accuracy (up to 12%) in depression detection. This may also be affected by duplicate data. A non-IID version of FL demonstrates higher levels of robustness due to non-IID settings. Using a decentralized approach gives a considerable edge for FL, which provides a huge benefit in data privacy and governance, where data is collected remotely from patients, without involving their physical movements, beyond the boundaries of the healthcare center. In addition, FL operates on a consensus solution where local models and their weights are transferred in collaboration without exchanging data. Figure 6-8 shows the working mode of operation of FL, with the sequence of steps involving a central server and several distributed clients who together participate in the decentralized mode of training.

FL in mental health can help us to detect and diagnose different diseases as illustrated in Figure 6-9. We can use different types of traditional ML models like decision trees, XGBoost, and random forest, as well as deep learning models like CNN, and base models to train both numerical, image, audio, and video data to detect symptoms related to depression, schizophrenia, suicide, bipolar disorder, and AHDH detection.

Federated Learning in Mental Health

Depression	Schizophrenia	Suicide	Violence	OCD	Bi-polar disorder	ADHD

❖ Tabular
  ➢ Decision Trees, XGBoost, RandomForest & linear SVM
  ➢ DeepMood (LSTM)
❖ Image (Resnet-18)
❖ Audio
  ➢ CNN
  ➢ GoogleNet, MobileNet V2 & ResNet-18
❖ Text
  ➢ LSTM & Bi-directional LSTM
  ➢ CNN
  ➢ Deep Feedforward Neural Networks
  ➢ BERT models
	❖ Tabular
	  ➢ Constac software for decentralized data analysis
	❖ Image
	  ➢ Deep Feedforward Neural Networks
		❖ Text
		  ➢ CNN & LSTM
			❖ Text
			  ➢ Multi-layer Perceptron (MLP) Network
				❖ Tabular
				  ➢ Bi-directional LSTM
					❖ Tabular
					  ➢ Deep and Cross Network (DCN)
						❖ Image
						  ➢ Deep Feed-forward Neural Network

Figure 6-9. *Types of ML models in FL environment to treat a variety of mental health problems (Khalil et al., 2024)*

Types of FL in Mental Health

FL can broadly be classified into three types from the standpoint of data partitioning and how the models get trained collaboratively. These types are **horizontal FL (HFL)**, **vertical FL (VFL)**, and **federated transfer learning (FTL)** (Sharma & Guleria, 2023).

Type of FL	Working methodology
HFL	Clients using the local dataset to train their respective local models have the same features, leading to homogeneous feature sets in the patient datasets of participating clients. This is the most frequently used data partitioning scheme.
VFL	Clients using the local dataset to train their respective local models have different features, leading to the heterogeneous feature sets in patient datasets of participating clients. This occurs because patient data comes in from different hospital centers and diagnostic clinics.

(continued)

Type of FL	Working methodology
FTL	Clients using the local dataset to train their respective models have different features and patient profiles. Transfer learning exploits the advantage of a pretrained model already trained (that happens during the training phase of one model on a similar dataset at one client) to solve a different problem for another client.

In the above two codes (Listings 6-1 and 6-2), we have seen how ML models play a part in prediction. However, these models do not show how to anonymize the sensitive parameters which a student may not want to disclose. Let's look at the next example which helps us to add noise to a model's predictions using Differential Privacy.

Listing 6-3. This is an example of building private models using Differential Privacy on mental health data consisting of sensitive information

We took mental health survey data containing attributes like age, gender, country, treatment, family_history, remote_work, work_type, mental_health_cons, phys_health_cons, phys_health_intrvw, mental_health_intrvw, and work_type to predict mental_vs_physical condition to understand which of the mental or physical condition an employer feels need care and attention.

1.   In the next step, we install the necessary libraries.

```
%pip install pandas
%pip install diffprivlib
```

2. After installation, we import the necessary libraries
   and models from diffprivlib.

```
import pandas as pd
from sklearn.model_selection import train_test_split
import matplotlib.pyplot as plt
from diffprivlib.models import GaussianNB,
RandomForestClassifier
```

3. We map the data attributes to different category
   labels, for example, assigning them to different age
   groups. Next, we create a Gaussian Naive Bayes
   Classifier model and fit the train data to experiment
   on nondifferentially private models.

```
X_train, X_test, y_train, y_test = train_test_split(df_
data, df_target, test_size=0.2)
clf = GaussianNB(random_state=0)
clf.fit(X_train, y_train)
print("Non Differentially private test
accuracy  %.2f%%" %
      (clf.score(X_test, y_test) * 100))
```

This yields

```
Non Differentially private test accuracy  41.98%
```

Figure 6-10. *Results of non-DP solution accuracy*

4. Our next step is to attach a set of bounds to
   GaussianNB and set epsilon (ϵ) =0.1 to test the
   accuracy of differentially private models.

```
bounds = ([25, 23, 20, 15, 10, 9, 8, 7, 6, 5, 4,3, 2,
1, 0, 0, 1], [200, 150, 100, 80, 70, 60, 50, 40, 30,
20, 16, 1000, 100000, 4500, 100, 20, 10])
clf0 = GaussianNB(epsilon=0.1, bounds=bounds, random_
state=0)
clf0.fit(X_train, y_train)
clf0.score(X_test, y_test)
print("Differentially private test accuracy
(epsilon=%.2f): %.2f%%" %
      (clf0.epsilon, clf0.score(X_test, y_test) * 100))
```

This yields

```
Differentially private test accuracy (epsilon=0.10): 28.81%
```

Figure 6-11. *Results of DP at epsilon =0.10*

5. We also run the same test on a series of epsilons
    (between 0.1 and 100) with the bounds set in the
    previous step.

```
epsilons = np.logspace(-1, 2, 100)
accuracy = []
for epsilon in epsilons:
    clf1 = GaussianNB(epsilon=epsilon, bounds=bounds,
    random_state=0)
    clf1.fit(X_train, y_train)
    accuracy.append(clf1.score(X_test, y_test))
print("Differentially private test accuracy
(epsilon=%.2f): %.2f%%" %
      (clf1.epsilon, clf1.score(X_test, y_test) * 100))
```

This yields where we see the accuracy improves at epsilon = 100.00

```
Differentially private test accuracy (epsilon=100.00): 39.51%
```

Figure 6-12. *Results of DP at epsilon =100.0*

6. Our next step is plot the graph and visualize

```
plt.semilogx(epsilons, accuracy, label="Differentially
private Naive Bayes", zorder=10)
plt.semilogx(epsilons, baseline * np.ones_
like(epsilons), dashes=[2,2], label="Non-private
baseline", zorder=5)
plt.xlabel("epsilon")
plt.ylabel("accuracy")
plt.ylim(-5, 1.5)
plt.xlim(epsilons[0], epsilons[-1])
plt.legend(loc=4)
```

This gives us

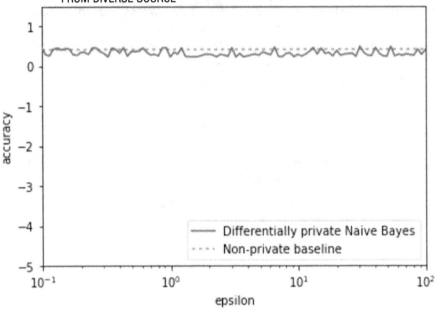

Figure 6-13. *Accuracy vs. privacy (epsilon=0.10) plot with Naive Bayes method*

7. We run another round of experiments with a new
   set of bounds and epsilons, which increases the
   accuracy level.

```
bounds = ([25, 23, 20, 15, 10, 9, 8, 7, 6, 5, 4,3, 2,
1, 0, 0, 1], [30, 40, 50, 55, 60, 65, 70, 75 , 80, 85,
90, 50, 40, 30, 20, 10, 100])
epsilons = np.logspace(-1, 2, 1000)
accuracy = []
for epsilon in epsilons:
    clf2 = GaussianNB(epsilon=epsilon, random_state=0)
    clf2.fit(X_train, y_train)
  accuracy.append(clf2.score(X_test, y_test)
```

8.   Next, we plot the results and visualize

```
plt.semilogx(epsilons, accuracy, label="Differentially
private Naive Bayes", zorder=10)
plt.semilogx(epsilons, baseline * np.ones_
like(epsilons), dashes=[2,2], label="Non-private
baseline", zorder=5)
plt.xlabel("epsilon")
plt.ylabel("accuracy")
plt.ylim(-5, 1.5)
plt.xlim(epsilons[0], epsilons[-1])
plt.legend(loc=4)
```

This yields

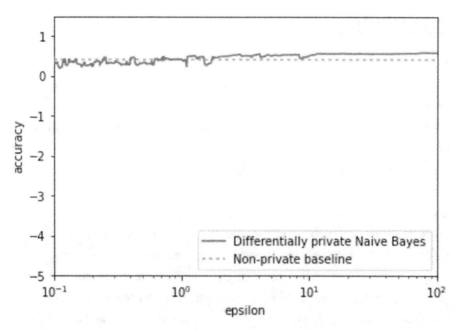

Figure 6-14. *Accuracy vs. privacy (epsilon= 100.0) plot with Naive
Bayes method*

```
Differentially private test accuracy (epsilon=100.00): 58.02%
```

Figure 6-15. *DP results at epsilon =100.0*

We have seen how adding Laplacian noise with a Naive Bayes Classifier yields DP models. Initially, the nonprivate model is seen to have a higher accuracy which diminishes adding noise. Further, on testing with noise boundaries, the accuracy has been seen to boost up. However, for the performance of this model, we should be aware that DP models with a privacy budget have lower accuracy compared to generic models with no noise. As developers, we need to choose the right trade-off between accuracy and privacy.

Now let us have the concluding thoughts from this chapter.

Conclusion

In this chapter, we have learned about the importance of privacy, security, and authenticity of mental health data collected by different wearables and apps to safeguard data from adversaries. This directly pointed us to the limitations of existing processes concerning FDA reviews, joint certification of software and hardware, absence of rigorous testing of embedded sensors, and calibration that directly impacts the data quality, leading to biased model predictions. With the current challenges on one hand and more and more digitization on the other hand, we got an understanding of how critical it is for us to protect virtual therapy sessions and do a complete threat analysis of mental health apps in the data transfer ecosystem. The chapter provided us with a complete overview of data protection privacy policies, standards, regulations, and data security measures with due emphasis on GDPR, HIPAA, and others, to protect against leakage of device data, personal data, user activity data, or leakage through external third parties through logging. We also studied all possibilities of data monetization to design an attack-resistant system. With a thorough insight into privacy threat modeling frameworks like LINDDUN, static, and dynamic analysis tools, along with privacy-preserving design principles on the cloud, we dived into system design considerations for mental health systems. We went further to understand privacy-enabled federated learning (FL) for mental health with different examples and the types of FL that can be used for mental health disease prediction and cure.

With a detailed background on data privacy and best practices for system design, we explore how "AI models from human-computer interaction textual data" can be used to predict mental distress in youth.

References

Alder, S. (2025). Healthcare Data Breach Statistics. *The HIPAA Journal.*
https://www.hipaajournal.com/healthcare-data-breach-statistics/

Alwakeel, A., Alwakeel, M., Zahra, S. R., Saleem, T. J., Hijji, M., Alwakeel, S. S., Alwakeel, A. M., & Alzorgi, S. (2023). Common Mental Disorders in Smart City Settings and Use of Multimodal Medical Sensor Fusion to Detect Them. *Diagnostics, 13*(6), 1082. https://doi.org/10.3390/diagnostics13061082

Bauer, M., Glenn, T., Geddes, J., Gitlin, M., Grof, P., Kessing, L. V., Monteith, S., Faurholt-Jepsen, M., Severus, E., & Whybrow, P. C. (2020). Smartphones in mental health: A critical review of background issues, current status and future concerns. *International Journal of Bipolar Disorders, 8*, 2. https://doi.org/10.1186/s40345-019-0164-x

Emily A. Dorner, Nancy L. Perkins, Jami Vibbert, & Jason T. Raylesberg. (2023). *FTC Announces $7.8 Million Fine as Part of Settlement With BetterHelp Regarding Health Information Privacy Practices | Enforcement Edge | Blogs.* Arnold & Porter. https://www.arnoldporter.com/en/perspectives/blogs/enforcement-edge/2023/03/ftc-announces-betterhelp-fine

Hunter, K. (2024). *NHS hack warning issued to everyone in Dumfries
and Galloway.* https://www.bbc.com/news/articles/cn00q1329420

Ienca, M., & Malgieri, G. (2022). Mental data protection and the
GDPR. *Journal of Law and the Biosciences, 9*(1), lsac006. https://doi.
org/10.1093/jlb/lsac006

Iwaya, L. H., Babar, M. A., Rashid, A., & Wijayarathna, C. (2022). On the
privacy of mental health apps. *Empirical Software Engineering, 28*(1), 2.
https://doi.org/10.1007/s10664-022-10236-0

Khalil, S. S., Tawfik, N. S., & Spruit, M. (2024). Exploring the potential
of federated learning in mental health research: A systematic literature
review. *Applied Intelligence, 54*(2), 1619–1636. https://doi.org/10.1007/
s10489-023-05095-1

Lyngaas, S. (2023). *Mental health startup exposes the personal data
of more than 3 million people | CNN Politics.* CNN. https://www.cnn.
com/2023/03/10/politics/cerebral-mental-health-privacy-data-
exposure/index.html

Martínez-Pérez, B., de la Torre-Díez, I., & López-Coronado,
M. (2014). Privacy and Security in Mobile Health Apps: A Review and
Recommendations. *Journal of Medical Systems, 39*(1), 181. https://doi.
org/10.1007/s10916-014-0181-3

Mukherjee, A. (2023, September 3). Data Security in the Age
of AI: ISO 27001 Compliance. *Medium.* https://medium.com/@
mukherjee.amitav/data-security-in-the-age-of-ai-iso-27001-
compliance-29f700f0f787

Settu, M. (2017). *AN APPROACH TO SECURE MENTAL HEALTH
DATA IN THE CLOUD USING END-TO-END ENCRYPTION
TECHNIQUE. 8,* 87–98.

Sharma, S., & Guleria, K. (2023). A comprehensive review on federated
learning based models for healthcare applications. *Artificial Intelligence in
Medicine, 146,* 102691. https://doi.org/10.1016/j.artmed.2023.102691

252

Sivan, R., & Zukarnain, Z. A. (2021). Security and Privacy in Cloud-Based E-Health System. *Symmetry, 13*(5), Article 5. https://doi.org/10.3390/sym13050742

U.S. Department of Health and Human Services. (2017). *HIPAA Privacy Rule and Sharing Information Related to Mental Health.* U.S. Department of Health and Human Services. https://www.hhs.gov/sites/default/files/hipaa-privacy-rule-and-sharing-info-related-to-mental-health.pdf

Wang, Z., Yang, Z., Azimi, I., & Rahmani, A. M. (2024). *Differential Private Federated Transfer Learning for Mental Health Monitoring in Everyday Settings: A Case Study on Stress Detection *Authors equally contributed.* https://arxiv.org/html/2402.10862v2

Xu, X., Peng, H., Bhuiyan, M. Z. A., Hao, Z., Liu, L., Sun, L., & He, L. (2022). Privacy-Preserving Federated Depression Detection From Multisource Mobile Health Data. *IEEE Transactions on Industrial Informatics, 18*(7), 4788–4797. IEEE Transactions on Industrial Informatics. https://doi.org/10.1109/TII.2021.3113708

SECTION 3

How AI Models Can Predict Different Mental States Of The Teenage Population?

## CHAPTER 7

AI Models from Human–Computer Interaction Textual Data Recorded to Predict Mental Distress in Youth

In this chapter, a comprehensive examination of various text processing algorithms for detecting mental stress is presented. The examination highlights the importance of textual data in different spheres of teenage youth and how this data can become a valuable tool for an alerting system to assess mental health risk. Readers understand how textual data, collected from applications and third-party sources, can be effectively processed at multiple stages, to predict different levels of risk. The chapter introduces diverse deep learning algorithms and explores their application in detecting mental distress in different scenarios. Equipping the readers

© Sharmistha Chatterjee, Azadeh Dindarian, Usha Rengaraju 2025
S. Chatterjee et al., *Revolutionizing Youth Mental Health with Ethical AI*,
https://doi.org/10.1007/979-8-8688-1186-9_7

with ML libraries, tools, and techniques, the chapter empowers them to efficiently train and predict mental distress, enabling a more data-driven and effective approach toward supporting mental health assessment and intervention.

In this chapter, these topics will be covered in the following sections:

- Usage and applications of textual data in mental health symptom detection

- Preprocessing of textual data

- Text generation and classification with ML models

- Prospects of NLP techniques in improving mental health diagnosis

Usage and Applications of Textual Data in Mental Health Symptom Detection

AI and **natural language processing (NLP)** have played a pivotal role in collecting and analyzing historical data from various sources to diagnose the patient's disease. As we have seen in Chapter 4, "Empowering Mental Well-Being Through Data-Driven Insights," data collection techniques and analysis tools have become more advanced to drive meaningful insights on patients' mental conditions from electronic medical records and social media posts. In addition, cell phones and wearables have made a remarkable accomplishment through mental health apps, equipped with ecological momentary assessment measurements and deployment of interventions based on the severity of devices.

Identifying emotions during various activities has helped to capture trends and patterns of a person's mental health through journaling, social activities, engagement with academic institutions, or through parent and student participation in psychometric tests. All of the data sources rely heavily on NLP, and even virtual assistant tools like chatbots are developed and are built on advanced algorithms of NLP to assist in real-time patient treatment, diagnosis, and monitoring. AI-powered chatbots come with advanced features to offer personalized recommendations and user-tailored responses to enhance emotional and mental well-being. Leveraging NLP by tracking patients' day-to-day life activities is trying to address the mental health challenges for the larger societal group at an affordable cost and reach.

NLP engines integrated with automated data and AI pipelines have made a tremendous impact in assessing the severity of mental health, prescribing recommendations, and connecting to emergency services whenever needed. The end-to-end design is backed by fine-tuned processing algorithms (text, audio, and video) to infer current conditions and learn/relearn new patient conditions, with continuous data ingestion and feedback capability. It helps to update any prediction changes and connect the youth and their patients to affiliated medical practitioners. The benefits of NLP are not only realized by youth and their parents; it manifests to a large extent to doctors and care providers as they can get real-time updates and act promptly to control the severity of the mental health conditions. The quality of mental health support improves as a result, yielding better infrastructure and operational management, leading to more room for research to better to improve further **cognitive behavioral therapy (CBT)** and other treatment or counseling services such as behavioral activation, interpersonal psychotherapy, selective serotonin reuptake inhibitors, and tricyclic antidepressants.

The importance of text data has even been better understood when data generated by sensors has been merged with other contextual information about patients regarding their mental status and social relationships. The results give better avenues for researchers to explore how data from multiple sensor integration have improved detection and treatment plans for patients over a period of time.

To understand the range of sources where textual data can be gathered and assembled to derive insights, let us look through Figure 7-1 which illustrates the percentage of data crawled from social media platforms, EHR, surveys, interviews, and other online forums before feeding them to the NLP engine (Zhang et al., 2022).

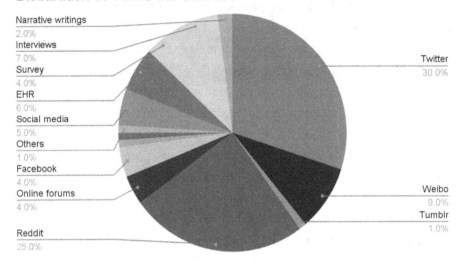

Distribution of Text Data Sources

Narrative writings 2.0%
Interviews 7.0%
Survey 4.0%
EHR 6.0%
Social media 5.0%
Others 1.0%
Facebook 4.0%
Online forums 4.0%
Reddit 25.0%
Twitter 30.0%
Weibo 9.0%
Tumblr 1.0%

Figure 7-1. *Diverse textual data sources to assess mental health using NLP (Zhang et al., 2022)*

Determining youth mental stress problems before has been carried out previously with data gathered from interviews and questionnaires which suffered from the limitations of reachability to larger societal groups within a certain community. Scalability and adequate privacy control measures remained an issue for many small startups as well as large mental health organizations, to design a fully efficient scalable real-time system that can learn from incoming text data and predict risk. Now many health organizations have started to adopt new collaboration methods, by allowing online group meetings to share information and seek help. This has paved the way for heavy-scale text data processing, from social media platforms using real-time data extraction techniques from Twitter and Reddit using application programming interfaces (APIs) (Tweedy and PRAW, respectively). Such APIs crawl the data from the social platforms based on keyword-based queries and feed to data preprocessing and algorithm (NLP) processing units to infer depression-related (positive) or standard posts (negative). Crawled data can be further labeled using information available from mental health community groups to extract the intended mental health problem data.

With suicide rates steadily rising among young people, we see an average of 10.5 suicide cases per 100,000 people (who.int, 2025). Since social media's rise to prominence, mental health data has been extracted to infer information, particularly from youth. Another mechanism of collecting anonymous contributions from social media platforms is to invite open dialogue-based conversations on socially defamed topics and encourage users to freely talk about their mental health problems and find support. Even open chat rooms, discussion rooms (recover your life, end his life), and TeenHelp.org, an Internet support forum, have been created for users to come and speak freely and the data used later on, to train psychological distress detection or self-harm detection models.

With advanced infrastructure and platforms being developed for mental health monitoring and the detection of depression and anxiety, **Twitter**, boasting over 300 million monthly active users, is widely recognized as a dominant social media tool. As Twitter encourages users to post their tweets or retweet, researchers have crawled the data using APIs to generate annotated datasets to help label users' depression status. Twitter has restrictions on sharing tweets publicly and tweet crawling has been limited over the recent past. Even many of the tweet identifiers can disappear before crawling the data.

Reddit is another well-known social media platform where users can publish posts and comments. Reddit offers the flexibility of grouping the posts into different subreddits based on the theme of the topics (i.e., depression and suicide). It also offers an open policy, where datasets are made available to the public which has resulted in the creation of a depression dataset named Reddit Self-reported Depression Diagnosis (RSDD) (georgetown-ir-lab.github.io, 2017). It contains posts from 9k depressed users and 100k control users. Researchers have even proposed an anorexia and self-harm detection algorithms from data available on the Reddit platform.

Apart from social media platforms, electronic health records (EHRs) serve as a rich source of secondary healthcare data, to extract patients' historical medical records, which include patients' profile information, medications, diagnosis history, and images. NLP techniques can be used to interpret narrative notes and diagnose EHR datasets for suicide screening for predicting other depression symptoms and mental health.

In addition, interviewing also serves as a data collection method where interviewing people helps to gather information on how they are feeling through a set of questions. Such responses can be used to detect mental illness after processing the linguistic information extracted from transcribed clinical interviews.

The intervention was comprised of therapeutic reminders delivered by SMS messages that were sent out 2 days, 7 days, 15 days, and monthly after hospital discharge. Each text message also provided a link to a mobile application that contained a questionnaire eliciting responses related to the patients' sources of help, evidence-based self-help strategies, and structured interview questions related to suicidal ideation, psychiatric symptoms, and satisfaction with care. Participants were also asked one unstructured, open-ended question related to their current mental state, "how are you feeling today?", and were encouraged to report on their progress since the hospitalization. Participants were able to enter responses to questionnaires up to once per day and were instructed to answer as often as they wished (Cook et al., 2016).

End users on mobile applications are often prompted with natural language-based queries to provide natural language-based answers in response to average nightly sleep (hours), sleep quality (0–100), anger rarely (0–100), changes in appetite (0–100), medication adherence (0–100) and WHO_5 well-being scale to assess their levels of stress, anxiety, or depression.

Text data processing became critically important at the COVID-19 pandemic when the lockdown created a negative impact on the mental health of undergraduate students. This has propelled researchers to think about how processing posts from students at schools and university communities can help us to understand the true conditions of their mental health. However, there remain certain challenges with the text data that need to be addressed.

Mental health-related problems are highly unstructured, unpredictable data composed of idioms, jargon, and dynamic topics, so it has become essential to adopt a scalable smart technique for retrieving the most valuable data features with a minimal set of dimensions that contains the maximum information. The process would execute in a sequence of steps and would help to boost the accuracy of systems for healthcare platforms designed for mental health.

Now let us look at what are the different sequences of operations involved in text data preprocessing to conclude varying mood conditions of youth and predict risk.

Preprocessing of Textual Data

Our first and foremost step is to understand how to make the text data ready to feed to NLP algorithms to receive appropriate outputs. The steps can be summarized in terms of the following steps as illustrated in Figure 7-2 (Amanat et al., 2022):

1. **Employ third-party APIs**: Use of third-party APIs with authentication tokens to download data, for example, Twitter APIs allow data crawling for seven days.

AI powered Depression Detector for Social Network Platforms

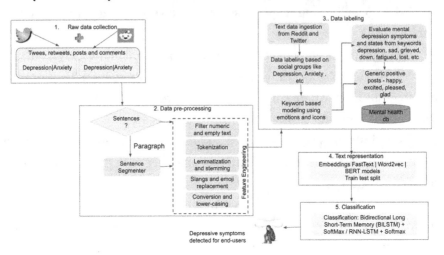

Figure 7-2. *Processing units for text data preprocessing and NLP algorithm (Amanat et al., 2022)*

2. **Removal of stop words**: The next step involves the removal of stop words from the data, where most frequent words like pronouns, prepositions, symbols (e.g., dates, #), conjunctions, articles (a, an, and the), and numerical characters are considered to be of little value that they would contribute anything to mental health information. The removal process is carried out by splitting the text into words and then eliminating words present in the list of stop words provided by the NLTK (Amartey Roland Teye, 2023).

3. **Tokenization**: The primary function of this step is to split or break the text into individual words, to make it easier for ML models to interpret long text, in terms of subwords instead of the original word. The

process helps to address out-of-vocabulary words,
by referring to smaller chunks of main words, that is,
the subwords.

4. **Lemmatization and stemming**: Both techniques
   aim to remove prefixes or suffixes present at the
   beginning or at the end of word tokens to generate
   the root word. The objective is to deduce the
   keyword and remove the redundancy, as the
   inflected words carry the same meaning, as the
   stem word. Lemmatization goes one step beyond
   by leveraging corpus for stop words and WordNet
   corpus for the concerned language to scan and
   explore the entire morphology of the neighboring
   words to interpret the word's part of speech. An
   example of lemmatization of "mice" is "mouse"
   which retains a normalized form of the word.

5. **Text and emoji replacement**: Stop word removal is
   then followed by replacing jargon and emojis with
   their factual text. Emojipedia is a critically important
   method in substituting emojis and icons to add
   more authenticity in mental health risk prediction.

6. **Expanding words to their generic meaningful
   forms**: This step may be represented in shorter
   forms like "b4," which has to be written as "before"
   or "2moro" which has to be spelt like "tomorrow."
   The meaningful representation of words then needs
   to be converted to lowercase characters.

7. **Data visualization**: Visualize the data for positive
   and negative sentiments as data visualization can be
   done through appropriate tools like WordCloud.

8. **Part-of-speech (POS) tagging**: This step involves the use of NLTK POS tagger to classify words into their parts of speech, a processing company known as **part-of-speech (POS) tagging**. These parts of speech include nouns, verbs, adjectives, adverbs, determiners, and conjunctions, to identify POS tags that impact to derive insights for anxiety and depression and eliminate POS that does not contribute to the identification of depression and anxiety. POS tagging isolates nouns, adjectives, and adverbs which plays a greater influence in determining the sentiments.

One primary challenge that may occur is using the right depression lexicon as it (Cha et al., 2022) can hold different meanings in social media (one such example is the word "isolation" which can refer to both emotional isolation and physical isolation). So it becomes essential to understand the context of the data and label it accordingly.

Now that we have gained an understanding of the different steps involved in text preprocessing, let us look at a practical example of how each of the processing steps changes textual data before passing the data to an ML engine.

In this part, we take a mental health dataset data and growing stress prediction. This dataset comprises various linguistic, psychological, and behavioral attributes of a very huge population.

Listing 7-1. This is an example of how a typical text processing functions translate a paragraph text to a proper text before sending it to an ML model. The input text considered here is given below

"I'm a very empathetic person and I always try 2 be kind and loving to people However, I'm constantly worried that I'm not doing it right, or that the things I enjoy and the way I think/act is offensive and not very caring.

I'm fine with changing how I behave if what I'm doing bothers people, but I got so overwhelmed and frustrated on what if I I am doing is offensive to someone. I feel unknowingly I may have hurt someone, and the thought makes me feel guilt and makes me feel that I can't enjoy my life very much. I want to be courteous to everyone and respect things that bother them and then not do/say those things, but I feel like I wearing myself out by constantly overthinking of myself."

Here, we have used NLTK library to process and understand human language data. NLTK has an easy-to-use interface to process more than 500 corpora and lexical resources such as WordNet. It also offers a suite of text processing libraries for classification, tokenization, stemming, tagging, parsing, and semantic reasoning and wrappers for processing industrial-strength NLP libraries. We have also used spaCy in the example below that heavily relies on providing software for production usage through in-built capabilities like tokenization, dependency parsing, and named entity recognition. spaCy has within it deep learning workflows that allow building relations with statistical models trained by machine learning libraries like TensorFlow, PyTorch, or MXNet.

1. In the first step, we install the necessary libraries.

```
%pip install nltk spacy
```

2. After installation, we import the necessary libraries and download the complete set of English stop words.

```
import nltk
import space
from nltk.corpus import stopwords
nltk.download('punkt')
from nltk.tokenize import sent_tokenize, word_tokenize
from nltk.stem import WordNetLemmatizer
nltk.download('averaged_perceptron_tagger')
```

```
from nltk.corpus import wordnet
nltk.download('averaged_perceptron_tagger')
nltk.download('stopwords')
stop_words = set(stopwords.words('english'))
print(stop_words)
```

3. Then, we convert the text to lowercase characters
   and remove numerical stop words from the text.

```
nlp = spacy.load("en_core_web_sm")
modified_str = ''.join([i for i in text if not
i.isdigit()])
doc = nlp(modified_str)
filtered_words = [token.text.lower() for token in doc
if not token.is_stop]
clean_text = ' '.join(filtered_words)
print("Text after Stopword Removal:", clean_text)
```

This gives output as

empathetic person try kind loving people, constantly
worried right, things enjoy way think / act offensive
caring . fine changing behave bothers people, got
overwhelmed frustrated offensive . feel unknowingly
hurt , thought makes feel guilt makes feel enjoy life .
want courteous respect things bother / things , feel
like wearing constantly overthinking

4. We pass the text for tokenization as

```
tokenized = sent_tokenize(clean_text)
wordsList = word_tokenize(clean_text)
print(wordsList)
```

This yields the following output for all
tokenized words:

```
['empathetic', 'person', 'try', 'kind', 'loving',
'people', ',', 'constantly', 'worried', 'right',
',', 'things', 'enjoy', 'way', 'think', '/', 'act',
'offensive', 'caring', '.', 'fine', 'changing',
'behave', 'bothers', 'people', ',', 'got',
'overwhelmed', 'frustrated', 'offensive', '.', 'feel',
'unknowingly', 'hurt', ',', 'thought', 'makes', 'feel',
'guilt', 'makes', 'feel', 'enjoy', 'life', '.', 'want',
'courteous', 'respect', 'things', 'bother', '/',
'things', ',', 'feel', 'like', 'wearing', 'constantly',
'overthinking', '.']
```

5. Our next action is to run through POS tagging engine
    for each of the tokens generated in the previous step.

```
for i in tokenized:
    wordsList = [w for w in wordsList if not w in
    stop_words]
    pos_tagged = nltk.pos_tag(wordsList)
    print(pos_tagged)
```

This yields output as

```
[('empathetic', 'JJ'), ('person', 'NN'), ('try',
'VB'), ('kind', 'NN'), ('loving', 'VBG'), ('people',
'NNS'), (',', ','), ('constantly', 'RB'), ('worried',
'VBD'), ('right', 'JJ'), (',', ','), ('things', 'NNS'),
('enjoy', 'VBP'), ('way', 'NN'),.....
```

6. In the next step, we try to get the root word by applying lemmatization.

```
lemmatizer = WordNetLemmatizer()
wordnet_tagged = list(map(lambda x: (x[0], pos_
tagger(x[1])), pos_tagged))
lemmatized_sentence = []
for word, tag in wordnet_tagged:
    if tag is None:
        # if there is no available tag, append the
        token as is
        lemmatized_sentence.append(word)
    else:
        # else use the tag to lemmatize the token
        lemmatized_sentence.append(lemmatizer.
        lemmatize(word, tag))
lemmatized_sentence = " ".join(lemmatized_sentence)
print("Lemmatized Sentence", lemmatized_sentence)
```

On running the above lines, we get the output as

Lemmatized Sentence empathetic person try kind love people, constantly worry right , thing enjoy way think/act offensive caring . fine change behave bother people, get overwhelmed frustrated offensive . feel unknowingly hurt, think make feel guilt make feel enjoy life. want courteous respect thing bother/thing, feel like wear constantly overthinking.

7. After employing lemmatization, we can go one step further to obtain the subjectivity and polarity of the Twitter reviews as follows:

```
from textblob import TextBlob
def getSubjectivity(review):
```

```
    return TextBlob(review).sentiment.subjectivity
def getPolarity(review):
    return TextBlob(review).sentiment.polarity

def analysis(score):
    if score < 0.10:
        return 'Negative'
    elif score == 0.10:
        return 'Neutral'
    else:
        return 'Positive'
```

In [15]:

```
subjectivity = getSubjectivity(lemmatized_sentence)
polarity = getPolarity(lemmatized_sentence)
print(subjectivity)
print(polarity)
pscore= analysis(polarity)
print(pscore)
```

This yields output as follows, denoting subjectivity as 0.54 and polarity as 0.25, and it's a positive sentiment.

0.5402380952380953

0.25023809523809526

Positive

The above output represents the classification of the sentence as opinionated as subjectivity is responsible for capturing whether a sentence is "opinionated" or "not opinionated." Further, the sentence polarity is also represented to be positive as in the above code we assume that a polarity score above 0.10 is positive.

Now having seen the complete text processing steps in the above example, let us also see how we can visualize the symptom keywords toward detecting mental health problems. The above example (Listing 7-1) does all the steps needed to complete text data preprocessing.

After data preprocessing, we should also gain an awareness of how to visualize the preprocessed and labeled text through a WordCloud to understand more the frequency of appearance of words and their polarity. So let us walk-through the use case of visualizing positive and negative sentiments using a WordCloud.

Listing 7-2. This is an example of how we can visualize a WordCloud for positive and negative sentiments using Twitter data

In this part, we take a social media Twitter dataset data to analyze human depression, by identifying words from depression and nondepression categories. The dataset is created using various tweets from Twitter users, and each tweet is annotated with a label of "0" or "1," where "0" signifies a stress-negative tweet and "1" signifies a stress-positive tweet. There are more than 20000 tweets in the dataset.

1. In the first step, we install the necessary libraries.

   ```
   %pip install wordcloud
   ```

2. After installation, we import the necessary libraries.

   ```
   from wordcloud import WordCloud, STOPWORDS
   import matplotlib.pyplot as plt
   import pandas as pd
   ```

3. Then, we load the Twitter mental health dataset and view the column entries.

   ```
   path = "Mental-Health-Twitter.csv"
   df = pd.read_csv(path).drop(columns=['Unnamed: 0',
   "post_created", "user_id"])
   ```

4. After loading the data, our next step is to segregate it based on the labels (0 or 1 based on whether stress is negative or positive).

```
X = df.post_text.values
y = df.label.values
df_pos = df[df['label'] == 1]
df_neg = df[df['label'] == 0]
print("Positive and Negative tweet shapes", df_pos.
shape, df_neg.shape, df.shape)
df_neg.head(5)
```

This yields the following output for the negative sentiments.

	post_id	post_text	followers	friends	favourites	statuses	retweets	label
10000	819457334271279105	MY ENEMY'S INVISIBLE , I DON'T KNOW HOW TO FIGHT	123	145	1068	23801	0	(
10001	818505546424647680	im gonna burn my house down into an ugly brack	123	145	1068	23801	0	(

5. Our next action is to tokenize the positive sentiment text (string), convert it to lowercase characters through an iteration process, and visualize it through a WordCloud.

```
comment_words = '' "
stopwords = set(STOPWORDS)
for val in df_pos.post_text:
    val = str(val)
    tokens = val.split()
    for i in range(len(tokens)):
        tokens[i] = tokens[i].lower()
    comment_words += " ".join(tokens)+" "
```

273

```
wordcloud = WordCloud(width = 800, height = 800,
                    background_color ='white',
                    stopwords = stopwords,
                    min_font_size = 10).
                    generate(comment_words)
```

This yields the following WordCloud for all positive
sentiments, where we see words like love, good,
treatment, and sleep getting priority in the display
along with mention of certain mental health
keywords like depression, migraine, and addiction.

Figure 7-3. *WordCloud demonstrates positive sentiment for
mental health*

6.   Our last step is to visualize the same WordCloud
     using negative sentiments and train the data to
     identify signs of depression from positive and
     negatively labeled tweets.

7.   We may further want to identify the features
     from tweets that are contributing positively to
     depression. We can do so by using **singular value
     decomposition (SVD),** where we first import
     the libraries as before and use TfidfVectorizer to
     vectorize/tokenize the text data, based on the
     relevancy of words present in the document.

```
from sklearn.decomposition import TruncatedSVD
from scipy import sparse as sp
vectorizer = TfidfVectorizer()
text_features_train = vectorizer.fit_
transform(df['post_text'])
text_features_train.shape
```

8.   Then, we need to employ truncated SVD to reduce
     the dimensions of high-dimensional data while
     retaining most of the original information. In
     contrast to normal SVD that only computes a k
     number of singular values (where k is a user-defined
     hyperparameter) to represent the output dimension
     or number of features, truncated SVD operates on
     term count/tf-idf matrices, returned by the vectorizers
     of textual data (often representing sparse matrices).

```
pca = TruncatedSVD(n_components=2)
features_components = pca.fit_transform(text_
features_train)
total_var = pca.explained_variance_ratio_.sum() * 100
```

275

```
print("Total explained variance", total_var)
df_features_components = pd.DataFrame(features_
components)
df_features_components = pd.concat([df_features_
components, df[['label']]], axis=1, ignore_index=True)
df_features_components.columns = ['SVD_comp_1', 'SVD_
comp_2', 'target']
df_features_components.describe(include='all')
```

We can see that TruncatedSVD output resembles
nearly the same as **principal component analysis
(PCA)**, with the only exception of performing well
on large sparse datasets which cannot be centered,
unlike PCA. This helps to save memory and
computing time.

This yields the following output as the total
explained variance and the two SVD components.

⇥ Total explained variance 0.7998760745928815

	SVD comp_1	SVD comp_2	target
count	2.000000e+04	20000.000000	20000.000000
mean	6.817227e-02	0.010666	0.500000
std	6.021349e-02	0.065779	0.500013
min	-6.196116e-13	-0.206869	0.000000
25%	1.538943e-02	-0.027662	0.000000
50%	4.486680e-02	0.004388	0.500000
75%	1.163199e-01	0.016844	1.000000
max	3.622653e-01	0.436240	1.000000

9. On further plotting with a graph, we get the
    key two principal identified components as
    illustrated below.

```
cmap = {0: 'red', 1: 'blue', 2: 'green'}
df_features_components.plot(kind='scatter', x='SVD
comp_1', y='SVD comp_2', c=[cmap.get(t, 'black') for t
in df_features_components['target']])
```

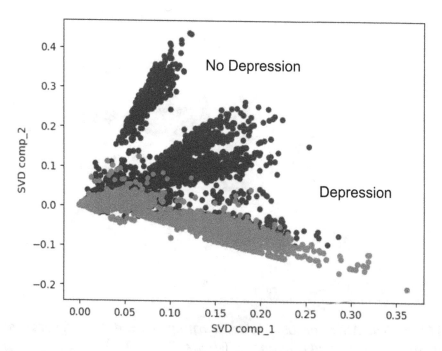

Figure 7-4. *Truncated SVD to extract high-dimensional features on
low dimension to detect depression vs. no depression*

In addition to using WordClouds, we can also visualize the positive
and negative sentiments for depression by using attention-based
LSTM. Figures 7-5 and 7-6 further illustrate positive and negative vibes
for mental depression or psychological stress, respectively, by depicting

a change in the intensity of the words, by deepening the color tone of the tokenized words. For example, negative words like "frustrating" and "tensed" are marked much darker than neutral words like "a lot" or positive words like "confident" as in Figure 7-5 (Winata et al., 2018). The change in the intensity of colors toward a darker shade happens due to a change in the weight of the attention layers for the corresponding words.

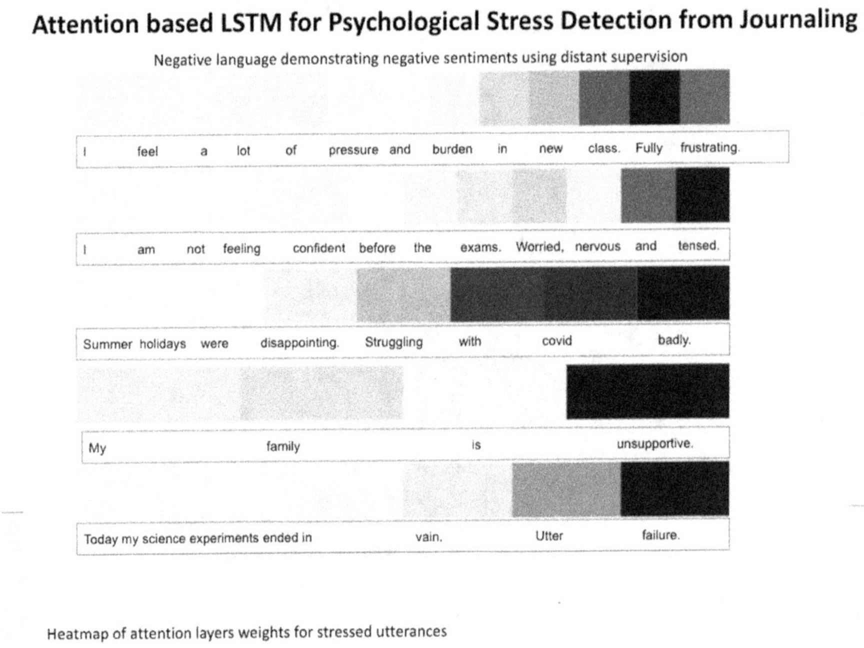

Attention based LSTM for Psychological Stress Detection from Journaling

Heatmap of attention layers weights for stressed utterances

Figure 7-5. *Attention-based deep learning of text data to generate a heatmap for stressed utterances (Winata et al., 2018)*

Attention based LSTM for Psychological Stress Detection from Journaling

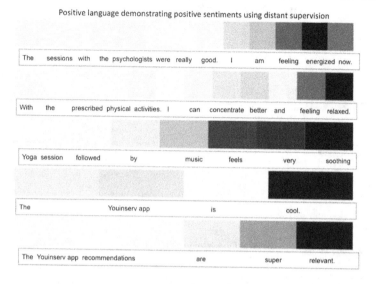

Positive language demonstrating positive sentiments using distant supervision

The sessions with the psychologists were really good. I am feeling energized now.

With the prescribed physical activities. I can concentrate better and feeling relaxed.

Yoga session followed by music feels very soothing

The Youinserv app is cool.

The Youinserv app recommendations are super relevant.

Heatmap of attention layers weights for unstressed utterances

Figure 7-6. *Attention-based deep learning of text data to generate a heatmap for unstressed utterances (Winata et al., 2018)*

In Figure 7-6, we find that the color tone becomes darker from neutral to more positive words (Winata et al., 2018). Such examples include "feeling relaxed," "very soothing," and "relevant." Neutral words like "sessions," "prescribed," and "Yoga session" are marked in light yellow to green shade.

In addition to looking at a lens of positive and negative sentiments separately, we can also use an x-y coordinate axis to rate all the words on four dimensions—pleasantness, unpleasantness, activation or arousal state, and deactivation state. The former mood state pleasant/unpleasant state can be further broken down to analyze activation/deactivation situations. Figure 7-7 further illustrates this using practical words related to depression and anxiety.

Representation of Depression Symptoms from Twitter Data on a Mood Activation Scale

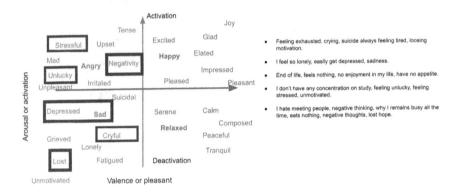

Figure 7-7. *Activation scale for pleasant or unpleasant moods*

We saw text preprocessing techniques along with data visualization, to show how to represent keywords from mental health problems or represent positive or negative polarized textual words.

Now let us take one step further to understand different types of traditional and deep learning models that can be used for assessing risk.

Text Generation and Classification with ML Models

Now that we have preprocessed the data, we need to fit the data into an ML model to gain predictive insights.

Training ML models with text data can be done using traditional and deep learning methods. Figure 7-8 illustrates the traditional text vectorization techniques using CountVector, TF-IDF, or hashing technique (Chowanda et al., 2023; Kurniadi et al., 2024) followed by training, for supervised or unsupervised textual data. It also represents different types of regression and classification algorithms that can be applied on text data or **singular value decomposition (SVD)**, truncated SVD, PCA, and clustering methodologies for unsupervised data.

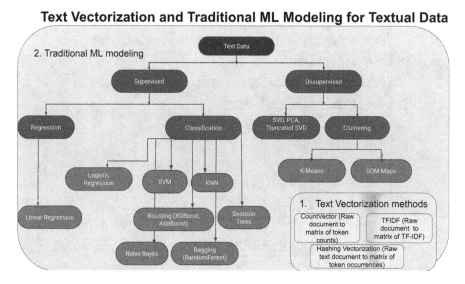

Figure 7-8. *Vectorization and prediction/classification ML models on text data (Amanat et al., 2022; Hu et al., 2022)*

The below code sample illustrated text vectorization techniques using CountVector, TF-IDF, and hashing. In this part, we have taken a Reddit thread dataset data to perform human stress detection.

Listing 7-3. This is an example of vectorization of text data followed by training an ML model

sentences = ['Feeling worried, even though you have a God who is ready to help you in any case', 'Why is my school so different? The others have passed why do i still have assignments and exams']

1. In the first step, we import the CountVector or related vectorization library and fit the sentences to retrieve the counts.

```
from sklearn.feature_extraction.text import
CountVectorizer
vectorizer = CountVectorizer()
vectorizer.fit(sentences)
vectorizer.vocabulary_
```

```
{'feeling': 9,
 'worried': 27,
 'even': 7,
 'though': 23,
 'you': 28,
 'actually': 0,
 'have': 11,
 'god': 10,
 'who': 25,
 'is': 14,
 'ready': 18,
 'to': 24,
 'help': 12,
 'in': 13,
 'any': 2,
 'case': 4,
 'why': 26,
 'my': 15,
 'school': 19,
 'so': 20,
 'different': 5,
 'the': 22,
 'others': 16,
 'passed': 17,
 'do': 6,
 'still': 21,
 'assignments': 3,
 'and': 1,
 'exams': 8}
```

Figure 7-9. *Vectorization to find the word count*

2. In the next step, we apply preprocessing of the text
   data and fit to any classification/regression model
   based on the use case. In this example, we have
   used LogisticRegression.

```
count_vectorizer = CountVectorizer(token_
pattern=r'\w{1,}',
                        ngram_range=(1, 2), stop_words =
stopwords,preprocessor=custom_preprocessor)
count_vectorizer.fit(X_train)
train_vectors = count_vectorizer.transform(X_train)
test_vectors = count_vectorizer.transform(X_val)
from sklearn.linear_model import LogisticRegression
clf = LogisticRegression(C=1.0)
scores = model_selection.cross_val_score(clf, train_
vectors, y_train, cv=5, scoring="f1")
clf.fit(train_vectors, y_train)
clf.score(test_vectors, y_val)
```

The above steps help to train the model and give a test score of 85.32%.
The vectorization technique can be extended to **TF-IDF** or using a **hashing**
mechanism as illustrated below.

Using TF-IDF:

```
tfidf_vectorizer = TfidfVectorizer(min_df=3,  max_features=None
,analyzer='word',token_pattern=r'\w{1,}',
ngram_range=(1, 3), use_idf=1,smooth_idf=1,sublinear_tf=1,
         stop_words = stopwords)
```

Using hashing:

```
hash_vectorizer = HashingVectorizer(n_features=10000,norm=None,
alternate_sign=False)
hash_vectorizer.fit(X_train)
```

As you can see in the code above, we have used "*n*-grams" feature, to extract a contiguous sequence of *n* words (instead of a single word to help to give the model more linguistic context). The n-gram size is configurable leading to unigram, bigram, or trigram settings that helps to determine the rate of accuracy by identifying the greatest positive predictive value (PPV) (Parikh et al., 2008).

Specificity is the metric that helps the model predict a true negative of each class available (Parikh et al., 2008).

Specificity = TN / TN + FP (true negative/true negative + false positive).

Sensitivity is a metric that helps to determine the model's ability to predict the true positives of each available class (Amartey Roland Teye, 2023).

Sensitivity = TP / TP+FN (true positive/true positive + false negative).

As there exists a close trade-off between specificity and sensitivity, it is often found that unigrams (single words) and bigrams (two words following a token) showed lower specificity but higher sensitivity, while trigrams have a higher specificity and lower sensitivity. The trade-off needs to be rightly chosen based on the use case and the text data processed.

Evaluation of the above metrics is essential in all of the three processes, CountVector, TF-IDF, and HashingVectorization. After tokenization of the data, the regular training methodology follows as illustrated in the CountVector technique. However, applying traditional ML models requires several steps to extract the key features of the data before applying supervised/unsupervised/recommended ML models, as shown in Figure 7-10. The figure illustrates a typical traditional ML pipeline that starts with feature engineering and dimensionality reduction modules, followed by exclusive ML engines that run clustering/classification or recommendation algorithms.

Teenage group Segmentation - Location based Analytics

An example to predict risk using clustering/PCA/SVD based on different segments of teenage population.

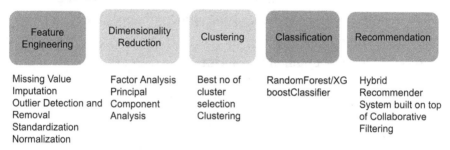

Feature Engineering	Dimensionality Reduction	Clustering	Classification	Recommendation
Missing Value Imputation Outlier Detection and Removal Standardization Normalization	Factor Analysis Principal Component Analysis	Best no of cluster selection Clustering	RandomForest/XG boostClassifier	Hybrid Recommender System built on top of Collaborative Filtering

Figure 7-10. *Activation scale for pleasant or unpleasant moods*

An important point of consideration is how these classification models exhibit varying performance metrics in different domain communities due to missing cross-domain features. Two notable examples are the demographic and linguistic information, which further instigates us to bring in cross-lingual method by ensembling multiple languages classification models. The kind of ensemble cross-lingual classification model takes maximum benefit of usage of words from English. Further, the cross-lingual classification model performance can be improved by using post behavior, image, sentiment, etc., that are available in the social media.

With a detailed overview of traditional ML models in place, let us do a deep dive into deep learning ML models to understand better how deep learning models can help in better efficient text data analysis and prediction.

Deep Learning-Based ML Models

Deep learning plays an important role in modeling textual data to create appropriate embeddings of the word vector. Some of the most popularly used models are FastText, word2vec, Bi-LSTM, and BERT which can be

employed for classification tasks to predict different categories of anxiety
or depression symptoms.

Figure 7-11 illustrates different types of static and contextual word
embeddings that help to predict similar types of symptoms in mental
health in the embeddings space. While pretrained word embeddings fall
in the static embedding category, contextual embeddings are obtained by
giving input to the pretrained transformed-based models.

Types of Text/Word Embeddings

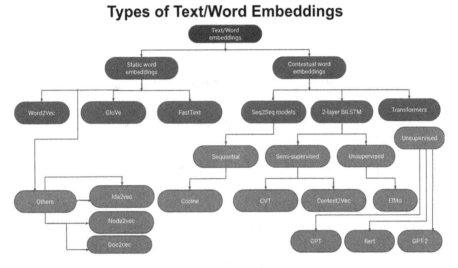

Figure 7-11. *Types of deep learning models with different word*
embeddings

Dynamic embedding considers all words in a given sentence to generate
the context and uses a process called attention to emphasize certain words
by scanning the neighboring words in the given sentence. An example
of an attention mechanism is *"The most inattentive student was always
scolded by teachers, so she became more demotivated to study,"* where all the
embeddings are dynamically generated from the pretrained model. The
attention mechanism is further illustrated in Figure 7-12, how certain words
in the sentence shown on the left side are overemphasized compared to its
neighbors to derive the context and sentiment of the sentence.

Now let us understand the underlying working philosophy of different types of static and dynamic word embeddings.

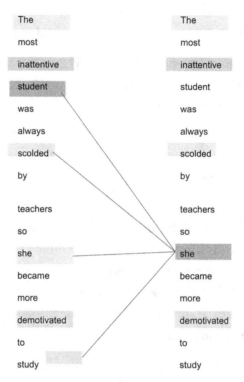

Figure 7-12. *Dynamic embeddings—attention mechanism of transformer model to show context dependency of neighboring words*

To begin with, let us first explore static word embeddings, which include Word2Vec, GloVe, and FastText.

Word2Vec

Word2Vec is an effective method of embedding, trained using neural networks, where vector math is applied to represent word representations (Jatnika et al., 2019). An example can be cited as "mental depression to disappointing performance to cheerfulness to excellent performance in class." The words "mental depression," "cheerfulness" and "disappointing," "excellent" are exactly a pair of opposites, whose relationships can be carved out among words for other neighboring words. The learning models that can be used to learn the word embeddings using Word2vec are

- **Continuous bag-of-words (CBOW) model**: Uses the current word context by learning the embedding to predict the next word

- **Continuous skip-gram model**: Uses the current word by learning the embedding to predict the surrounding words around the current word.

These models rely on learning models based on the nearby context words, and the configured distance or window of neighboring words helps to determine the context of the word. Training these models to create high-quality word embeddings from high-dimensional data can be done quickly. Although these models represent larger corpora of text with billions of words, they suffer from the limitation of small context and absence of global occurrence.

GloVe

Word2Vec has been extended to generate the **Global Vectors for Word Representation (GloVe)** (Jeffrey et al., 2014) algorithm, using word-context matrix factorization techniques. It represents a matrix of (words × context) cooccurrence data, that is derived by counting the frequency of appearance of a specific word in a particular "context" for each "word." For example, the sentences "A poor performer is under mental depression" and "A top performer is cheerful" would yield a matrix as given in Table 7-1.

Table 7-1. GloVe—word-context matrix representation

	Words											
Context	A	poor	performer	is	under	mental	depression	A	top	performer	is	cheerful
A	0	1	2	2	1	1	1	2	1	2	2	1
poor	1	0	1	1	1	1	1	1	1	1	1	1
performer	2	1	0	2	1	1	1	2	1	2	2	1
is	2	1	2	0	1	1	1	2	1	2	2	1
under	1	1	1	1	0	1	1	1	1	1	1	1
mental	1	1	1	1	1	0	1	1	1	1	1	1
depression	1	1	1	1	1	1	0	1	1	1	1	1
A	2	1	2	2	1	1	1	0	1	2	2	1
top	1	1	1	1	1	1	1	1	0	1	1	1
performer	2	1	2	2	1	1	1	2	1	0	2	1
is	2	1	2	2	1	1	1	2	1	2	0	1
cheerful	1	1	1	1	1	1	1	1	1	1	1	0

Table 7-1 denotes the common word association between two sentences, where a count of 2 shows the words "a" and "performer" are common in both sentences. Due to the combinatorial nature of the matrix, there are many contexts, and the factorization of the matrix would yield a lower-dimensional matrix, where each row represents a vector representation for the corresponding word. GloVe leads to more effective word embeddings, by building a word context or word cooccurrence matrix, by leveraging the entire corpus instead of defining a window to limit the local context.

FastText

FastText is also a Word2Vec extension which considers each word composed of n-characters (Bojanowski et al., 2017). Overall, these character n-grams sum up to build the word and illustrated for the word vector "orange":

"<or", "ora", "oran", "orang", "orange" "orange>", "ran", "rang", "range" "range>", "ang", "ange", "ange>", "nge", "nge>", "ge", "ge>"

The most distinguishing feature of FastText compared to the other two methods is the use of n-grams. Instead of extracting complete words from the training data, FastText chooses vectors for individual words and the n-grams within each word. FastText experiences a considerable rise in the amount of computation in the training time, as the means of the target word vector and its n-gram component vector leads to an increase in the volume of the data.

FastText's character n-gram addresses the "out-of-vocabulary" problem by breaking down the word in terms of the smallest components, like the word "aquarium" that can be broken down into "aq/aqu/qua/uar/ari/riu/ium/um." Though the interpretation of the meaning of n-character subword is not instantaneous, it can be detected using the common root like in case of "aquarium" and "Aquarius," using the common root "aqua."

Seq2seq Models

Seq2seq models are designed using neural networks with two key components: an encoder and a decoder, to receive an input sequence, process it, and generate an output sequence. It can be largely employed in designing chatbots and processing tasks that involve sequential data. The combined encoder-decoder architecture is trained on a given input sequence to maximize the conditional probabilities of the target sequence.

The trained input data from the encoder is passed to the last state of its recurrent layer. This acts as an initial state for the first recurrent layer for the decoder. The decoder feeds in the last state of the encoder processes in subsequent layers and sends it to an output-dense layer, which then generates a one-hot vector representing the target word as illustrated in Figure 7-13. The figure demonstrates the processing steps involved in the encoder-decoder layers, which can then be summarized as follows:

1. **Encoder input layer**: Takes an input sequence and moves it to the embedding layer.

2. **Encoder embedding layer**: Receives the input sequence and converts each word to a fixed-size vector.

3. **Encoder first LSTM layer**: The generated vector from the word is then passed from its output to the next layer using CuDNNLSTM layer or LSTM because or BiLSTM.

4. **Encoder second LSTM layer**: The same flow continues and the last layer's states are passed on to the first LSTM layer of the decoder.

5. **Decoder input layer**: Receives the input from the encoder and moves it to the embedding layer.

6. **Decoder embedding layer**: It then converts each received word to a fixed-size vector.

7. **Decoder first LSTM layer**: The process of receiving a vector and moving to the next layer continues, and the last layer is initialized to be the last state of the LSTM layer from the decoder.

8. **Decoder second LSTM layer**: On processing the output from previous layers, the output is then moved to a dense layer.

9. **Decoder dense layer (output layer)**: It is responsible for generating the final hot target vector, once it receives the output from the previous layer.

The greatest advantage of this kind of architecture is that it supports the mapping of sequences of different lengths to each other when the inputs and outputs are not correlated and their lengths vary. Apart from chatbots, this kind of architecture can be used for mental health during speech recognition (Morgan et al., 2021), question answering tasks, and time series applications.

Encoder Decoder Architecture for Mental Health Chatbots

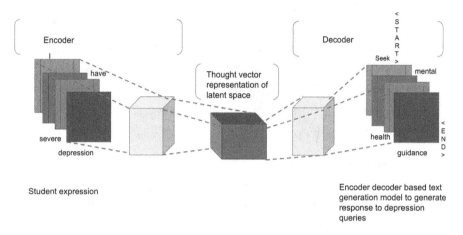

Figure 7-13. *Generative AI to design mental health chatbots using seq2seq models*

Now that we have an overview of how dynamic contextual embeddings play a pivotal role in seq2seq models, let us do a deep dive in doing text classification with the second type of seq2seq models using an LSTM network.

LSTM Models

In contrast to encoder-decoder models, a stacked LSTM or RNN model produces output sequences and input sequences, of the same length. Encoder-decoder sequence models can also be designed using LSTM networks. Let us now do a walkthrough of a simple bidirectional LSTM network to understand how it can classify and input text data to depression positive or negative.

Listing 7-4. This is an example of text classification using a
bidirectional LSTM network

1. In the first step, we import the necessary libraries
   from keras.layers that will be used to create
   the sequential network, provide as input the
   embeddings, and measure the efficiency of the
   model predictions.

```
from keras.models import Sequential
from keras.layers import Embedding, Conv1D,
MaxPooling1D, Bidirectional, LSTM, Dense, Dropout
from keras.metrics import Precision, Recall
from keras.optimizers.legacy import SGD
from keras.optimizers import RMSprop
from keras import datasets
from keras.callbacks import LearningRateScheduler
from keras.callbacks import History
from keras import losses
import tensorflow as tf
```

2. In the next step, we concentrate on the tokenization
   process to tokenize the input text into sequences of
   integers and then pad each sequence to the same
   length, to transform the text into a sequence of
   integers.

```
def tokenize_pad_sequences(text):
    tokenizer = Tokenizer(num_words=max_words,
    lower=True, split=' ')
    tokenizer.fit_on_texts(text)
    X = tokenizer.texts_to_sequences(text)
```

```
X = pad_sequences(X, padding='post',
maxlen=max_len)
return X, tokenizer
```

3. Once we are ready with the padded sequences,
    we define the parameters like vocabulary and
    embedding size, number of epochs for the training
    process, learning rate of the neural network, and
    momentum. This would serve as an input to
    regulate the convergence process by controlling the
    stochastic gradient descent (SGD) optimization
    process. We also build the model by specifying the
    embedding format, size, and the neural network
    layer architecture.

```
vocab_size = 5000
embedding_size = 32
epochs=20
learning_rate = 0.1
decay_rate = learning_rate / epochs
momentum = 0.8
sgd = SGD(lr=learning_rate, momentum=momentum,
decay=decay_rate, nesterov=False)
model= Sequential()
model.add(Embedding(vocab_size, embedding_size, input_
length=max_len))
model.add(Conv1D(filters=32, kernel_size=3,
padding='same', activation='relu'))
model.add(MaxPooling1D(pool_size=2))
model.add(Bidirectional(LSTM(32)))
model.add(Dropout(0.4))
model.add(Dense(2, activation='softmax'))
```

```
tf.keras.utils.plot_model(model, show_shapes=True)
print(model.summary())
model.compile(loss='categorical_crossentropy',
optimizer=sgd,
                metrics=['accuracy', Precision(),
Recall()])
batch_size = 64
history = model.fit(X_train, y_train,
                        validation_data=(X_val, y_val),
                        batch_size=batch_size,
                        epochs=epochs, verbose=1)
```

4. Finally, we want to evaluate the model performance
    by logging and plotting the accuracy metrics.

```
loss, accuracy, precision, recall = model.evaluate(X_
test, y_test, verbose=0)
print('Accuracy  : {:.4f}'.format(accuracy))
```

This yields the following accuracy graph along with the training and
validation loss.

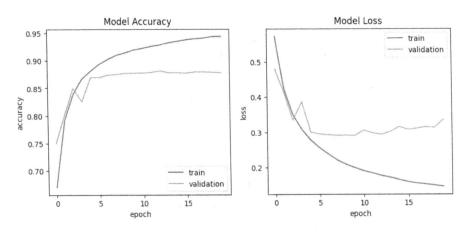

Figure 7-14. *Model accuracy and loss metrics for the training and
validation process*

Transformers

Transformer models are another kind of dynamic contextual word embedding model which operates on a sequence of tokens or other structured data. It uses a technique called self-attention that moves the data through a series of layers of feedforward neural networks. Using transformer models, mental healthcare apps can aid user's journaling activities by completing sentences as given below:

Mental health app: Good morning, how are you?

User: I am doing good.

Mental health app: Did you go to school?

User: Yes, today is the science exam.

Mental health app: Are you feeling nervous or tense?

In this way, the transformer models can be embedded within mobile apps to predict the next three or four consecutive words, by keeping track of the context of what is being written by the user. The transformer has four main components:

- **Tokenization**: Breaking the long input text into smaller parts in terms of words, to feed to an ML model that it can easily understand. The process of breaking to subwords helps to address out-of-vocabulary words, as the smaller chunks will have pretrained embedding stored. Padding follows tokenization where all input text is taken to be of equal length, with CLS token denoting the start of a sentence and SEP indicating the end of a sequence.

- **Embedding:** Conversion process by which a model converts the token into a vector representation, to represent its meaning in the context of the given sentence. It also generates a sequence of vectors in output corresponding to the input text.

- **Positional encoding:** Process by which we can enhance the meaning of the embedding by adding a sequence of predefined vectors to the embedding vectors of the words. This process not only helps to generate a unique vector for each input sentence but also adds more information to similar sentences having the same words through the specification of the order in the different output vectors. For example, the vectors corresponding to the words "Mental," "depression," "is," "impacting," "Diana's," "academics," and "." become enriched with the modified vectors when the positions are added like "Mental (1)," "depression (2)," "is (3)," "impacting (4)," "Diana's (5)," "academics (6)," and ". (7)." It is further illustrated in Figure 7-15 the steps involved in generating a unique vector corresponding to the sentence to represent all necessary information on the meaning and order of the words.

A Transformer Model Architecture showing text data processing

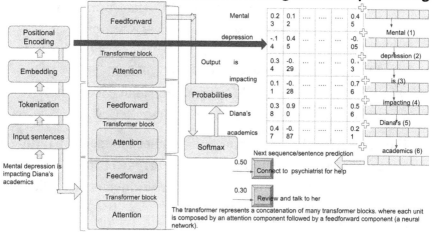

Figure 7-15. *Mental depression text processing and next sequence generation using a transformer model*

- **Transformer block:** To predict the next word in the sequence, we feed the data obtained in the previous step to a large neural network, which comprises two key components—attention and feedforward. We have illustrated the concept of attention in Figure 7-12 and how attention impacts different words in a sentence in Figures 7-5 and 7-6. We need to repeat the two key components, what is called a transformer several times to help language models understand and propagate the context in subsequent blocks. The primary function of the transformer is to assign the highest score to the words that are most likely to appear in the next sentence. It does so by the process of attention where the most important token in a sentence is evaluated compared to others and a weighted score is assigned to the token.

- **Softmax**: The softmax is the next processing unit after the transformer where it receives the transformer output scores for the words. This layer is responsible for converting these scores to probabilities where the highest score receives the highest probability. The transformer assigns the highest probability of 0.50 to "Connect" and a probability of 0.30 to "Review."

The transformer-based training and enhanced models can further help us generate future sentences to provide support as a chatbot. Now let us explore different components of a typical transformer model, called BERT.

BERT Model

With more research on NLP and advancement in this field, deep learning language models like **Bidirectional Encoder Representations from Transformers (BERT)** (Hu et al., 2022; Vaj, 2023) started to find huge applications to improve efficiency. The BERT model has been fine-tuned in domains related to biology, data science, and medicine, of which some of the healthcare-related BERT models are specified below.

- **BERT-Base**: This is a BERT model at its base, which has been trained on large text data and can be fine-tuned for specific mental health tasks.

- **BERTweet**: This is a BERT model that has been largely trained on Twitter data, to derive insights on mental health-related conversations on social media.

- **BioBERT**: This is a BERT model that has been pretrained on biomedical text, consisting of mental health-related research papers and clinical notes.

- **ClinicalBERT**: This is a pretrained BERT model obtained from clinical text that consists of electronic health records (EHRs) and clinical notes that can help in mental health diagnosis and treatment.

- **SciBERT**: This is a pretrained BERT model obtained from scientific text, like mental health research papers, which can be further reused for mental health research and analysis.

- **PubMedBERT**: This is a pretrained BERT model obtained from PubMed abstracts which can be further used for mental health research by digging deep into the research papers.

- **HealthBERT**: This is a pretrained BERT model obtained from health-related text, part of which also includes mental health-related articles, which can help in mental health-related content on the web.

- **BlueBERT, MedBERT**: This is a pretrained BERT model trained on a huge amount of biomedical text, like mental health research papers, articles, and clinical notes.

- **RoBERTa**: This is a pretrained BERT-based model optimized on a few NLP tasks, on mental health, that has demonstrated remarkable benchmark results.

One unique feature of BERT is that it employs a bidirectional view of words in a sentence to properly analyze the relationships and improve the context. It has often been used to analyze Twitter data to predict sentiments. These data contain depression and anxiety-related posts and have often been fed by researchers and data scientists to the BERT model. It efficiently preserves the contextual and semantic meaning of the words immediately preceding it and following it, among all words and subwords present in the entire tweet history. Another approach used by NLP researchers is to transfer the knowledge from a large pretrained model (BERT) to a smaller model to boost performance and accuracy, a process known as distillation. When BERT is used along with **Bidirectional Long Short-Term Memory (Bi-LSTM)**, it can boost the classification accuracy of depression- and anxiety-related posts on Twitter up to 98%. The process of transferring the required domain knowledge to a smaller network (Distiled_BERT) through distillation helps in mental health-related problem identification, as the context remains specific to the healthcare environment.

BERT can function in two steps: pretraining and fine-tuning. The pretraining stage involves training the model on unlabeled data over pretraining tasks. In the second stage of fine-tuning, the initialization of the BERT model is done using a set of pretrained parameters, where all of the parameters are further fine-tuned, leveraging the labeled data from the downstream tasks. The downstream task is equipped with fine-tuned models, initialized with the same pretrained parameters. BERT can help us in many tasks like **next sentence prediction (NSP)**, **question answering (QA), and natural language inference (NLI)** when used in mental health apps, question answering, psychiatric test screening, and summarization processes.

Some of the other transformer models that use bidirectional contextual representations are BERT, ELMo, and OpenAI GPT.

Among the different types of NLP techniques available, the single key challenge in all types of prediction tasks is the appropriate selection and implementation of deep learning-based classifier models that perform accurately for sentiment analysis. Even standard classifiers like SVM, logistic regression, random forest, and AdaBoost do not consider the semantic context. Social media data contains a lot of linguistic features and also suffers from the limitation of identifying different aspects of mental health like schizotypy, social anxiety, eating disorders, generalized anxiety, obsessive-compulsive disorder, apathy, alcohol abuse, or impulsivity. Elastic net regularizer yielded better results for identifying different forms of anxiety and depression but failed for alcohol abuse or impulsivity.

Deep learning classifier models, such as CNN or XGBoost, when applied along with the generic text prepossessing techniques, failed to capture the semantic and syntactic meanings of a word from large volumes of data (Amanat et al., 2022). The inability to maintain long-range dependency further substantiates the fact that transformer modes like BERT can play a significant role in mental health sentiment prediction and proactive servicing. Researchers have often applied classical feature analysis methods, using deep learning methods like CNN, an LSTM, and an LSTM-CNN merged to predict the risk of suicides in online forums. The results though yield better results than traditional classifier models and LSTM, they often take a hit on the performance metrics, as they lack long-term dependency and representation of words along with the desired context (Amanat et al., 2022).

Now that we have gained a fair understanding of how NLP can play an important role in proactive mental health problem identification, we even know how to build the individual microservices that constitute the relevant components of the real-time mental health diagnosis framework. Hence, we need to stitch the individual pieces together to design the end-to-end system to build a deployable framework that runs in the cloud. Figure 7-16 illustrates the different components involved from data aggregation to text data processing, data labeling, and creation of embeddings to create the classification model (Malgaroli et al., 2023). The classification model can be tuned using hyperparameter tuning to improve the model performance. The architecture further demonstrates the model-serving framework that can help serve the model. We also have an incremental retraining mechanism that gets triggered when the feedback mechanism suggests a drop in prediction accuracy beyond a certain threshold and we observe data or model drift.

ML Model End-to-End Training Pipeline at Cloud with Text Data

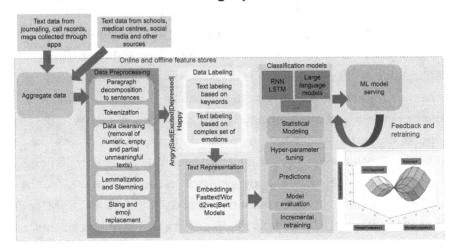

Figure 7-16. *End-to-end processing engines to predict mental depression using text data (Malgaroli et al., 2023)*

Now let us investigate the potential prospects of improving mental health diagnosis systems using NLP-based techniques.

Prospects of NLP Techniques in Improving Mental Health Diagnosis

We see how social media data (such as Facebook, Twitter, and Instagram) and data obtained from other public and online forums and different channels can be used to predict mental health. However, the challenge remains as these channels and platforms remain accessible to only a few and this leads to disparities in understanding the concerns arising among racial/ethnic minorities and young adults. Even the major research and analysis on depression has used social media data where tweets are in English. Other country-specific languages like Japanese, Korean, French, and Italian can also be used to study the nature of mental health conditions among young adults and undergraduate students.

More than often, population groups such as racial/ethnic minorities, young adults, and individuals from lower socioeconomic strata may be completely absent from recording their thoughts and feelings in the online channels and hence remain excluded from receiving the AI-driven insights and proactive servicing that can be obtained through the mental health framework. As a result, they face barriers to mental health services which remain confined to a selected few. In such scenarios we have to look for alternative ways to a sustainable lower-cost, faster, and alternative mental health risk assessment techniques where NLP can play a formidable role. Improving our health system surveillance methods can leverage the power of NLP to help us capture depression and suicide risks broadly. Among low-cost interventions, we can implement text messaging systems, within vulnerable communities specially among high-risk youth. When youth and adolescents section of our population are provided more timely treatment before exams and social events, then some of the challenges get addressed. However, the entire ecosystem around the youth population should be more proactive in assisting with communications between schools, families, youth services, and law enforcement bodies and estimate the progress, based on the intervention and assistance. In addition, such low-cost solutions can bring in more inclusivity on the usage of the mental health framework.

Now let us summarize what we have learned in this chapter.

Conclusion

In this chapter, we have learned about how different sources of textual data can help us in predicting mental health conditions. We focused on data from different social media platforms, EHR records, and online forums and on interviewed answers to understand their role in helping to generate insights on mental health-related problems and user sentiments, along with how data can help in generating future predictions. In this respect, we gained a detailed understanding of the different steps involved in text data processing before feeding the data to traditional or deep learning-based model engines, with practical examples using codebase. This gave us a better way to comprehend how social media data can be put into real-world implementations in the mental health context. We also got an insight into the different models available in the space of NLP both for the traditional and deep learning worlds and how they can be used effectively in the mental health problem domain. In the context of deep learning models, the chapter provides us with more focused attention on the importance of transformer models in the world of NLP, their mode of operation, and how they can better derive insights and a higher level of accuracy. With a solid background in ML algorithms, we also got an overview of how text processing and ML engines and their respective components can be deployed and run in a cloud environment to serve users through the model-serving framework. In the concluding part of this chapter, we looked at the limitations of the current data and explored the potential prospects of NLP to make it more inclusive and generate more value for the youth section of the population.

With detailed background text data processing and its role in proactive servicing for mental health-related problems, we explore how AI models from human–computer interaction, recorded speech, and video data can be used to predict mental distress in youth.

References

Amanat, A., Rizwan, M., Javed, A. R., Abdelhaq, M., Alsaqour, R., Pandya, S., & Uddin, M. (2022). Deep Learning for Depression Detection from Textual Data. *Electronics*, *11*(5), Article 5. https://doi.org/10.3390/electronics11050676

Amartey Roland Teye. (2023). *Natural Language Processing forMental Health Diagnostics*. 10.13140/RG.2.2.35011.21283

Bojanowski, P., Grave, E., Joulin, A., & Mikolov, T. (2017). Enriching Word Vectors with Subword Information. *Transactions of the Association for Computational Linguistics, 5*, 135–146. https://doi.org/10.1162/tacl_a_00051

Cha, J., Kim, S., & Park, E. (2022). A lexicon-based approach to examine depression detection in social media: The case of Twitter and university community. *Humanities and Social Sciences Communications, 9*(1), 1–10. https://doi.org/10.1057/s41599-022-01313-2

Chowanda, A., Andangsari, E. W., Yesmaya, V., Chen, T.-K., & Fang, H.-L. (2023). BERT-BiLSTM Architecture to Modelling Depression Recognition for Indonesian Text from English Social Media. *2023 4th International Conference on Artificial Intelligence and Data Sciences (AiDAS)*, 54–59. https://doi.org/10.1109/AiDAS60501.2023.10284707

Cook, B. L., Progovac, A. M., Chen, P., Mullin, B., Hou, S., & Baca-Garcia, E. (2016). Novel Use of Natural Language Processing (NLP) to Predict Suicidal Ideation and Psychiatric Symptoms in a Text-Based Mental Health Intervention in Madrid. *Computational and Mathematical Methods in Medicine, 2016*, 8708434. https://doi.org/10.1155/2016/8708434

georgetown-ir-lab.github.io. (2017). *Reddit Self-reported Depression Diagnosis (RSDD) dataset.* https://georgetown-ir-lab.github.io/emnlp17-depression/

Hu, Y., Ding, J., Dou, Z., & Chang, H. (2022). Short-Text Classification Detector: A Bert-Based Mental Approach. *Computational Intelligence and Neuroscience, 2022*(1), 8660828. https://doi.org/10.1155/2022/8660828

Jatnika, D., Bijaksana, M. A., & Suryani, A. A. (2019). Word2Vec Model Analysis for Semantic Similarities in English Words. *Procedia Computer Science, 157*, 160–167. https://doi.org/10.1016/j.procs.2019.08.153

Jeffrey, P., Socher, R., & Manning, C. D. (2014). Glove: Global vectors for word representation. *2014 Conference on Empirical Methods in Natural Language Processing (EMNLP)*, 1532–1543.

Kurniadi, F. I., Paramita, N. L. P. S. P., Sihotang, E. F. A., Anggreainy, M. S., & Zhang, R. (2024). BERT and RoBERTa Models for Enhanced Detection of Depression in Social Media Text. *Procedia Computer Science*, *245*, 202–209. https://doi.org/10.1016/j.procs.2024.10.244

Malgaroli, M., Hull, T. D., Zech, J. M., & Althoff, T. (2023). Natural language processing for mental health interventions: A systematic review and research framework. *Translational Psychiatry*, *13*(1), 1–17. https://doi.org/10.1038/s41398-023-02592-2

Morgan, S. E., Diederen, K., Vértes, P. E., Ip, S. H. Y., Wang, B., Thompson, B., Demjaha, A., De Micheli, A., Oliver, D., Liakata, M., Fusar-Poli, P., Spencer, T. J., & McGuire, P. (2021). Natural Language Processing markers in first episode psychosis and people at clinical high-risk. *Translational Psychiatry*, *11*(1), 1–9. https://doi.org/10.1038/s41398-021-01722-y

Parikh, R., Mathai, A., Parikh, S., Chandra Sekhar, G., & Thomas, R. (2008). Understanding and using sensitivity, specificity and predictive values. *Indian Journal of Ophthalmology*, *56*(1), 45–50.

Vaj, T. (2023, May 1). 10 BERT Recommended for mental health context. *Medium*. https://vtiya.medium.com/10-bert-recommended-for-mental-health-context-2b94cfa33514

who.int. (2025). *Suicide—India*. https://www.who.int/india/health-topics/suicide

Winata, G. I., Kampman, O. P., & Fung, P. (2018). Attention-Based LSTM for Psychological Stress Detection from Spoken Language Using Distant Supervision. *2018 IEEE International Conference on Acoustics, Speech and Signal Processing (ICASSP)*, 6204–6208. https://doi.org/10.1109/ICASSP.2018.8461990

Zhang, T., Schoene, A. M., Ji, S., & Ananiadou, S. (2022). Natural language processing applied to mental illness detection: A narrative review. *Npj Digital Medicine*, *5*(1), 1–13. https://doi.org/10.1038/s41746-022-00589-7

CHAPTER 8

AI Models from Human–Computer Interaction Speech and Video Data Recorded to Predict Mental Distress in Youth

This chapter provides an in-depth study of algorithms related to speech, audio, and video data to proactively detect mental anxiety and depression. The chapter emphasizes collecting the data from devices and providing know-how for merging textual and audio data in determining mental distress and risk levels. With an understanding of deep learning algorithms, this chapter delves into different metrics to assess mental health conditions and provide recommendations for treatment and

© Sharmistha Chatterjee, Azadeh Dindarian, Usha Rengaraju 2025
S. Chatterjee et al., *Revolutionizing Youth Mental Health with Ethical AI*,
https://doi.org/10.1007/979-8-8688-1186-9_8

support. This chapter introduces readers to diverse mental health parameters for measuring mental health conditions by illustrating with examples. This chapter also discusses leveraging synthetic data generation to predict mental distress accurately. By combining advanced processing techniques and deep learning, this chapter opens new avenues for prior data processing to enable accurate and sensitive detection of mental distress in today's data-driven world. The readers at the same time are made aware of the best practices of selecting data and training the deep learning models to avoid misuse and inaccurate prediction.

In this chapter, these topics will be covered in the following sections:

- Why depression classification is important

- Textual data and speech/audio data in depressive symptom detection

- Video data and deep learning algorithms in depressive symptom detection

- Knowledge of best practices to build scalable depression models

Why Depression Classification Is Important

Depression affects patients with negative thoughts and feelings, and their psychological level is much different than normal people due to the generated brain signals. They even have fewer feeling-good hormones, such as serotonin and oxytocin, released from their brain. They are prone to more use of negative words and words of rejection, negative expressions that exemplify their mental state, like sadness, stress, demotivation, or dissatisfaction. Mental stress and anxiety can lead to irregular menstrual cycles in females. Depressive disorder patients often find problems in making eye contact and are prone to making less frequent

mouth movements. They even suffer from normal physical movements
restricting their physical activities. Research results demonstrate EEG
(electroencephalography) signals, **NMR (nuclear magnetic resonance)**
signals, and other audio or image signals show different characteristics
for a depressed patient as compared to a normal patient. It becomes
even more difficult when psychologists are unable to predict depression
symptoms, their severity, and their occurrence episodes. This puts
forward a stronger need for healthcare systems to develop and equip
mental healthcare professionals with accurate, automatic, and accessible
technology that detects and classifies depression for ease of treatment by
psychologists and the well-being of patients.

Depression and anxiety impact social functions like employability
and different dimensions of patients' lives, including social interactions
and everyday routine activities. The mode of operation of social functions
and interactions is guided by people's behavior, their physical and mental
fitness, and marital status, as they directly influence how they interpret and
respond to each other and to their surroundings. Deviation from social
functions results in maladaptive behaviors and impaired interactions
resulting in social dysfunction that impacts the individual, their well-
being, the community, and society by and large. This is more common
in younger adults who experience higher rates of anxiety than the aged
elderly population. In a nutshell, social dysfunction leads to undesirable
consequences within society, which needs to be addressed to control
the spread and provide the right tools to individuals and their families.
The need of the hour is an on-time diagnosis of psychological disorders/
dysfunctions, along with continuous monitoring of emotional, cognitive,
and sociodemographic features (like education and life stage challenges at
different ages) and other external contributing factors (like environmental
or societal challenges like COVID-19). This can lead to improved social
engagement, interaction, and adaptability by identifying the cause and
magnitude of depression proactively.

Table 8-1. *Depression-related symptoms at varying levels of intensity*

Sl. no.	Type of depression	None	Mild	Moderate	Severe	Very severe
1.	Anxiety	0	1	2	3	4
2.	Tensity	0	1	2	3	4
3.	Funk	0	1	2	3	4
4.	Insomnia	0	1	2	3	4
5.	Memory or attention disorder	0	1	2	3	4
6.	Depression	0	1	2	3	4
7.	Muscle impairment	0	1	2	3	4
8.	Sensory system symptoms	0	1	2	3	4
9.	Cardiovascular system symptoms	0	1	2	3	4
10.	Gastrointestinal symptoms	0	1	2	3	4
11.	Genitourinary symptoms	0	1	2	3	4
12.	Respiratory symptoms	0	1	2	3	4
13.	Autonomic symptoms	0	1	2	3	4
14.	Abnormal behavior during talks	0	1	2	3	4

With a growing concern about social dysfunctions impacting adolescents and youth, depression detection and classification become an important step toward proper treatment, medication, tracking, and monitoring progress. Depression classification requires text, speech, and video data, where data recorded individually or in combination can aid in classification mechanisms. To aid in the depression classification process, we need text, speech, video (Ashraf et al., 2024), and sensor data that can capture the symptoms listed in Table 8-1. These symptoms manifest in varying levels of seriousness based on the level of depression. Hence, the signals of the symptoms need to be captured and fed to train ML learning models that will craft out the most impactful features and finally predict the probability of the individual falling into any of the specified categories of the mental health disease.

To enhance depression-related knowledge and classify the pattern of depression, human behavior and surroundings need to be closely monitored so that real-time data can be fed to AI engines to continuously predict the probability of depression in the near future. To accurately classify depression propensity, we need to evaluate the following conditions (Ding et al., 2020):

- **Severity of the disease: Depression** can be broadly classified into mild, moderate, and severe based on the symptoms and how it affects individuals with different conditions. Mild depression manifests through a loss of interest in daily or social life without incurring too many disturbances in normal life or impacting social interactions or learning abilities. On a higher scale, moderate depression leads to emotional swings (ups and downs), where disturbance of the mental state leads to thinking abilities slowing down actions to respond in daily life. In situations when left untreated, moderate depression increases, resulting in major depression that creates suicidal tendencies. While mild depression can be treated with self-psychological indications and a counselor's help, moderate to severe depression should not be left untreated and necessitates hospital visits and professional treatment in due time.

- **Number of episodic occurrences of the disease**:
Depression can also be classified based on the number
of episodes of the disease, where episodic occurrences
can range from a single occurrence to recurrent
occurrences. An individual can experience single-onset
depression, where the individual has experienced
depression once in a lifetime, as an effective treatment
could prevent the reoccurrence of the disease. Though
this is very rare, recurrent depressive disorders result in
the emergence of repeated symptoms as the treatment
could not have probably fully remitted the disease. This
kind of reoccurrence is mostly common and needs
basic treatments that can mitigate the severity and
reduce the reoccurrence period.

- **Age-level depression**: Depression can also be
classified according to age levels, such as children and
adolescents' emotional imbalance, adult depression,
and senile depression. Depressive symptoms appearing
at varying age levels are different, based on the life
stage challenges and family and social conditions.
It has been observed that the onset of depression
happens more in adults than in the young population.

Depression Detection Deep Learning Models

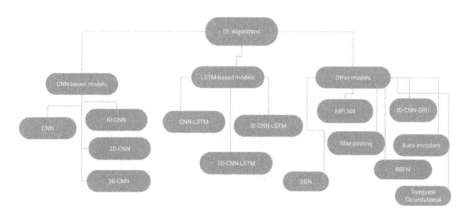

Figure 8-1. *Deep learning-based classification methods for depression detection (Bolhasani, 2021)*

Having an understanding of how depression classification is affected by its severity, number of episodic occurrences, and age of the individual, let us first understand different depression detection and classification mechanisms using deep learning techniques. Figure 8-1 illustrates the varied categories of deep learning methods, broadly classified into CNN-based methods, LSTM-based methods, and other methods.

Convolutional neural networks (CNN)-based methods: These types of deep learning methods help us to predict depression status (positive or negative) by monitoring user activity or speech or video data. A CNN-based network is composed of the following layers (Tufail et al., 2023):

- **Convolution layer**: It is the first layer and serves as the building block of the neural network. It is capable of detecting any type of input, that is, text features and audio features. The input is convolved with the filter of the same size to obtain the output, with the help of filter kernels, which use a backpropagation algorithm to update the weights.

315

- **Max-pooling layer**: The pooling layer is situated in between two convolution layers (convolution layer and fully connected layer), which functions to decrease the size of the input by removing unnecessary information. By this, it helps to control the overfitting problems in neural networks.

- **ReLU as an activation function**: Rectified linear unit functions to bring in nonlinearity by replacing all negative data with zeros. The main objective of this layer is to keep relevant information and pass it to the next layer. There are different types of activation functions, such as SoftMax, ReLU, tanH, and Sigmoid, to control the varying types of nonlinearity.

- **Fully connected layer**: A single or two fully connected layers appear before the output classification layer, to classify and update the results.

- **Batch normalization**: The goal of batch normalization layers is to generate a more stable output, by normalizing the output of previous layers.

- **Dropout layer**: Dropout is a layer that helps to control the overfitting of the model output and enables a faster training rate by randomly dropping a few values of the neural network.

CNN can be used in different dimensions on different sets of data to classify depression (Wang et al., 2022). This type of neural network can encode temporal and spectral features from the voice signals or static facial or physical expression features from the video frames to predict different types of mental disorders. For example, a one-dimensional convolutional neural network (1D CNN) can be used to detect depression on datasets containing measurements of motor activity. We can use multiple models

316

with different time segments and different functions to train and classify the occurrence of depression, like **ADHD (attention deficit hyperactivity disorder)** or **ASD (autism spectrum disorder)**.

Using a wrist-worn accelerometer can be used to record user activity and detect the peak amplitude of user movement accelerations. The data recorded at different time intervals, in time-lapses of one day for seven days at an interval of one minute, can yield a transient voltage signal proportional to the rate of acceleration. The preprocessed data when fed to a two-dimensional convolutional network (2D CNN), CNN) will generate a pixel of the image corresponding to each record, recorded per minute of user activity. Training 2D CNN models using data from a mobile device or a smartwatch can predict depression in real time with over 75% accuracy. It can also determine psychological stress level, by learning a person's breathing patterns from thermal images. This kind of real-time depression detection using 2D CNN has a low computational cost, can assist the patients proactively with necessary resources, and helps them with future follow-ups. 2D CNN plays an important role in inferring any change in motor activity to identify episodic occurrences of depression that can help to classify patients according to the level of the disease.

3D CNN Modeling of Brain Images to predict Depression in Youth

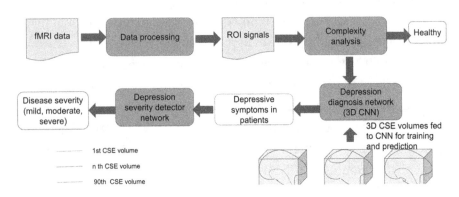

Figure 8-2. *3D CNN modeling for depression detection using fMRI data*

CNN-based complex architecture can aid in mental task and workload classification on receiving data from EEG signals (Tufail et al., 2023). EEG signals capture varying neuronal changes from cognitive actions and mental and motor tasks of users when they are subjected to standardized stress tests and workload tasks and submit questionnaires provided by psychologists. CNNs help to classify motor, cerebral, and behavioral disorders when presented with input signals in the form of 2D or 3D time-frequency representations. 1D CNN can be used when input data is fed in from single-channel EEGs while 3D CNN can be used for multichannel EEGs.

Further, the 3D CNN architecture can be leveraged to identify patterns from **functional magnetic resonance imaging (fMRI)** and **structural magnetic resonance imaging (sMRI)** data between brain regions. Such 3D pattern recognition helps us to detect physiologically useful information primarily useful for the pathophysiology of **late-life depression (LLD)** occurrences (Lin et al., 2023).

Even **resting-state functional magnetic resonance imaging (rs-fMRI)** is also helpful for researchers in identifying neurobiological mechanisms that signify adolescent depression. sMRI plays a formidable role in identifying youth suffering from major depressive disorder by nailing the brain's subregions. This in turn helps to predict adolescent suicide risk and improve suicide prevention among high-risk patient adolescent populations. Figure 8-2 illustrates the process of feeding fMRI data to predict depression.

It uses functional and structural information of the brain, to detect severe, moderate, and mild depression incidents by leveraging a nonlinear technique called **cross-sample entropy (CSE)** (Lin et al., 2023). CSE volumes in the left inferior parietal lobule, left parahippocampal gyrus, and left postcentral gyrus can be used to train 3D CNN models, to diagnose and predict the severity of diseases such as ADHD. The use of 3D CNN models takes into consideration the temporal (the multiscale entropy analysis) and the spatial (i.e., the 3D approach) features of the fMRI data. 3D CNN can also be used to extract structured and unstructured EHR data, **electroencephalogram (EEG)** features, spectral changes of EEG hemispheric asymmetry, and spatial and temporal changes to differentiate between different mental stress and anxiety conditions.

CSE is a technique to evaluate the degree of asynchrony or dissimilarity of two data series obtained from brain imaging. This metric does not show considerable change where the cross-approximate entropy does not change and shows no dependency on directions. It further enables us to reveal how the brain regions are functionally connected to the data series and can be simultaneously revealed.

Among other deep learning methods, the **multilayer perceptron (MLP)** neural network operates in a forward direction, where each node transmits data to its forward node using a backpropagation algorithm. The difference between MLP and **radial basis function (RBF)** is RBF consists of one hidden layer in contrast to MLP, which has several hidden layers. An MLP modeling framework with its hidden network in a black

box model can access the emotional states of any individual based on
education, age, and social and cognitive skills, along with the individual's
behavior, to access the covariate relationships of these factors. This
can help us to identify primary contributing factors and predict social
dysfunction to assist with systematic interventions. Further, using neural
networks paves the way for detailed analysis of how different symptoms
are interlinked and connected with observed behaviors. The nonlinear
relationships of the neural network help us to evaluate the risk probability
that an individual showing abnormal behavior can demonstrate suicidal
tendencies. Among feedforward neural networks, **deep belief networks
(DBN)** can assist in discovering strong relationships between rs-fMRI and
sMRI data. DBN consists of multiple layers of hidden units and increasing
the depth of DBN can aid in improving the diagnostic classification of
mental disorders (Su et al., 2020). After extracting the features from brain
morphometry, the DBN can pinpoint the differences between the different
classes—the frontal, temporal, parietal, and insular cortices—and in
some subcortical regions. Once the brain anatomy is identified using DBN
including the corpus callosum, putamen, and cerebellum, it becomes
easier to detect mental health disorders like childhood schizophrenia,
major depressive disorders, and bipolar symptoms.

Radial basis functions are a special type of feedforward neural network
that is faster to train and is composed of three layers: an input layer, a
hidden layer, and the output layer. The radial basis functions neural
network delegates the most important functionality to a hidden layer
for any computation-related tasks, and output obtained from this layer
is multiplied by the neuron's weight and then summed up to generate
the output. The output layer is assigned for any kind of task related to
classification or regression. This type of NN can be used in the diagnosis
of **bipolar disorder (BD)**, a mental illness where patients experience
episodes of depression, psychosis, changes in mood state, and manias
(IBM, 2023; Luján et al., 2022).

Among other DL-based algorithms, **autoencoders (AE)** also emerge
as a powerful **automatic depression detection (ADD)** tool as they are
capable of extracting highly relevant vocal and visual features (Ding et al.,
2020; Selvaraj & Mohandoss, 2024).

Classifier based Stacked Deep Auto Encoder Models

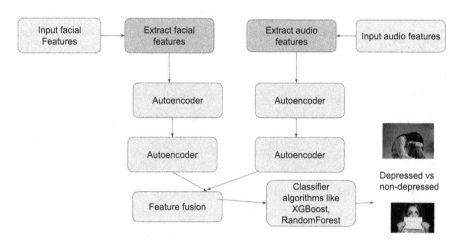

Figure 8-3. *Depression identification with stacked deep autoencoder*
models (Kumar et al., 2020)

AEs run unsupervised ML algorithms to extract low-dimensional
features using nonlinear dimensionality reduction mechanisms. This type
of deep neural network uses unsupervised learning to reconstruct the
input signal in order to reduce the reconstruction error (the difference
between the original input vector and the reconstructed output vector).
The training triggers a comparison between the input and the output of the
network to make them close to each other, through the use of an encoder
and decoder network.

In an AE, the responsibility of the encoder is to map the input vector
(the features) into a hidden representation (the latent representation)
of the input with lower dimensionality. Then, the decoder component

of AE can perform the reconstruction transformation on the hidden representation. This results in a reconstructed version of the input signal where the hidden style and characterization of the original input are viewed as a reduced feature vector, keeping the most important and relevant relationships of the original input features. For example, speech-based adversarial AEs can be used for emotion recognition, to extract higher dimensional features from audio samples and then pass the extracted features to adversarial AE to represent the extracted hidden information. Figure 8-3 illustrates how an AE model functions to extract image and speech data.

Autoencoders provide remarkable benefits in sensing-based **human activity recognition (HAR)**, which can enable faster and more accurate model predictions for identifying mental depression (Sardari et al., 2022; Thapa et al., 2023).

- It can use a semi-supervised model, without any labeled data or alterations in the pretrained model/classifier, to identify human activity and detect mental health conditions.

- **Adversarial autoencoder (AAE)** comes with more efficacy and robustness, equipped to learn recent changes in human activities.

- No manual feature extraction is required, as it can automatically deduce structure and learn spatiotemporal characteristics from sensor data that is unprocessed and unstructured.

- AE can be used effectively as the state-of-the-art model, across different platforms and domains to proactively predict mental health problems.

- AE can be leveraged to generate synthetic data to generate more samples for the underrepresented group among a set of depressed and nondepressed individuals.

Autoencoders can be further combined with LSTM networks to diagnose mental depression from facial expressions and speech responses. LSTM can play a significant role in modeling temporal information of the responses and yield more accurate results in detecting bipolar and unipolar disorders.

In addition, hybrid neural networks like LSTM and CNN can play a significant part in the detection of depression from images and videos. At first, the CNN models extract the face appearance features. At the next step, the deep-learned appearance features are assembled with audio and face shape features to feed to an LSTM to capture long-term sequence features.

Further, neural networks can be leveraged to predict cognitive functioning (HYPO) by feeding on human behavior and interactions to deduce chances of social dysfunction. Using cognitive functioning, different mental processes like memory, attention, and problem-solving abilities can be evaluated to determine social adaptability. To predict social functioning on a continual basis, the level of anxiety and depression needs to be assessed using any of the neural networks like MLP or other models running on a scalable framework that can ingest data in real time. Any real-time data ingestion, modeling, and monitoring framework supported by a 360-degree patient treatment and care service can leverage data insights and take remedial actions to promote healthy social functioning.

In Chapter 7, "AI Models from Human-Computer Interaction Textual Data Recorded to Predict Mental Distress in Youth," we studied how the classification of text data has helped in predicting stress and anxiety. Now let us understand how text and speech data can help to build AI-based models to aid in depression classification.

Textual Data and Speech/Audio Data in Mental Health Symptom Detection

Research studies have explored theories for speech-based depression detection, which have discovered that neurophysiological changes can occur during depression see (Ding et al., 2020). The pressing need is the use of physiological signals, including four basic rhythm waves of EEG signals, which leverage processing technology like **fast Fourier transform (FFT)** and ML models like SVM to identify depression tones.

Neurophysiological changes impact laryngeal control (i.e., the behavior of the vocal folds) to a certain extent, and the detection becomes feasible using deep learning neural networks such as a **CNN (convolution neural network)**, which can determine the base structure to demonstrate the potential of depression detection by leveraging the audio modality of speech. Depression detection becomes feasible with the design of an automatic depression detection framework that can rely on the acoustic characteristics of a given language to diagnose degrees of mental illness.

One such research study examined 153 patients with major depressive disorder and health counterparts without current or past mental depression. One hundred sixty-five people were introduced to the framework to detect depression through voice recording using their smartphones (Kim et al., 2023). The framework recorded the voices of the participants (who spoke in the Korean language) while reading textual sentences that were predecided. The procedure relied on text-dependent read speech tasks. The procedure may include traditional ML models that can do feature engineering on the acoustic features, an ML model selected that will have a certain number of parameters to train and classify log-Mel spectrograms using deep convolutional neural network (CNN), or models that can train and classify log-Mel spectrograms by leveraging pretrained networks available online. Among the three approaches, automatic depression detection on speech data yields the highest accuracy of 78.14%

on a CNN network, showing deep-learned acoustic characteristics are capable of giving better performance than using traditional approaches that rely on pretrained models .

Data Ingestion Preprocessing of Textual and Speech Data

One major advantage of developing a framework that feeds in speech data from mobile phones is that smartphone-based speech recording is easily accessible, and patients who suffer from major depressive disorder demonstrate consistency in mood states that capture the depressive statuses well during reading text-dependent sentences. This happens as text-dependent read speech helps in expressing the reader's behavior in a controlled manner over the text of the same content and length. This in turn powers up speech and affective constraints, reducing and removing acoustic variability to a greater extent in comparison with the spontaneous mode of text utterances, providing a solid ground-truth base of data for speech analysis. However, in order to ensure quality in speech recordings, the recording needs to be carried out in a quiet room, and the microphone needs to be positioned roughly within a distance of 30 cm from the participants.

Depression detection in clinical settings demands data aggregation, which follows a standard approach to drive consistent outcomes in an automated environment. As stated above, text-dependent read speech helps to infer voice characteristics by including every sound that represents normal speech. In addition, we also need to focus our attention on varying acoustic qualities from larger datasets that can help to identify depression symptoms from variations in acoustic characteristics. Multiple past research studies have used a convolutional neural network (CNN) layer and a dense layer to model the depressive state; some studies have started with local regional languages to extrapolate the inference

for the larger scale of the population. Figure 8-4 shows an automated
depression detection framework that predicts the depressive states from
text-read speech data. It further demonstrates how text and speech data
are simultaneously fed to their respective preprocessing units to feature
engineer and extract the dominant features. Speech processing uses an
important visual metric called a log-Mel spectrogram that represents
sound frequencies over time to prominently distinguish subtle sound
features. Here, frequencies are mapped to the Mel scale (resembling
human hearing) before getting them represented on a logarithmic scale.

While log-Mel spectrogram is successful in extracting the speech
samples, tokenization helps to split the text sentences before passing it to
the speech and text deep learning model inputs for training.

We have a speech data-based depression detection model and a BERT-
based CNN model to classify depression from text data. The outputs from
both models are then concatenated to generate the predictive probability
score of the final status of the depression.

Speech and Text Data Integration generating Depression Detection Model

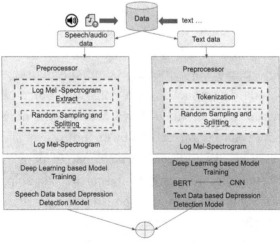

Figure 8-4. *Deep learning-based ML model for depression detection
using speech and text data (Park & Moon, 2022)*

326

In order to capture speech/audio from a user, a smartphone can be used, where speech captured during reading can be divided into three tasks:

- Vowel recognition

- Digit recognition

- Passage reading and interpretation (predefined text-based sentences)

Figure 8-5 illustrates the different stages of speech capture and analysis of two different groups of patients, control and test groups. The analysis process involves feeding the data to any of the CNN (VGG16, VGG19, ResNet50 Inception-v3, MobilieNet-v2, VGGish, YAMNet) models to derive the differences in acoustic features that help to distinguish between normal and major depressive disorder patients. CNN models demonstrate consistently better performance in several research studies than conventional ML models, as CNN models are capable of extracting acoustic features in a broader spectrum. The above-stated CNN-based deep models are primarily used for image, speech analysis, or object detection.

- **VGG-16/VGG-19**: "VGG" refers to the Visual Geometry Group of the University of Oxford, while the "16" refers to the network's 16 or 19 layers that have weights, suitable for image classification.

- **ResNet50**: Residual networks with 50 layers, suitable for image classification on large datasets.

- **Inception-v3**: Pretrained convolutional neural network of 48 layers deep used for image analysis and object detection.

- **MobilieNet-v2**: Mobile-friendly convolutional neural networks(CNNs), with 53-layer deep.

- **VGGish**: Deep learning model based on the VGG
  (Visual Geometry Group) architecture, developed by
  Google for audio feature extraction.

- **YAMNet**: Yet another multiscale convolutional neural
  network, trained to recognize and classify over 500
  audio event classes directly from raw audio waveforms.

The audio passage read by individuals before feeding the data
into the depression detector, as shown in Figure 8-5, should contain
balanced proportions of consonants, vowels, and digits. It is generally
a recommended practice to use neutral expressions standardized with
regional/English language phrases which removes any kind of unwanted
bias and can be treated as a reliable protocol. It is also a recommended
best practice to record the individual's age in years, a number of education
years, and other lifestyle attributes like nonalcoholic, nonsmoker, and
HAM-Dd recorded at weekly/biweekly intervals (a scoring factor in
depression analysis signifying 10–13 mild; 14–17 mild to moderate; >17
moderate to severe depression) and self-assessment records like PHQ-9e.

Speech Analysis for Analysing Mental Health Disorder

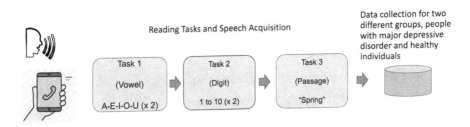

Figure 8-5. *Speech input from smartphones for detecting major
depressive disorder (Kim et al., 2023)*

While lifestyle attributes and self-assessment records help to correlate the prediction results. It helps in explaining how depression can be associated among the youth population based on family, education history, and daily routines. However, there is scope for improvements in the prediction accuracy of models. CNN models can further improve the classification accuracy based on the optimization of CNN architecture and with a wider spectrum of text-dependent speech datasets (Wang et al., 2022). A rich collection of datasets once incorporated in audio-based depression detection studies can better assist better in detecting anomalies and identifying trends based on lifestyle conditions.

In the above section, we have seen how the speech data can be fed into the system and processed data can be preprocessed before feeding into the deep learning network. Now let's take a closer look at the relevant extracted features, their meanings, and how they can play a significant role in identifying depression in speech data.

Depression Detection Using Relevant Audio/Speech Features

In this subsection, we will study the different speech features that are extracted in the preprocessing before feeding them into the deep learning model.

To process the audio frames in a deep learning model, often a fixed-length frame can be chosen with augmented samples, where each frame needs to be transformed into channels of size 64, which can yield an LM spectrogram with a size of 64×200 pixels (Bolhasani, 2021; Du et al., 2023). These image-like patches can further be converted to a spectrogram (mel_db) by using a Mel scale. The LM spectrogram is one of the popular adopted mechanisms for processing audio applications that use deep learning. The spectrogram necessitates the input of a 2D image representation of an audio signal with complex features, which are

processed downstream to extract the most important features.
The extracted frame-level-based 2D LM spectrograms are further fed to a
2D CNN for training and prediction.

Conventional ML models are capable of extracting the three most
popular acoustic features like **Mel-frequency cepstral coefficients
(MFCCs), extended Geneva Minimalistic Acoustic Parameter Set
(eGeMAPS), Interspeech Computational Paralinguistics Challenge
(COMPARE),** and **Linear Prediction Coefficients (LPC)** from baseline
audio signals. MFCC features represent the characteristics of a sound,
such as its timbre, while eGeMAPS is a basic standard acoustic parameter
set that plays an important role in automatic voice analysis, such as
paralinguistic or clinical speech analysis. While analyzing patients'
physiological conditions, this feature set is instrumental in detecting
physiological changes during the generation of voice. It is primarily
composed of the following parameters (Min et al., 2023):

- **Frequency-related parameters** like pitch; jitter;
  formant 1, 2, and 3 frequency (center frequency of first,
  second, and third formant); and formant 1 bandwidth
  of the first format

- **Energy/amplitude parameters** like shimmer (peak
  amplitude difference between consecutive periods),
  loudness (signal intensity from an auditory source),
  and **harmonics-to-noise ratio (HNR)**

- **Spectral (balance) parameters** like alpha ratio (energy
  sum from 50 to 1000 Hz and 1 to 5 kHz), Hammarberg
  index (strongest energy peak in the 0–2 kHz region to
  the strongest peak in the 2–5 kHz region), spectral slope
  0–500 Hz and 500–1500 Hz, energy ratios of spectral
  harmonic peak at the first, second, and third formant's
  center frequency to the energy of the spectral peak, and
  analyzing the harmonic difference (like H1-H2, H1-H3)

These features are highly impactful in comparing the effectiveness of the deep spectral representations model to identify emotional changes through speech. MFCC is a highly influential feature in speech recognition systems as it exhibits short-term power spectral features in acoustic speech signals. It is capable of demonstrating any change in the human vocal tractFCC which can detect for depression.

Let us now see with a practical example how features can be extracted from speech data and then used for training and prediction.

Listing 8-1. In this code example, we use audio .wav datasets to extract short-term and mid-term audio features. Here, we have used the pyAudioAnalysis library to extract the features and plot it. We also learn about how to run regression models to analyze audio/speech data when the emotional state is not a discrete class but a real-valued measurement

1. In the first step, we install the necessary libraries.

```
!git clone https://github.com/tyiannak/
pyAudioAnalysis.git
!pip3 install -r ./pyAudioAnalysis/requirements.txt
!pip3 install -e pyAudioAnalysis/.
```

2. After installation, we import the necessary libraries and load the .wav file.

```
from pyAudioAnalysis import audioBasicIO
from pyAudioAnalysis import ShortTermFeatures
import matplotlib.pyplot as plt
from pyAudioAnalysis import MidTermFeatures
[Fs, x] = audioBasicIO.read_audio_file("test.wav")
```

3. In the next step, we analyze the short-term features
    which split the input signal into short-term
    windows (frames) and compute several features for
    each frame.

```
F, f_names = ShortTermFeatures.feature_extraction(x,
Fs, 0.050*Fs, 0.025*Fs)
plt.subplot(2,1,1); plt.plot(F[0,:]); plt.xlabel('Frame
no'); plt.ylabel(f_names[0])
plt.subplot(2,1,2); plt.plot(F[1,:]); plt.xlabel('Frame
no'); plt.ylabel(f_names[1]); plt.show()
```

4. Followed by this, we also do mid-term feature
    analysis by specifying parameters in relation to
    feature creation, the mid-term window size (mw),
    the mid-term step size (ms), the short-term window
    size (sw), and the short-term step size (ss).

```
[Fs, x] = audioBasicIO.read_audio_file("test.wav")
F, s_names, f_names = MidTermFeatures.mid_feature_
extraction(x, Fs, 1.0*Fs, 1.0*Fs, 0.050*Fs, 0.050*Fs)
plt.subplot(2,1,1); plt.plot(F[8,:]); plt.xlabel('Frame
no'); plt.ylabel(f_names[8])
plt.subplot(2,1,1); plt.plot(F[0,:]); plt.xlabel('Frame
no'); plt.ylabel(f_names[0])
plt.subplot(2,1,2); plt.plot(F[1,:]); plt.xlabel('Frame
no'); plt.ylabel(f_names[1]); plt.show()
plt.subplot(2,1,1); plt.plot(F[2,:]); plt.xlabel('Frame
no'); plt.ylabel(f_names[2])
plt.subplot(2,1,2); plt.plot(F[3,:]); plt.xlabel('Frame
no'); plt.ylabel(f_names[3]); plt.show()
```

5. On plotting the mid-term features, we get the
    following outputs of the audio features plotted
    against the frame numbers.

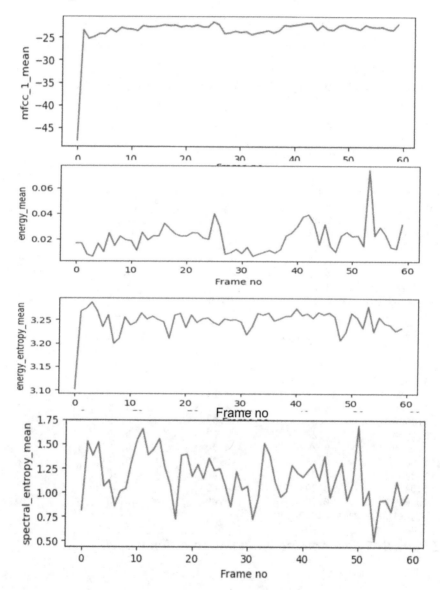

Figure 8-6. *Audio features with respective frame numbers*

6. We can also train a regression model, using all the
   .wav files from the speechEmotion folder.

```
aT.feature_extraction_train_
regression("pyAudioAnalysis/pyAudioAnalysis/
data/speechEmotion/", 1, 1, aT.shortTermWindow,
aT.shortTermStep, "svm", "pyAudioAnalysis/
pyAudioAnalysis/data/svmSpeechEmotion", False)
```

7. The training yields arousal and valence as the
   predominant features.

```
Regression task arousal
Param           MSE           T-MSE          R-MSE
0.0010          0.10          0.19           0.34
0.0050          0.08          0.12           0.35
0.0100          0.08          0.11           0.35          best
0.0500          0.11          0.09           0.33
0.1000          0.10          0.08           0.35
0.2500          0.11          0.09           0.33
0.5000          0.10          0.09           0.34
1.0000          0.12          0.09           0.33
5.0000          0.10          0.09           0.35
10.0000         0.10          0.09           0.36
Selected params: 0.01000
Regression task valence
Param           MSE           T-MSE          R-MSE
0.0010          0.30          0.37           0.35
0.0050          0.31          0.24           0.36
0.0100          0.28          0.17           0.35
0.0500          0.27          0.09           0.35
0.1000          0.28          0.09           0.39
0.2500          0.29          0.09           0.35
0.5000          0.33          0.09           0.37
1.0000          0.27          0.09           0.36
5.0000          0.29          0.09           0.36
10.0000         0.26          0.09           0.38          best
Selected params: 10.00000
```

Figure 8-7. *Error metrics from the regression model*

8. Next on the regression model, we can query individual .wav files to yield the probability of valence and arousal emotions.

```
!python3 pyAudioAnalysis/pyAudioAnalysis/audioAnalysis.
py regressionFile -i pyAudioAnalysis/pyAudioAnalysis/
data/speechEmotion/anger01.wav --model svm --regression
pyAudioAnalysis/pyAudioAnalysis/data/svmSpeechEmotion
```

```
!python3 pyAudioAnalysis/pyAudioAnalysis/audioAnalysis.
py regressionFile -i pyAudioAnalysis/pyAudioAnalysis/
data/speechEmotion/38.wav --model svm --regression
pyAudioAnalysis/pyAudioAnalysis/data/svmSpeechEmotion
```

We see that the first file has more arousal and negative valence emotion compared to the second file.

1st file

```
arousal    0.700
valence   -0.350
```

2nd file

```
arousal    0.012
valence    0.229
```

We just studied how different speech features like MFCCs and energy intensities can be extracted from audio files; now let us understand the role of COMPARE feature in speech recognition tasks.

On the other hand, COMPARE can be used to feature various new recognition tasks like social signals that involve laughter, sadness, and fillers. It is also a useful metric in detecting conflicts in dyadic group discussions and unusual communications due to pervasive developmental disorders and other emotional changes. COMPARE mostly operates on less descriptive speech features that capture spectral, cepstral, prosodic, and voicing-related dynamic information from the audio signal. Though deep learning models perform better in extracting the most relevant speech features, other ML models like SVM, **linear discriminant analysis (LDA)**, **k-nearest neighbor (KNN)**, and **random forest (RF)** can also assist in depression analysis. In addition to COMPARE, we can also use the metric LPC, which represents a set of coefficients to describe speech signals.

To further understand the above plots from Figure 8-6, we need to understand how the metrics vary among depressive and nondepressive patients. Entropy gives us a measurement of the uncertainty of the signal source, proving the underlying fact that greater randomness occurs due to larger entropy, resulting in a more scattered probability distribution. In Figure 8-6, we see the spectral entropy being plotted; we need to compare the same with nondepressive patients. The higher spectrogram entropy signifies nondepressed speech, which makes the speech richer and more complex in spectral structure. As depressive patients exhibit lower spectral entropy with a flattened pattern in the spectrum, the above plots confirm nondepressive speech. In addition, short-term energy represents the average energy (or amplitude change) of a frame of speech signal, which will become higher for common people because of the absence of slurring and pauses commonly found in depressive patients.

Line spectral frequencies (LSFs) also play a role in discovering depression from speech. LSFs are an important component that captures any effect occurring in the interconnected tube model of the human vocal tract, representing the state of the vocal tract as completely opened or closed. Mathematically, they represent quantized line spectral pairs (LSPs) that are used to represent linear prediction coefficients (LPCs).

Figure 8-8 illustrates how acoustic parameters like LPC and MFCC function through speech production to speech perception in identifying depression (Du et al., 2023).

Speech Production and Perception to extract Depression

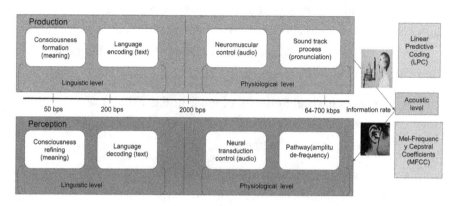

Figure 8-8. *LPC and MFCC model in closed-loop speech chain to extract depression (Du et al., 2023)*

It lays down linguistic and physiological aspects both for production and perception to evaluate LPC and MFCC from speech information (emanated between 0 and 700 kbps) to recognize depression. While speech production concentrates on neuromuscular control and soundtrack progress, speech perception concentrates on neuro transduction (conversion of the sound wave to an electrical impulse that can be interpreted by the brain) and amplitude frequency to infer the results. This is due to the fact that phonetic differences are substantial in the case of depressed patients and are better pronounced in the speech production and generation process.

At first, at the linguistic level, the message is encoded to text to enable sound generation from the vocal tract. The next step involves the interpretation of the language's acoustic properties at the physiological level to extract the essential features of speech necessary for its understanding at the physiological level. Following this, the text is decoded to its meaning at the linguistic level. As speech length remains variable, after feature extraction, a segmentation and fusion method may be applied to further extract intra- and inter-segment features to prevent any cropping and redundant information. In the end, the LPC and MFCC metrics are computed to evaluate the extracted features to determine vocal tract changes in patients and classify them as true positive cases of depression.

Overall, the entire process consists of three stages as illustrated in Figure 8-9 (Du et al., 2023; Wang et al., 2022).

- **Preprocessing**: Removal of noise and dividing the entire speech into segments of specified intervals and marking their sequences.

- **Intra-segment features extraction**: From the LPC and MFCC features, use a 1D CNN to extract the intra-segment high-level depressive features.

- **Inter-segment features extraction**: Use an LSTM-based classification technique to fuse segment features of each subject on a time domain to compute the correlation information between segments.

We also see two standalone ML models in place, where the generation model leverages extracted phonetic features (raw speech, speech with/without noise) from the speech generation process, whereas the perception model extracts perception characteristics from the speech perception process.

After initial steps of preprocessing, the extracted features corresponding to different segments from a single subject are mixed and are passed downstream to a 1D CNN model. 1D CNN plays a key role in convoluting all the frequencies coming from LPC and MFCC as they contain different magnitudes of dimensions of both time and frequency domains. This process helps to increase the model's sensitivity to the frequency domain. Further, 1D maximum pooling of CNNs enables to recording of short-term temporal dynamic variations and frequency correlations to evaluate depression-related features. Batch normalization and dropout can improve training speed and control overfitting. In the output, before generating the final predicted output, dense layers extract certain features and reduce the dimensions of the output. However, we need to be careful of class imbalance and maintain a 1:1 ratio of depressed and nondepressed segments in the training dataset.

Figure 8-9 demonstrates that inter-segment feature extraction is done by LSTM after intra-segment feature extraction that captures both the short- and long-term temporal correlation between the segments at the personal level. This method relies on concatenating the outputs of LPC and MFCC for every feature to assemble into a 32-dimension segment feature that is passed to an LSTM of two dense layers which helps to reduce the dimension and produce the classification. The LSTM, with its recurrent layers, are used to process variable-length speech inputs to yield the classification result at the output.

Depressive speech is marked by differences in behavior in speech, like lower speech rate, less pitch variability, pitch tone, rhythm, stress, voice quality, articulation, intonation, Mel spectrograms, and more self-referential speech. We can look at the amplitude of the differences existing between normal and depressive speech to understand the severity of depression. Figure 8-10 illustrates computing the depression index score by feeding in self-assessment results to the deep morning model predictions. It helps in quantifying the depression intensity and recommends remedial actions based on the severity of the disease.

Researchers have concluded that the audio CNN model is equipped with enhanced learning ability and higher learning feature parameters and exhibits lower loss in comparison with textual CNN. In addition, it also produces better accuracy results, (shooting up to an accuracy of 98% with a minimum 0.1% loss) making it a better choice for speech-related depression detection.

339

3 - Stage Process with 1D CNN and LSTM to identify Depression

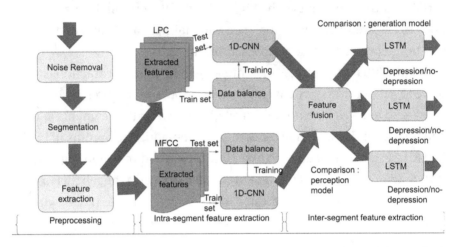

Figure 8-9. *1D CNN and LSTM deep learning-based models unified
to identify depression using LPC and MFCC features (Du et al., 2023;
Wang et al., 2022)*

LSTM models, in comparison with Bi-LSTM, exhibit lower accuracy
and a lower learning rate. Though researchers have found the training
accuracy and validation accuracy of Bi-LSTM are high, the Bi-LSTM model
is unable to retain information on audio and text features for a long period.
Even the LSTM model is found to lose the past data on any addition of new
information to the model.

One drawback of the Bi-LSTM model is that it requires more
preprocessing time; however, during training, an increase in the number of
epochs fastens the training time and results in reduced loss. With the high
prediction accuracy of audio CNN and Bi-LSTM, a combination of both
could be an ideal choice for faster learning and prediction. We can also use
the textual CNN model (accuracy in the range of 92% and loss of 0.2%) or
hybrid models like the textual CNN model in combination with LSTM or
Bi-LSTM.

Attention mechanism was introduced in LSTM and Bi-LSTM networks for detecting depression in speech-based data (Wang et al., 2022). As each frame in speech demonstrates varying emotions, it is important to give more importance to frames that exhibit full emotions through the attention mechanism and ignore frames that have poor emotional characteristics.

Therefore, we introduced an attentional mechanism in a Bi-GRU for the detection of depression.

The figure is similar to Figure 8-9 in terms of speech data preprocessing and feature extraction. Figure 8-10 also shows that in addition to feature extraction, statistical modeling can also be applied before feeding to the CNN model for training. FBank feature has been extracted to estimate the impact of the speech recognition system. This procedure is undertaken by processing each frame through a window function and Fourier transformation, to emulate the process of perceiving sound through human ears.

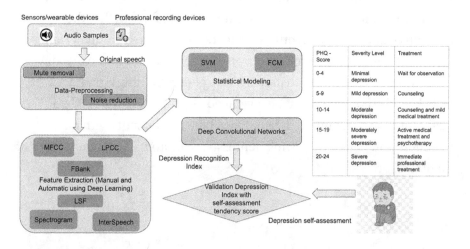

Figure 8-10. Speech preprocessing, statistical analysis, and self-assessment to detect mental health disease severity (Wang et al., 2021)

We have seen how CNNs are used for training in text and speech data containing depressive symptoms. Let us now consider the role of CNNs in generating synthetic data and removing bias in depression detection algorithms.

Synthetic Data Generation Using CNNs as GANs

CNNs are also used in EEG data augmentation using generative algorithms to facilitate unbiased classification algorithms. CNNs can be trained with time series datasets using **generative adversarial networks (GAN)** to generate synthetic data of the voice/speech of underrepresented groups. This in turn can ensure enough data representation of varying depressive symptoms and mental states across the population group in consideration with varying voice pitch, intensity, pronunciation, and other differentiating attributes. Synthetic data for patients who have had qualitative and quantitative evaluations can be fed to retrain the diagnostic model.

GANs comprise two neural networks: a generator producing synthetic data from random noise and a discriminator judging whether the presented data is real or synthetic. The training process gradually shifts the distribution of data produced by the generator toward the real data distribution. Figure 8-11 illustrates how a generator discriminator-based GAN architecture can be leveraged using CNNs to aid in data augmentation to maintain data balance between patients with no disorders and patients showing depressive symptoms of different types of mental disorders.

Synthetic Mental Health Data Generation with GAN

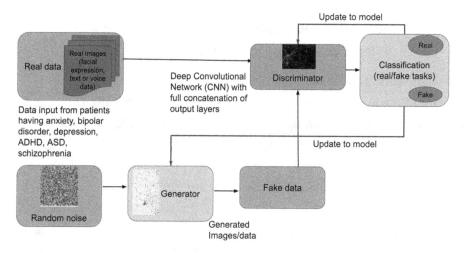

Figure 8-11. *The discriminator generator-based neural network to aid in synthetic data generation for mental health*

- The generator is responsible for generating fake images with random noise data fed into it.

- Over time, the generator starts producing realistic data and images.

- The discriminator is responsible for evaluating the difference between real and fake images/data produced by the generator and predicts the data/image to be real (1) or fake (0). The real data fed into the discriminator makes it easier for the discriminator to differentiate between real and fake data.

- The generator continues its learning in a feedback loop from the output generated by the discriminator.

- The process of continuous learning helps the generator to generate images/data at the output, which is very realistic and makes it difficult for the discriminator to differentiate between real and fake data.

This kind of adversarial network using GAN enables automated depression detection systems to generate synthetic **computed tomography (CT)** and **magnetic resonance imaging (MRI)** to help clinicians identify patients who are suffering from depression and those having a high likelihood of falling into depression. CNN-based GANs can further be used in generating time series synthetic data and image-to-image translation, such as synthesizing CT images from an MRI image.

Let us now see where a GAN fits into a depression data processing and training pipeline through Figure 8-12 (Wang et al., 2022).

Depression Detection from EEG Signals

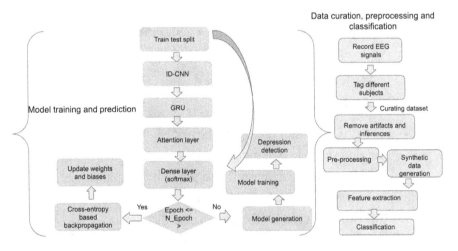

Figure 8-12. *EEG signal recording, data preprocessing, and model training to evaluate depression (Wang et al., 2022)*

To incorporate synthetic data generation in a data processing and training pipeline, we show in Figure 8-12 (the labels marked in pink) how data recorded through EEG signals are first tagged/annotated on different subjects and curated to form a consolidated dataset for depression detection. The dataset is further processed to remove noises and not needed artifacts, normalized and validated to evaluate the percentage of representations from minority datasets. The dataset from minority classes (rare forms of depression) is then passed through a GAN to generate more data samples to have a fully equal representation of depression and no-depression classes.

1D CNN in combination with GRU or LSTM can yield more accurate results when fed in with demographic data of the individuals taken in the study, such as age, gender, ethnicity, and socioeconomic status. Such models can be effective in classifying obsessive-compulsive disorders, addiction disorders, disorders linked to trauma and stress, mood disorders, schizophrenia, and anxiety disorders.

Further, Figure 8-12 (the labels marked in blue) shows train-test data splitting for the training and testing phase. Here, we have used a 1D CNN and a GRU followed by processing through an attention network to aid in the classification of depression and nondepression instances. As part of neural network training, the back propagation-based training method updates the weights and biases continuously till the model converges.

By this time, we have got a fair understanding of how different deep models play a critical role in depression detection from speech data. It's now time to study the role of deep learning models in identifying depression from video data, such as facial expressions, bodily movements, and activities.

Video Data and Deep Learning Algorithms in Depressive Symptoms Detection

Videos from social media also play an important part in detecting depression; one prime example is when audio and video features are extracted from the vlogs (video logs) on YouTube. In the first step, it is advisable to annotate the videos into two different types depression and nondepression, by analyzing the vlogs and then building an ML model that learns the audio-video and visual features and then helps to predict depressed individuals present in social media proactively to help them with early proactive treatment.

CNN-based deep learning models have enhanced depression detection methods when they fuse acoustic, visual, and textual modalities, as such models are capable of providing more comprehensive information on different stages of depression than static images.

Video Data Annotation

As stated above, video data annotation is an important step in marking ground-truth video data, to label videos of individuals who are truly suffering from depression (Min et al., 2023). Let us understand the sequence of operations needed in labeling the video data.

- At first, YouTube Data API can be used to search for videos exhibiting signs of depression as well as no depression.

- YouTube DL can be used to download the videos of both classes.

- The next step would be to analyze if the videos have a "vlog" format, to segregate videos where the individual speaks directly to the camera.

- Video data annotators would then label videos as having signs of depression if there are instances when the speaker shares their depression-related feelings and symptoms during the time of video recording. Such instances may include discussing their symptoms, mood swings, and their current medications.

- To avoid mislabeling, an agreed number of annotators need to be employed instead of a single annotator and a statistic termed Cohen's Kappa (κ) needs to be computed to measure the degree of agreement between two annotators who are classifying the videos into depression and no-depression categories as the labeling process is subjective.

- The mean kappa score computed from the given set of annotators from filtered vlogs justifies the labeling and completes the final annotation into two classes.

After annotation of video data comes the next important step of detecting the facial expressions using AI and ML algorithms (Min et al., 2023). Now let us delve into how to detect facial expressions using an open source Python library that uses a pretrained face expression recognition model to recognize emotional states from the vloggers.

Detecting Facial Expression

In this subsection, we demonstrate an example of a facial expression using an open source **facial expression recognition (FER)** library which takes a frame from an image (frame) as an input and passes it through the detector model to predict the emotion of the detected face (i.e., angry, disgust, fear, happy, neutral, sad, and surprise). The different emotion scores are evaluated and extracted for each frame in the vlog.

Following this, the model computes a mean score for each video feature by aggregating it into the video level. The video extraction process should be able to predict different stages of depression for early-stage intervention by including diverse data sources. Further, the frame-wise emotion and mood recognition process should have a robust labeling technique to label overlapping videos correctly to enhance intercoder reliability. This would help to increase the model performance by aligning the edge cases of overlapped video frames.

We need to combine the video feature extraction process with other audio features and visual features that can be obtained from a vlog. Further, we only need to consider frames where a single face is encountered. Before going to the details of the code, let's explore how we can open a live video feed and leverage a script to capture an image frame and later use that for detecting emotion.

- In order to capture images using a webcam, we can leverage (`https://github.com/kevinam99/capturing-images-from-webcam-using-opencv-python`) the file webcam-capture-v1.01.py to run the command as follows:

  python3 webcam-capture-v1.01.py

- Once the webcam starts running, the picture needs to be brought closer to save in the webcam frame.

- Once the object is in the right frame, press the key "s" to save a picture, or alternatively press "q" to exit.

- The key "s" helps to save the picture; you will get a view of the saved image in the same directory as that of the program, after getting converted to grayscale and then resized to 28×28 size.

Once the image is saved as "jpg," let us look at how to use that image to detect emotions.

Listing 8-2. In the following listing, we will try to understand how
a single image (from an extracted video frame) can be processed
through pretrained libraries to detect emotions

1. In the first step, we need to install the necessary
    libraries of pretrained facial expression along with
    tensorflow, opencv, and ffmeg.

    ```
    pip install fer
    pip install tensorflow>=1.7 opencv-contrib-
    python==3.3.0.9
    pip install ffmpeg moviepy
    ```

2. In the next step, we import the installed libraries,
    load the image which needs to be recognized, and
    pass it through the detector model to identify the
    emotions.

    ```
    from fer import FER
    import cv2
    img = cv2.imread("sampleimg.jpg")
    detector = FER()
    detector.detect_emotions(img)
    ```

3. As an output, it yields the image shown in
    Figure 8-13, centered around a rectangle with the
    different emotions and their probabilities.

    ```
    [{'box': [277, 90, 48, 63], 'emotions': {'angry': 0.02,
    'disgust': 0.0, 'fear': 0.05, 'happy': 0.16, 'neutral':
    0.09, 'sad': 0.27, 'surprise': 0.41}]
    ```

angry:
disgust:
sad:
fear
happy: 0.16
neutral:
sad:
surprise:0.41

Figure 8-13. *Emotion detection using a pretrained face expression recognition model*

In addition to open source libraries, visual indicators have also been used extensively to establish a linkage between nonverbal behavior and depression. **Facial action coding framework (FACS)** is one of such standard frameworks that relies on facial appearance to taxonomize human facial developments and classify physical articulations (see Ashraf et al., 2024). The process incorporates facial action units that record different actions of an individual's muscles (including hands, shoulders, and head movements) and can be decoded by psychologists for treatments.

Facial expression recognition algorithms play a significant role in autism, neurodegenerative detection, and predicting psychotic disorders. It is designed using CNN, which is used to build a **squeeze-and-excitation**

network (SENet) a combination of squeeze-and-excitation networks. This kind of architecture, in conjunction with ResNet, helps in capturing the most important emotional features from the face and ignoring less useful features and the residual neural network.

The role of conventional and deep learning-based algorithms is not limited to facial expression detection; but they also play a formidable role in recognizing video objects and human activities. Let us now discuss the different methodologies that can be used to detect youth depression.

Recognizing Video Objects and Human Activities to Identify Depression

In this section, let us delve into existing video frame-based detection algorithms that are illustrated in Figure 8-14.

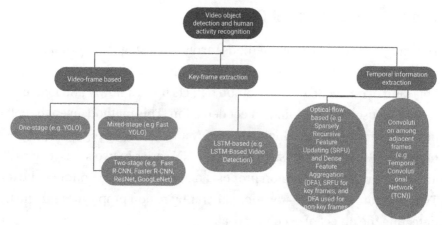

Video Object and Human-activity Recognition using Deep Learning

Figure 8-14. *Different types of video object and human activity detection algorithms*

Video Frame-Based Object Detection Algorithms

Video detection algorithms break up the video into individual frames, where each frame is treated as an image to detect depression by correlating different adjacent frames (Boesch, 2024). This kind of algorithm uses LSTM-based DL models.

One-Stage Video Object Detection

This kind of video object detection algorithm operates at a high speed to detect a single frame in order to facilitate real-time video detection. **You Only Look Once (YOLO)**, **Single Shot MultiBox Detector (SSD)**, and RetinaNet fall under this category. YOLO treats the object classification as a regression algorithm by configuring the image size to a specific resized object and then performing a nonlinear mapping between image features and neural network parameters to detect the image or video. SSD makes use of one single shot to detect multiple objects within the same image.

Two-Stage Video Object Detection

Two-stage object detection methods involve two-stage processing: the first one for feature extraction, followed by classification in the second stage. This kind of object detection method detects frames slowly and hence cannot be used for real-time object detection. This method, even though it has a slower speed, has a higher accuracy and leverages **regional proposal network (RPN)** methods such as the R-CNN series in two shots, to classify an object by defining a region proposal network to create boundary boxes. Two shots are used here—one for generating region proposals and one for detecting the object of each proposal.

Mixed-Stage Video Object Detection

The method of mixed-stage object detection involves both one-stage detection and two-stage detection, or other video detections, in scenarios where either of the one-stage or two-stage detection methods are not sufficient in recognizing the videos. Examples include minimum delay video object detection that leverages both one-stage and two-stage image detectors simultaneously and can perform in real-time speed.

It computes the distribution variation or observed changes of the video sequences statistically by using **cumulative sum (CUSUM)**. Following this, it assembles the feature map sampling values of the video sequence to aggregate the small deviations of the video sequences into fluctuations. A CNN detector is used to process frame-by-frame to filter out inaccurate boxes and boost the prediction accuracy.

Single, Two, Multi-stage Video Detection

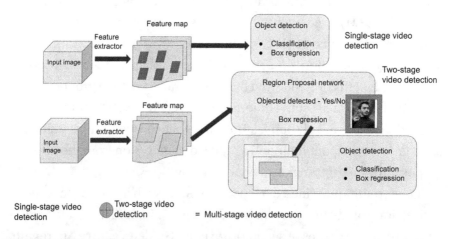

Figure 8-15. *Different stages of video object detection*

Figure 8-15 illustrates single, two, and multistage video object detection mechanisms (encord.com, 2023).

Key Frame-Based Video Detection

The key frame-based video detection (Seq-NMS) method attempts to find the most relevant and accurate region proposals from a video. The process concentrates on selecting the largest score of a frame sequence from a video, followed by rescoring the sequence and eliminating the overlapped region proposals. In addition, methods like **Scale-Time Lattice (ST-Lattice)** operate by propagating the key frame to non-key frames. This method can be used to recursively iterate and identify non-key frames from the middle of two key frames till all non-key frames have been identified.

Conventional ML techniques like **support vector machine (SVM)** can also be used to detect features from a video frame, after selecting the key frame, and by further tracking the non-key frames

Temporal Information-Based Video Detection

Temporal information-based video detection uses **long short-term memory (LSTM)** or convolution networks on multiple frames in sequential order, such as 3D convolution, to extract more relevant features frame by frame in a time sequence. In the initial step, a video is divided into equal-length clips, followed by the generation of tube proposals for each clip based on 3D convolutional network (ConvNet) features. This method uses 3D convolution features to recognize and localize different clips, facilitating spatiotemporal action video detection from the tube proposals by linking them using network flow.

Other traditional video detection methods, such as those that use optical flow (predicting movement between two images extracted from two consecutive frames of a video), are Flow-Guided **Feature Aggregation (FGFA)** and **Multimodal Alignment Aggregation and Distillation (MEGA)**. These methods use feature aggregation algorithms to consolidate frame features from key frames lying at varying distances. These include assembling adjacent frame features and long-term frame features, to boost the quality of extracted features from the objects.

Association Long Short-Term Memory (Association LSTM)

LSTM alone has limitations in modeling object association between consecutive frames to represent different detected objects. Hence, researchers have come up with another variant of LSTM termed as associative LSTM which uses its long short-term memory for learning features. It relies on temporal information processing capability to process multiple video frames simultaneously. It operates by regressing and classifying directly on object locations and categories, in addition to evaluating and associating high-quality features for output object representations. Here, the objective function tries to minimize the object regression error and association error to find the association between objects in two consecutive frames. This method helps to recover missing objects in the present frame and correct mislabeled objects at the same time by applying temporal consistency and semantic label validation across object locations and categories.

Apart from using deep learning models like CNNs or LSTMs, traditional ML classification models like **eXtreme Gradient Boosting (XGBoost)** or **Gradient Boosting Decision Tree (GBDT)** can also be used to build depression detection models. The XGBoost algorithm does feature selection by splitting the tree to maximum depth and then using tree pruning functionality to prune the tree backwards and remove splits beyond which there is no increase of additional information that would aid in selecting the right features. These classification algorithms have their inbuilt feature selection methods. However, in addition to those, we can also use **principal component analysis (PCA)** in extracting features from the face, head position, eyes, and mouth for depression detection. As data scientists, we need to be conscious of selecting the right number of feature sets, without which our model would face the problem of underfitting because fewer features are selected, or the model would undergo overfitting when more features are selected.

Now let us look at a practical example of how to recognize human
activities and identify their patterns using PCA.

Listing 8-3. Here, we have taken an activity recognition dataset
and performed human activity recognition. The dataset was created
by recording 30 participants performing activities of daily living
(ADL) while wearing a waist-mounted smartphone with embedded
inertial sensors. The objective is to classify activities into one of the
six activities performed, namely, WALKING, WALKING_UPSTAIRS,
WALKING_DOWNSTAIRS, SITTING, STANDING, and LAYING. The
dataset consists of more than 7000 data points

*This task can be viewed as a tabular classification task where we can
leverage tabular techniques like decision models. First, we begin by loading
and understanding the distribution of the data.*

*After this, we use the labelencoder to convert the string labels into integer
classes so that it can be fed into the models for training. Then, we move on
to use principal component analysis(PCA). This is because the original data
has too many features which can hamper the model training. So, by using
PCA, we project the higher dimension data into lower dimension so as to
decrease the number of features. Finally, we use the random forest classifier,
a tree-based algorithm from sklearn to make the predictions.*

1. In the first step, we install the necessary libraries.

   ```
   %pip install -q kaggle
   ```

2. After installation, we import the necessary libraries
   and load the dataset.

   ```
   import pandas as pd
   from sklearn.preprocessing import OneHotEncoder,
   LabelEncoder
   ```

```
from sklearn.decomposition import PCA
from sklearn.ensemble import RandomForestClassifier
from sklearn.ensemble import GradientBoostingClassifier
!kaggle datasets download -d uciml/human-activity-
recognition-with-smartphones
!unzip /content/human-activity-recognition-with-
smartphones.zip
df = pd.read_csv('train.csv')
df.head()
```

3. Then, we segregate the input X and target Y columns
   for train dataset and then label encode to convert
   the strings into integers.

```
X_train = df.iloc[:,0:len(df.columns)-1]
Y_train = df.iloc[:,-1]
le = LabelEncoder()
Y_train = le.fit_transform(Y_train)
le_name_mapping = dict(zip(le.classes_,
le.transform(le.classes_)))
print(le_name_mapping)
```

4. Our next step is to apply PCA transformation to
   reduce the features on a higher dimension space by
   projecting it on a lower dimension space.

```
pca = PCA(0.95)
pca.fit(X_train)
train_x_pca = pca.transform(X_train)
```

5. We then first apply use tabular algorithm like
   RandomForestClassifier.

```
clf = RandomForestClassifier()
clf.fit(train_x_pca, Y_train)
clf.score(train_x_pca,Y_train)
```

6. Then, we apply the next tabular algorithm
    GradientBoostingClassifier.

```
clf = GradientBoostingClassifier()
clf.fit(train_x_pca, Y_train)
clf.score(train_x_pca,Y_train)
```

We see the predicted output after fitting, the RandomForestClassifier
on the reduced PCA feature set scores 1.0, while the GradientBoosting
Classifier model scores 0.97 (Pampouchidou et al., 2016). Now both
these models can be further used for prediction on the reduced/PCA
transformed feature set. The principal components of the feature set are
given by Figure 8-16.

Dimensionality Reduction with PCA for Activity Recognition Dataset

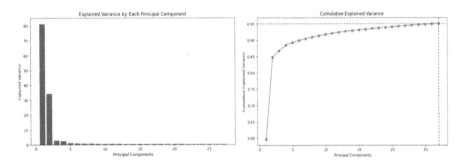

Figure 8-16. *PCA plot demonstrating the key features from user daily
activities*

Let us now take another example to understand facial emotion
detection from a medium-scaled pretrained model.

Listing 8-4. Here, we have taken a popular facial emotion dataset
Affectnet and performed emotion recognition to detect mood
changes and patterns from the facial images. The dataset consists

of images for eight different emotional classes, namely, anger,
contempt, disgust, fear, happy, neutral, sad, and surprise with more
than 1500 images for each emotion

*We will treat this as a simple image classification model where we will use
a popular EfficientNetV2 pretrained on the ImageNet dataset. EfficientNetV2
is a very popular medium-sized model which has proved its proficiency on
various benchmarks. We will also use the KerasCV library which makes the
popular models accessible in one place and provides ease of access. Keras
library is easy to use for dataset preparation and fine-tuning the model.*

1. In the first step, we install the necessary libraries
    and unzip the dataset that we would need to use.

    ```
    %pip install -q kaggle
    !kaggle datasets download -d vfomenko/young-
    affectnet-hq
    !unzip -qq /content/young-affectnet-hq.zip /data
    ```

2. After installation, we import the keras libraries, load
    the image dataset into keras, and start building the
    deep learning model pipeline.

    ```
    import keras
    import keras_cv
    import numpy as np
    train_ds = keras.utils.image_dataset_from_directory(
                directory="data/",
                labels='inferred',
                label_mode='categorical',
                batch_size=32,
                image_size=(224, 224))

    backbone = keras_cv.models.EfficientNetV2Backbone.
    from_preset(
    ```

```
        "efficientnetv2_b0_imagenet"
)
model = keras_cv.models.ImageClassifier(
    backbone=backbone,
    num_classes=8,
    activation="softmax",
)
model.compile(
    loss='categorical_crossentropy',
    optimizer=keras.optimizers.Adam(learning_
    rate=1e-5),
    metrics=['accuracy']
)
model.fit(
    train_ds,
    epochs=3,
)
```

3. Then, we take an unknown input image and
    start making predictions, which yields the output
    emotion of surprise.

```
ds = train_ds.take(1)
for im,_ in ds:
  img = im
img
import matplotlib.pyplot as plt
plt.imshow(img[0].numpy().astype(int))
emotions = ['anger', 'contempt', 'disgust', 'fear',
'happy', 'neutral', 'sad', 'surprise']
pred = model.predict(img)
print(emotions[np.argmax(pred[0])])
```

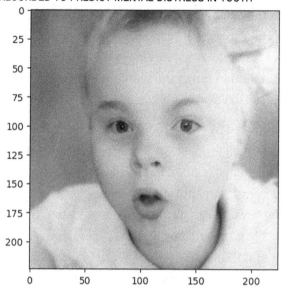

Figure 8-17. *Predicted emotion through ImageClassifier*

We are now knowledgeable of the fact that, same as images, videos can also be used to analyze at an individual frame level to extract human emotions to identify depression. Let us now get an overview of how curvelet transformation and local binary patterns play a role in detecting depression.

Detecting Depression Using Video-Based, Curvelet Transform, and Local Binary Patterns

We have illustrated through Figure 8-18 the steps involved in depression identification using curvature information of a human face. These are summarized below through the following steps (Pampouchidou et al., 2016).

- Segregate the first frame of the video to initialize the face region.

- Track the remaining portion of the video and extract relevant features using the **Kanade-TomasiLucas (KLT) tracker**, which leverages the spatial intensity information to direct the search for the position that produces the best match.

- It is followed by processing the extended face region using curvelet transform for every frame to generate a pseudo-image (CurveFace). This helps to concentrate on the curvature information of an image, helping to identify different facial expressions from their curves (e.g., mouth corners are angled down in a sad expression). In addition, individual-specific biometrics (such as different shapes of facial features and their symmetries), as well as occlusions (such as facial hair, and eye-glasses), can be also discovered.

- The last step involves computing the **Local Binary Pattern descriptor (LBP)** for each individual CurveFace to generate the feature vector for the frame-based classification. Local Gabor Binary Patterns in Three Orthogonal Planes further generate frame-based transformed versions (different planes in time (XZ and YZ)) by consolidating the window of frames.

- The generated planes (XZ and YZ) are processed row-wise and column-wise in sequence (the first row is combined with the plane corresponding to the first column; the second row is combined with the second column), which helps to preserve motion information across. This method relies on short feature vectors in frame-based classification to determine the nearest neighbors that classify the video into different categories of depression.

Video-based Depression Detector

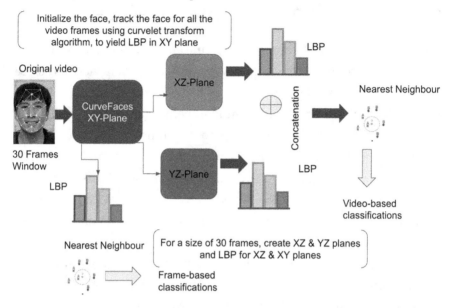

LBP - **Local Binary Pattern** descriptor

Figure 8-18. *Depression detection using curvelet transform and local binary patterns (Pampouchidou et al., 2016)*

After gaining insights on video depression detection techniques, it is time for us to revisit the key learnings for this chapter and also get a know-how of using best practices in building ML models to identify depression.

The best practices would help us to know the promises as well as negative angles of training computationally intensive deep learning models from images and videos to detect depression. It will help us to be cautious of the environmental impacts before training large-scale models.

Knowledge of Best Data and AI Practices to Build Scalable Depression Models

The performance of different deep learning algorithms cited above strongly depends on increased data scale that would help yield large-scale models (Varoquaux et al., 2025). However, the increase in scale comes with the increased cost of training models with larger datasets using GPUs to complete training faster. This cost becomes reasonably higher when the model achieves a reasonable level of accuracy, and the model may show diminishing returns even after a given point (Varoquaux et al., 2025). For example, adding 567 times more computes would add a 5% error rate on ImageNet. As cost becomes a concerning factor, researchers are investigating ways to re-architect the neural networks to lower the computational intensity, which would result in increased performance. Having more efficiency in computational infrastructure and faster learning speed will result in producing less CO_2 emissions and heating. Training and deploying AI models demands large quantities of raw materials for manufacturing computing hardware. In addition, the amount of energy required to power infrastructure and water to cool ever-growing datacenters leads to large quantities of CO_2 emissions during their training process. For example, LLMs can emit up to 550 tonnes; more so, natural language processing (NLP) models on GPUs can lead to approximately 600,000 lb of CO_2 emissions. The contributing factors lead to unsustainable practices that vary differently based on the deployment choices made (e.g., type of GPU used, batch size, precision, etc.).

Moreover, vision and video-based algorithms do not always perform better in certain contexts with scale. Hence, DL models lead to unsustainable environmental conditions without improving their performance after the models saturate. We need to be conscious of the fact that more varied instances of ground-truth depression data are needed for large-scale models; however, the data needs to have surveillance, in the absence of which would drive unethical and unauditable data practices.

Let us now take a glance at the nuances in selecting the right model, suitable for identifying depression in a trustworthy manner.

Right Model Selection

With the rise of LLMs getting used in the mental health industry, the selection of right model makes a difference not only in identifying the right depression or providing suggestions to the patients but also helps to develop environmentally friendly sustainable solutions.

As data scientists of mental depression detector models, we need to be aware of the differences existing in model performance between models within a similar size class as the same difference persists between models of different sizes. This would help us to select the right model and make the right choices for selecting the hyper-parameters during training. Further, the use of fine-tuned LLMs (GPT-3.5 Turbo or LLaMA2-7B) for depression detection is more centered around assembling domain-specific expert knowledge as opposed to pulling in general-purpose knowledge to fully explore the in-context learning ability of the LLMs. Hence as data engineers and mental healthcare providers or experts, there's a greater need to feed in authentic data sources and have regulatory and auditory checks that can safeguard against ethical questions that may arise later. For example, fine-tuning LLMs, using only users' social media data is not an appropriate recommendation to build a trustworthy tool for proactively assessing mental health risks, as there's a strong need for additional data sources to support the prime indicators of mental health gathered through social media data see (Varoquaux et al., 2025).

Even in some scenarios, LLM's after-instruction tuning does not match the model's expected prediction accuracy, which has remained an area of concern in areas related to moral question answering, in mental health. LLMs demonstrate serious limitations in completing clinical questionnaires for depression and anxiety (Varoquaux et al., 2025).

With the pros and cons in mind, let us investigate certain key factors in consideration during training deep learning models for depression detection.

Key Considerations for Training Large-Scale Deep Learning Models

With the limitations around training video and vision-based large-scale models, researchers have highlighted the prime factors listed below to trade off the benefits achieved using predictions of transformer-based models.

Quality vs. Quantity of Large Datasets

As stated above, a large data size does not always improve the model performance by ensuring complete coverage in model predictions. Researchers have observed that sometimes image or text datasets are not widely representative of the entire population under consideration. To eliminate the chance of having disproportional representations in datasets, depression datasets must include equivalent classes from different regions, countries, and communities to constitute different types of facial and body expressions from people with different physical attributes. Along with this, privacy protection should exist against predatory data-gathering practices in each of these regions, countries, or communities.

Interpretability of Models for Complex Datasets

While deep learning transformer-based models are far capable of assessing depression context and background using self-attention mechanisms and deep layers, as designers of mental health predictive AI systems, we should be able to design interpretable frameworks for large-scale generative models to explain why the model generated the predictions and attribute them to the right input data sources to justify processing of the sequential data through its layers. Traditional ML models perform better when they have larger instances of preestablished features; they fail due to complexities present in depression expressions and symptoms in speech, text, and video data. On the other hand, DL models also fail to decipher emotional nuances from the deeper context of messages, speech, facial expressions, and body movements in many scenarios when one data source is not supported by the other.

Adaptability and Generalization

LLM-based models may hallucinate when trained on certain datasets and evaluated on others. This may happen due to the wide dynamic nature of depressive behavior and physical attributes that are inherently present in the user's journaling activities, speech, facial and other physical movements, and gestures. As models lose accuracy and relevancy, all data, AI, and analytics professionals centered around the platform should come together to understand and explain the skewness of data specific to any linguistic population and raise an alert if the model shows drift over other cultural segments of the population.

Scalability and Real-Time Analysis

A mental health platform should have enough computational resources to handle new incoming data (text, video, image, or speech) and detect changes in model performance or drifts in real time. At the same time, the platform should be scalable to trigger new learning, kick off retraining, or fine-tune existing models.

Integration of Domain-Specific Knowledge

The platform, people, and process of mental health should take into consideration the constant discovery and learning of mental health knowledge specific to certain topics such as different categories of mental

health problems and integrate it into the platform. The data integration
will come along with large-scale computational models to increase the
performance of detecting depression (Varoquaux et al., 2025).

Conclusion

In this chapter, we have learned about how depression impacts individuals
and the mental and physical side effects of depression. We also learned
about different tools, techniques, and signals to capture information,
including speech and video data which prompted us to learn about the
advantages on the need of depression classification. Here in this context,
we came to know about different types of symptoms and how their
intensities can drive us to take remedial actions. In the next section, we
delved into different types of deep learning models and how they assist in
predicting depression with varying use cases. As we concentrate on the
impact of depression in youth, we take a step further to understand how
deep learning-based methods can be used when we have text and speech
data, in scenarios when we can combine journaling/social media data
with questionnaires filled by individuals. While we studied different neural
network architectures like CNN, autoencoders, and others, we also got an
understanding of the role of adversarial networks in generating synthetic
datasets for underrepresented classes. In this chapter, we got into the
details of step-by-step speech and video data feature extraction processes
and how this can aid in predicting depression proactively. We saw practical
examples of how facial expressions can help locate depressions, along
with video-based object recognition and activity detection methods. This
chapter concludes by emphasizing the right practices of data collection
and training large-scale models to prevent bias and hallucination, which
can help the AI-based mental health detection system to be used as a
unified framework for everyone.

Now after having a background of depression intensity and level, let us move into the next chapter, Chapter 9, "AI Models for Detecting Different Types and Levels of Risks Associated," to learn more about the different steps involved in processing data to get detailed insights on the level of depression.

References

Ashraf, A., Gunawan, T. S., Bob Subhan Riza, Haryanto, E. V., & Janin, Z. (2024). On the review of image and video-based depression detection using machine learning. *ResearchGate*, *19*(3), 1677–1684. https://doi.org/10.11591/ijeecs.v19.i3.pp1677-1684

Boesch, G. (2024, October 4). *Object Detection: The Definitive 2025 Guide*. Viso.Ai. https://viso.ai/deep-learning/object-detection/

Bolhasani, H. (2021). *Depression Diagnosis by Deep Learning Using EEG Signals: A Systematic Review* (No. 2021070028). Preprints. https://doi.org/10.20944/preprints202107.0028.v1

Ding, Y., Chen, X., Fu, Q., & Zhong, S. (2020). A Depression Recognition Method for College Students Using Deep Integrated Support Vector Algorithm. *IEEE Access*, *8*, 75616–75629. https://doi.org/10.1109/ACCESS.2020.2987523

Du, M., Liu, S., Wang, T., Zhang, W., Ke, Y., Chen, L., & Ming, D. (2023). Depression recognition using a proposed speech chain model fusing speech production and perception features. *Journal of Affective Disorders*, *323*, 299–308. https://doi.org/10.1016/j.jad.2022.11.060

encord.com. (2023). *Object Detection: Models, Use Cases, Examples*. https://encord.com/blog/object-detection/

IBM. (2023). *What Is an Autoencoder? | IBM*. https://www.ibm.com/think/topics/autoencoder

Kim, A. Y., Jang, E. H., Lee, S.-H., Choi, K.-Y., Park, J. G., & Shin, H.-C. (2023). Automatic Depression Detection Using Smartphone-Based Text-Dependent Speech Signals: Deep Convolutional Neural Network Approach. *Journal of Medical Internet Research, 25*(1), e34474. https://doi.org/10.2196/34474

Kumar, R., Nagar, S. K., & Shrivastava, A. (2020). Depression Detection Using Stacked Autoencoder from Facial Features and NLP. *SMART MOVES JOURNAL IJOSTHE, 7*(1), Article 1. https://doi.org/10.24113/ojssports.v7i1.115

Lin, C., Lee, S.-H., Huang, C.-M., Chen, G.-Y., Chang, W., Liu, H.-L., Ng, S.-H., Lee, T. M.-C., & Wu, S.-C. (2023). Automatic diagnosis of late-life depression by 3D convolutional neural networks and cross-sample Entropy analysis from resting-state fMRI. *Brain Imaging and Behavior, 17*(1), 125–135. https://doi.org/10.1007/s11682-022-00748-0

Luján, M. Á., Torres, A. M., Borja, A. L., Santos, J. L., & Sotos, J. M. (2022). High-Precise Bipolar Disorder Detection by Using Radial Basis Functions Based Neural Network. *Electronics, 11*(3), Article 3. https://doi.org/10.3390/electronics11030343

Min, K., Yoon, J., Kang, M., Lee, D., Park, E., & Han, J. (2023). Detecting depression on video logs using audiovisual features. *Humanities and Social Sciences Communications, 10*(1), 1–8. https://doi.org/10.1057/s41599-023-02313-6

Pampouchidou, A., Marias, K., Tsiknakis, M., Simos, P., Yang, F., Lemaitre, G., & Meriaudeau, F. (2016). Video-based depression detection using local Curvelet binary patterns in pairwise orthogonal planes. *2016 38th Annual International Conference of the IEEE Engineering in Medicine and Biology Society (EMBC)*, 3835–3838. 2016 38th Annual International Conference of the IEEE Engineering in Medicine and Biology Society (EMBC). https://doi.org/10.1109/EMBC.2016.7591564

Park, J., & Moon, N. (2022). Design and Implementation of Attention Depression Detection Model Based on Multimodal Analysis. *Sustainability, 14*(6), Article 6. https://doi.org/10.3390/su14063569

Sardari, S., Nakisa, B., Rastgoo, M. N., & Eklund, P. (2022). Audio based depression detection using Convolutional Autoencoder. *Expert Systems with Applications, 189*, 116076. https://doi.org/10.1016/j.eswa.2021.116076

Selvaraj, A., & Mohandoss, L. (2024). Enhancing Depression Detection: A Stacked Ensemble Model with Feature Selection and RF Feature Importance Analysis Using NHANES Data. *Applied Sciences, 14*(16), Article 16. https://doi.org/10.3390/app14167366

Su, C., Xu, Z., Pathak, J., & Wang, F. (2020). Deep learning in mental health outcome research: A scoping review. *Translational Psychiatry, 10*, 116. https://doi.org/10.1038/s41398-020-0780-3

Thapa, K., Seo, Y., Yang, S.-H., & Kim, K. (2023). Semi-Supervised Adversarial Auto-Encoder to Expedite Human Activity Recognition. *Sensors, 23*(2), Article 2. https://doi.org/10.3390/s23020683

Tufail, H., Cheema, S. M., Ali, M., Pires, I. M., & Garcia, N. M. (2023). Depression Detection with Convolutional Neural Networks: A Step Towards Improved Mental Health Care. *Procedia Computer Science, 224*, 544–549. https://doi.org/10.1016/j.procs.2023.09.079

Varoquaux, G., Luccioni, A. S., & Whittaker, M. (2025). *Hype, Sustainability, and the Price of the Bigger-is-Better Paradigm in AI* (No. arXiv:2409.14160). arXiv. https://doi.org/10.48550/arXiv.2409.14160

Wang, H., Liu, Y., Zhen, X., & Tu, X. (2021). Depression Speech Recognition With a Three-Dimensional Convolutional Network. *Frontiers in Human Neuroscience, 15*. https://doi.org/10.3389/fnhum.2021.713823

Wang, Z., Ma, Z., Liu, W., An, Z., & Huang, F. (2022). A Depression Diagnosis Method Based on the Hybrid Neural Network and Attention Mechanism. *Brain Sciences, 12*(7), 834. https://doi.org/10.3390/brainsci12070834

## CHAPTER 9

AI Models for Detecting Different Types and Levels of Risks Associated

This chapter provides an area to delve deeper into intelligent algorithms to detect patterns that can help in the early diagnosis of different types of disorders, classify the severity of the problems, and the actions for alleviating those concerns. This chapter also drives us to detect risks from patterns and markers by bringing to readers' attention how different data sources can identify biomarkers leading to early detection of diseases. Readers are increasingly aware of the fact that mental data can be leveraged in proactive detection and early addressing of problems with timely interventions from caregivers. The primary objective of this chapter is to empower mental health-associated stakeholders with AI modeling tools and techniques. In addition, this chapter further extends its boundary by looking through the lenses of the essential plugins and components necessary for a risk framework under different scenarios. It explores the thorough use of diverse datasets and models brain scans and behavioral patterns. This chapter further allows us to evaluate GenAI and its role in assessing the severity of risks.

© Sharmistha Chatterjee, Azadeh Dindarian, Usha Rengaraju 2025
S. Chatterjee et al., *Revolutionizing Youth Mental Health with Ethical AI*,
https://doi.org/10.1007/979-8-8688-1186-9_9

In this chapter, these topics will be covered in the following sections:

- Principles of a risk assessment framework

- Evaluation metrics for depression, use cases, and actions

- Role of GenAI in risk assessment

With the above discussion areas in mind, let us delve into the consequences of not managing varying levels of risks and providing a strategy to categorize risks and plan for suitable proactive actions so that mental health problems do not become more severe.

Principles of a Risk Assessment Framework

In this section, let us first look at how different categories of mental health problems pose threats and make it necessary to have a risk assessment framework for mental health. This helps us to understand the severity of the problem for varying age groups, gender, and other needs of the population before bringing in AI to predict and address those problems (see Skianis et al., 2025).

Risk assessment and management is an important dynamic process to let information flow, monitor individual conditions in lieu of the change in social environmental conditions, enable information formulation, and undertake prompt decisions to help patients and patient families. The clinicians, clinical service teams, service providers, public health professionals, and policymakers need to constantly monitor patients' mental health conditions, the functioning of the framework, and the ecosystem to provide appropriate guidance and support.

The Ministry of Health Guidelines for Clinical Risk Assessment and Management in Mental Health Services (1998) in New Zealand had set the below-mentioned guidelines relevant to the context of mental health (Matthewson, 2002):

374

- Risk measurement governs the likelihood of occurrence of an adverse event or events (like self-harm or doing harm to others), arising out of mental health problems.

- Risk assessment involves listing the key factors, behaviors, or environmental conditions causing the illness during any stipulated period of time.

- Risk management involves summarizing and organizing the risk data to provide an information base, by prioritizing the leading factors causing the risk and monitoring the factors from time to time to infer the state of progression of the risk.

- The risk management framework also aims to minimize the likelihood of the occurrence of adverse events by providing safe, meaningful, and effective care to the patients, in a manner that promotes the patient's mental well-being.

- Risk measurement and management should function proactively by categorizing risk based on the type of illness, historic past actions, and factors that could increase the level of the risk.

Let us now explore the key factors to consider for designing a risk framework.

Designing Risk Framework with Key Factors for Consideration

With the guidelines of risk assessment and management in place, to design and scale the framework, we have to synthesize data and AI into the framework. First and foremost, we need certain metrics that need constant monitoring, evaluation, and interpretation to derive benefits and maximum value from the framework.

Previous frameworks of mental health lacked research, leading to no root cause analysis of individuals suffering from social isolation, loneliness, loss of social connections, and economic hardship. In addition, it failed to take into consideration a detailed understanding of environmental factors and different scenarios and behavioral circumstances of individuals due to the presence of diverse subgroups that lead to positive mental health conditions. The role of the framework starts when a patient first comes in contact with a service, avails of the service, or opts out of the subscribed service and avails of any other alternative mental healthcare services. The most popular risk metrics in the context of mental health involve

- Propensity to self-harm (psychological and/or physical), commit suicide, or demonstrate risks of harm to service/care providers, family members, or to any other people

- Increase in illness and tendency of addiction to drugs or alcohol

- Risk of being vulnerable and likelihood of exploitation, with threat of abuse or violence from outside

- Risk of incurring violence (physical, emotional, sexual) from others or getting indulged in similar acts of violence

- The threat of intimidating someone, especially dependents and children

- The tendency toward property damage through nuisance, hasty, and careless behavior, including rash driving

The underlying risk factors can be modeled using certain existing frameworks, like the bioecological model's systems framework, which uses ten predictors to assess mental health risk for an individual. The adolescent risk assessment framework under our current scope and discussion should take into consideration microsystem risk (parenting behaviors and family environmental context variables) and exosystem risk (mother and father cultural orientation toward the individual, siblings, and extended family) to infer the severity of risk in adolescent mental health (Siddique & Lee, 2024). Further factors for evaluation include ***externalizing*** and ***internalizing*** behavior that can be either self-reported by children or reported by parents (Rothenberg et al., 2023). While externalizing behavior demonstrates how an individual behaves outside to the external world and includes behavior like lying, truancy, vandalism, bullying, and drug and alcohol use, internalizing behavior includes how an individual perceives oneself in their mind. Hence, internalizing behavior reflects core feelings of an individual and feelings such as loneliness, self-consciousness, nervousness, sadness, and anxiety, which are self-reported by parents and captured through 33 and 31 questions, respectively, comes under this. Figure 9-1 illustrates the most important internalizing and externalizing behavior features where the degree of importance of the feature incurs an influence on the risk intensity.

A. Internalizing Behavior Feature Importances

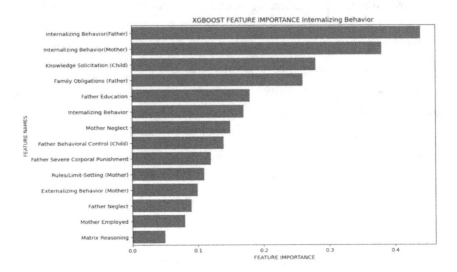

B. Externalizing Behavior Feature Importances

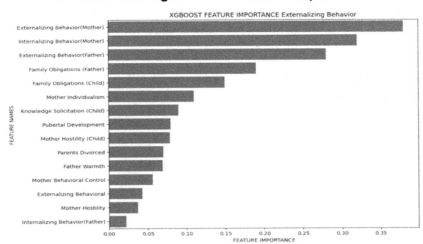

Figure 9-1. *Internalizing and externalizing behavior characteristics of youth to determine the level of risk*

378

From a clinical perspective, it demands the involvement of psychiatrists and psychological care providers to undertake a thorough mental state examination when patients demonstrate or desire to perform the above-mentioned threatening actions. Even situations leading to mood fluctuations with low or elevated mood conditions, signs of hostile behavior, or inability to control anger deserve medical attention and the need for evaluating the aforesaid risks. In scenarios where patients show suspicious behavior and an intention to harm others, any kind of illusion, fantasy, or hallucination behavior raises threat. These signs, along with instances of low self-control and noticeable idiosyncrasies in behavior, create a red flag to assess the behavioral risk and develop a mitigation plan. The risk framework should also take into consideration family history, the patient's past behavior history, illness, and current social, economic, and other environmental factors that can impact their social and personal relationships and disturb their mental state. Such factors might include immigration status, any change in legal status or life-changing events, accommodation, arrest or criminal charges, academic performance, family disputes, or any other threat arising out of social, economic, or political factors. In addition, issues related to isolation due to hospital discharge or any treatment may cause sufficient mental distress and must be brought within the purview of the risk evaluation framework.

Another mental health model is the socioecological model (SEM) founded by the School of Public Health, Minnesota, which takes into consideration societal, policy, communities, organizations, interpersonal, and individual happiness factors, as this complex range of social changes and nested environmental conditions helps youth and adolescents to react to different stress conditions in life and impacts mental health and well-being (Michaels, C., et al., 2020). This school of public health gives great importance to proactively assessing and treating so that children and families thrive mentally, socially, and emotionally, with and without diagnoses.

With respect to the youth population, life-changing events between the course of 16 and 29 play a significant role in leading to depression in many situations. Hence, the depression risk assessment should consider any parameters in these formative years that are related to developing autonomy in family relations, becoming financially independent, getting engaged as a couple, owning residential properties, and becoming responsible as a family leader. All these engagements and participation in society bring about increased responsibility and duty as a full citizen, causing increased pressure and anxiety, which need further exploration.

Mental Health Risk Severities

Researchers have found that the majority of mental health disorders become visible to children and youth before they are 14 years old. Failure to detect the problems leads to serious consequences for the child or youth's development processes, impacting their mental, physical, and social health in the long term. They become prone to anxiety and depression, leading to risk factors in their lifestyle and eating habits. As a result, they suffer from adverse health conditions, violent behavior, and negative social and family relations and cannot grow up to live to their full potential. The risks of acts of violence toward others, chances of getting harmed, abused or exploited by others, and suicide cases among youth are on the rise as a result of mental illness.

On the other hand, if such severities are dealt with early, with early detection and treatment, it will boost children's and adolescents' quality of life, with positive results on their academic performance, physical well-being, and social relationships. Additionally, it can also help them to build resilience to tackle any situation when they grow as adults.

Further, any kind of anxiety or depression symptoms have been seen to worsen during natural calamities like the COVID-19 pandemic (Abdel-Bakky et al., 2021). The COVID-19 lockdown and social distancing policies completely disturbed children's and adolescents' social lives by limiting their social interactions at school. They became addicted to excessive mobile phones, online gaming, movies, laptops, and television screens at times, which resulted in youth engaging in negative behaviors such as drinking or smoking. The considerable decrease in nongaming leisure activities and lack of physical exercise demonstrated a close relation between health-related risk behavior and gaming levels (Kim et al., 2024). This correlation became higher for high-risk gamers who demonstrated increased levels of health-related risk behaviors than regular students. Though it impacted both males and females, the high-risk female group exhibited high-risk conditions as compared to their male counterparts.

Unhealthy practices such as drinking, smoking, stress, depression, and improper eating habits were on the rise owing to isolation and resultant stress during COVID-19. An increase in depression and anxiety also led to deteriorating academic performance at schools (studied by the **Korea Creative Content Agency (KOCCA, 2022)**).

As gaming turned out to be one of the strong factors in mental health threats, let us explore how diverse gaming participants could be impacted negatively based on their age, gender type, time, and expertise level in gaming.

Gaming as a Mental Health Risk Evaluation Criteria

In this subsection, we will take a look at how different parameters pertaining to engagement in gaming can increase the risk of mental health. Figure 9-2 illustrates varying characteristics of gaming participants that can influence different intensities of mental stress. For example, participants of high school with three to five years of gaming experience, spending more than three hours on mobile games, show signs of severe addictions and possess an increased likelihood of mental stress, anxiety, and depression. Similarly, gaming can help students alleviate stress when they spend less than an hour as a recreation between different academic sessions. It has further been observed in studies that middle school students suffer from higher levels of fatigue, improper eating habits, and physical inactivity, whereas higher signs of depression were common among high school students. The risk behavior varies based on the combination of factors listed below.

Gaming Participants' Characteristics with Risk of Mental Stress

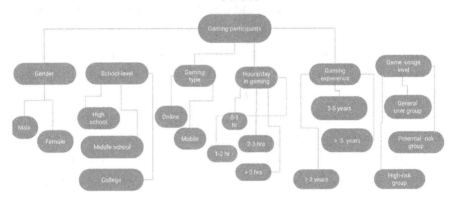

Figure 9-2. Risk category evaluation using characteristics of participants in gaming

However, gaming tools not only serve as an effective mechanism to substantiate the mental health risk but also offer unique advantages by providing a digital platform to improve attention span and memory and encourage positive behavior change. The pros and cons should be consciously weighed out for mild, moderate, or severe risk-prone individuals in the youth population to assist treatment for mental illness, including anxiety, depression, or posttraumatic stress disorder.

As designers of mental health risk evaluation systems, we should be conscious of the risk scoring procedures and effective design strategies to consider negative perceptions of risk. The design and operating procedure of the risk assessment framework should involve a human-in-the-loop to audit the propensity of the risk scores to take immediate appropriate actions for the longer-term benefit of the children and youth population.

Let us now discuss a suitable risk assessment framework.

Tools and Techniques for Risk Assessment Framework

A risk evaluation framework should not rely on a single data source to justify the level of risk and raise an alert. It should rather use aggregated sources and weigh them based on the recency of information, the number of data points received in a specified time interval, along with feedback information received from human-in-the-loop to compute an aggregated and expected risk score. Figure 9-3 illustrates a risk evaluation framework with diverse information sources.

Further to illustrate Figure 9-3, when we have two or more sources available for a single participant, like clinical assessment scores, call records history, and gaming and journaling behavioral characteristics, then we should formulate a weighted score as stated below.

Figure 9-3. *Risk category evaluation techniques among participants*

We can execute independent ML models to predict the intensity of depression from each one of the call history records, gaming, journaling, and clinical assessments. However, clinical assessments are always verified by psychiatrists and can be given a higher weight when compared to others. Again, if the participant has been active in day-to-day journaling activities, the ML model used to predict depression severity from journaling can be given a higher score. Hence, if ML_1, ML_2,...,ML_n are the n independent ML models corresponding to the individual predictions of the above depression model and if W_1, W_2,..., W_n are their corresponding scores, then the effective depression intensity can be given by

$$\frac{W_1 * ML_1 + W_2 * ML_2 + ... W_n * ML_n}{W_1 + W_2 + ... + W_n}$$

where $W_1 > W_2$ (weight given to model predicting depression intensity based on clinical assessments) and $W_2 > W_3$ (weight given to model predicting depression intensity based on recency of data from regular journaling activities). This kind of ensemble-based modeling can be used for mental depression and anxiety classification tasks (Siddique & Lee, 2024).

Formulating Risk Classification with ML Models

In this subsection below, we give three code snippets (voting, averaging, and weighted averaging), to demonstrate how each classifier ML model quantifies a risk and how we can mitigate bias by using a consolidated ensemble of models (Qasrawi et al., 2022). It also helps us to understand how feature selection from different classification models can output a specific risk probability to quantify risk as mild, medium, or severe, but the final decision can be corrected or influenced by bringing in all the classifier models involved. In all these examples, the set of imports for the python libraries will be common. We will see with the below code snippets how voting classifier is used as an ensembling technique to aggregate results from DecisionTreeClassifier, KNeighborsClassifier, LogisticRegression, and MLPClassifier (Rahman et al., 2023).

```
from sklearn.ensemble import VotingClassifier
from sklearn.tree import DecisionTreeClassifier
from sklearn.neighbors import KNeighborsClassifier
from sklearn.linear_model import LogisticRegression
from sklearn.neural_network import MLPClassifier
```

1. Voting is a kind of ensembling technique, which relies on the criteria of the max voting method that can be used for risk classification problems. Here, we see in the given code snippet below that multiple models are used to make classification predictions for each data point, which are treated as a "vote." The predictions which we get from the majority of the models are taken as a final vote.

```
model1 = LogisticRegression(random_state=3)
model2 = tree.DecisionTreeClassifier(random_state=3)
model3 = KNeighborsClassifier(random_state=3)
model = VotingClassifier(estimators=[('lr', model1),
('dt', model2), (knn, model3)], voting='hard')
model.fit(x_train,y_train)
model.score(x_test,y_test)
```

2. We can also use the averaging technique to average out the predictions from all the models and use it to make the final prediction through the final probability.

```
model1 = tree.DecisionTreeClassifier(random_state=3)
model2 = KNeighborsClassifier(random_state=3)
model3= MLPClassifier(random_state=3, max_iter=300)
model1.fit(x_train,y_train)
model2.fit(x_train,y_train)
model3.fit(x_train,y_train)
pred1=model1.predict_proba(x_test)
pred2=model2.predict_proba(x_test)
pred3=model3.predict_proba(x_test)
finalpred=(pred1+pred2+pred3)/3
```

3. Each model is assigned different weights to justify the importance of each model for prediction. This method is more useful when one of the models has human-in-the-loop or has recent data to account for its importance. For example, here we have given 40% weight to the multilayer perceptron classifier which has the most recent data as a result of tracking in regular journaling activities. Following

this, we have 35% weight given to the clinical
assessment survey risk classification model, which
was completed a month earlier.

```
model1 = tree.DecisionTreeClassifier(random_state=3)
model2 = KNeighborsClassifier(random_state=3)
model3= MLPClassifier(random_state=3, max_iter=300)
model1.fit(x_train,y_train)
model2.fit(x_train,y_train)
model3.fit(x_train,y_train)
pred1=model1.predict_proba(x_test)
pred2=model2.predict_proba(x_test)
pred3=model3.predict_proba(x_test)
finalpred=(pred1*0.25+pred2*0.35+pred3*0.4)
```

The above code samples demonstrate how ensembled ML models
help to evaluate a risk score based on which different proactive actions can
be taken.

In the above subsection, we have learned how ML models can classify
depression intensities based on ensembling, giving the highly relevant
ML model to high level of importance. With that background in mind, we
should also try to explore what are the depression-identifying metrics that
can be collected from the students, youth, and their parents through a set
of questionnaires.

The risk model of depression identification must be designed with
recommendations from psychiatrists, care providers, and other support
based on the symptoms and depression type. Hence, as designers for a
risk-based system, we must take into consideration the type of depression
that an individual is undergoing and use journaling activities, descriptions
from Twitter text, or expressions expressed using unstructured interviews.
This kind of text data, when passed through transformer text classification

models, can identify the category of depression, which will be helpful further in weighing the risk and connecting to the proper support system through alerts and notifications.

Category of Depression Symptoms

The risk categorization of depression is highly interlinked with the depression symptoms under consideration, as a teenager under major depressive order would likely have moderate to high severity compared to someone under situational depression. A situational depression might be triggered by stressful academic pressure or ongoing exams that may likely disappear after the situation is over. Figure 9-4 illustrates how input processed data when fed to transformer models is capable of differentiating between bipolar and neuro-cognitive symptoms and can further distinguish between Bipolar-I and Bipolar-II symptoms or symptoms pertaining to situational or seasonal conditions that are leading to depression (Qasrawi et al., 2023). Let us illustrate further with a few examples.

A statement given by a youth on job loss stating "I am feeling full of guilt on losing my job. I have to take care of my sick parents and I am completely lost what to do", would classify this as a situational depression. On the other hand, an adolescent youth uttered, "I feel extremely tired and exhausted these days, I don't feel hungry in mornings and evenings, even in the week daya my sleep is disturbed at night, causing me to feel sleep during the day time", shows signs of dysthymia.

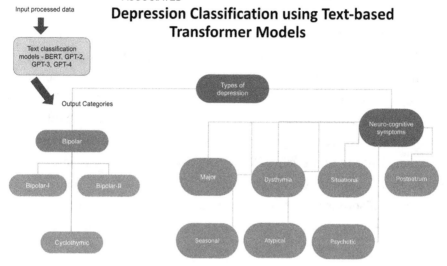

Figure 9-4. Classification of depression causing symptoms

Text-based classification models can help us to classify bipolar
depressive symptoms by identifying keywords and phrases that recognize
substantial changes in mood, energy levels, and behavior. A depression
risk framework would need to be served input from the classification
model on signs of Bipolar-I vs. Bipolar-II, to enable it to classify risks and
raise alerts by stating its risk level.

Further, adolescents suffering from Bipolar-I disorder experience
more severe mood highs (mania) with lesser chances of having depressive
episodes. If they have depressive symptoms, feelings of low mood, lack of
interest in daily or fun activities, and feelings of worthlessness are strong
signals for the classification models to detect mood and behavior changes.
A depression-based risk framework should be able to classify Bipolar-I as
a severe form of depression when compared to Bipolar-II, as mania makes
youth invincible, prone to hallucinations, and prompts them to participate
in risky behaviors like reckless driving, gambling, or indulgence in alcohol
or drugs. In addition, the risk framework should be responsible for raising
high flags and alerts to facilitate hospitalization.

Another mental disorder of high concern is major depressive disorder,
that is positively correlated with a large number of other mental health
conditions such as panic disorder, agoraphobia, generalized anxiety
disorder, post-traumatic stress disorder, obsessive-compulsive disorder,
and separation anxiety disorder. It becomes so serious that patients
diagnosed with major depression have an 8.2 times higher probability
of showing signs and being diagnosed with generalized anxiety than
someone without depression. Hence, the risk framework should be
pretrained with models having symptoms and intensities of major
depressive disorder, so as to proactively send alerts and intervene to
combat the seriousness of the disease.

Depression Symptoms

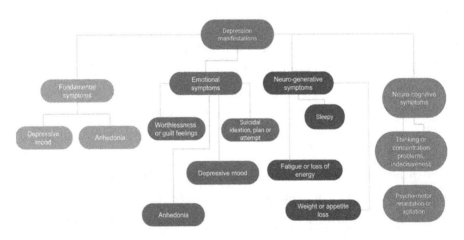

Figure 9-5. *Classification of depression causing symptoms*

On the other hand, adolescents suffering from Bipolar-II experience
a less severe high (hypomania); however, they also experience depressive
episodes. Here, the symptoms may make children and youth feel overly
excited, energetic, and happy, which may cause them to take prompt
decisions and risks. They may end up doing things without informing
parents, such as taking part in risky outdoor activities or getting more

violent in sports. As we have seen, Figure 9-4 helps us to recognize different types of depression at a high level. Figure 9-5 goes one step beyond to break down the depressive symptoms into different types like fundamental, emotional, neuro-generative, and neuro-cognitive symptoms, which will link to the type identified through Figure 9-4 and severity (severe, moderate, low, and mild) of depression.

As language serves as an important feature in identifying depression, let us explore how the use of language varies in different depression categories. Language has a close association with mental health as the sum of scores for each psychiatric disorder and the top ten text features associated with depression severity include total count of words, negative emotions, focus on present, verbs, auxiliary verbs, adverbs, tone, analytic, six-letter words, and leisure words. In addition, age and gender also have a role in influencing mental health symptoms.

Now let us look at a few examples of how language can serve as an important tool for AI models for measuring depression.

Language As a Tool for Identifying Depression Symptom

Social media text (e.g., Twitter) can play a pivotal role in recognizing depression symptoms and intensity. The use of elevated first-person singular pronouns is a dominant feature in depressive mood expression, which represents a high probability of self-focused attention, obsessive-compulsive disorder, anxiety, eating disorders, and schizophrenia. However, in certain cases, first-person singular pronouns are not a single source of evidence for mental health problems and can be marked by other disorders like comorbidity. Research studies have discovered third-person plural pronouns (they, them) are common for those who suffer from Schizophrenia as evidenced by discussion forum posts. Social media texts having statements like "I have PTSD" demonstrate self-confession of depression symptoms, which could additionally be enriched by treatment options.

A specific research study conducted on Twitter data (https://pmc. ncbi.nlm.nih.gov/articles/PMC8956571/) from over 1000 individuals had responses from nine different self-report mental health-related questionnaires. Twitter users with obsessive-compulsive symptoms were found to follow more accounts, while a high number of followers were noticed for people with eating disorders. The highest volumes of tweets were noticed among users with generalized anxiety. Twitter users with high levels of depression, apathy, impulsivity, obsessive-compulsive disorder, and schizotypy were also high on the insomnia index and were more prone to tweeting at night. Further, all kinds of mental health problems, except alcohol abuse and eating disorders, show a high affinity to tweet and post replies. Impulsivity, followed by apathy, exhibits the strongest correlation with the insomnia index, further relating the factor that higher impulsivity leads to more sleep disturbances. Sleep disturbances are more commonly recorded among users who possess suicidal thoughts, in contrast to users who show nonsuicidal mental illness and report anger/aggression or higher appetite tendency (Cook et al., 2016).

A positive benefit of expressing depression through language is that it can lower the adverse effects of stress. This kind of self-disclosure on social platforms also leads to increased life satisfaction and lower depression and controlled suicides among social media extrovert users than those who refrain from seeking social support and engaging. We see that there lies an opportunity if we can educate different support groups of mental health and friends, family, and content moderators about the types of comments and posts they can share when someone is experiencing mental health problems. Despite the role of language, we should be cognizant of the roles of support networks in lowering the seriousness of mental illness. Hence, such factors should also be considered in mental health risk models.

However, the main challenge comes as the segregation or categorization of the type of mental disorder from the language remains a real problem, as depressive symptoms overlap and strongly correlate with each other. Depression symptoms such as insomnia, psychomotor agitation/retardation, and depressed mood are the most related symptoms with relation to more than 28% of other symptoms in the network. Another challenge comes from social media research dominance among English-speaking countries and participants. This restricts mental health inference for online participants from other cultures, such as users from India and China who maintain a healthy presence on online platforms to post their queries. If we try to extrapolate models that are trained using the English language, we would see it leads to systematic biases and would yield inaccurate results in their predictions.

Let us explore what could be the alternative text-based solutions that can leverage NLP to mitigate biases.

Mitigating Biases with NLP in Accessing Risk

NLP-based solutions can play a significant role in extracting suicide risk themes from Facebook, Twitter, and Instagram platforms and textual information available from communication messages between patients and their families with physicians over secure messaging platforms (Zhang et al., 2022). It further plays an extensive role in reducing disparities among under-represented groups owing to the presence of skewed datasets in English. NLP-based techniques, when applied to low-cost text-based messaging solutions, can help to assess the suicide risk for racial/ethnic minorities, young adults, and individuals from lower socioeconomic strata where extensive suicide risk monitoring solutions are absent. NLP, along with risk assessment, can help to capture the sentiments and effects of risk interventions among vulnerable communities that exhibit emerging psychiatric issues among high-risk youth. This allows care provider services to assist in communication at schools and involve families, youth mental health services, and law enforcement bodies to provide needed treatment before aggravating the severity of the disease, leading to poor academic and social conditions.

While we have explored the type, symptoms, and level of depression
and the role played by text-based classification models, let us also
understand how interpretability can play an important role in conveying
the predictable chances of that symptom with underlying causal factors
leading to the occurrence of that specific symptom.

Bringing in Interpretability in Identifying Depression Symptoms

ML/AI models that come with interpretability to justify the causes and
the importance of the factors in the level of importance become critical in
explaining to psychiatrists and care providers to enable them to provide
the right counseling to the patients and their family members. If we can
combine the causes along with the level of attribution of the causes, its
dependency, and the interplay of other multiple related factors, then it
becomes appropriate to serve as an explanation of the exact condition the
patient is undergoing, to offer the right treatment.

Research studies conducted in samples indicate certain factors emerge
as most important for mental depression. Figure 9-6 further illustrates
that body mass index, number of sports activities per week, grade point
average (GPA), sedentary hours, and age play the most significant role in
increasing depression prediction accuracy when these features are used in
training the classifiers. With research bringing in extensive monitoring for
promoting psychological well-being in schools, colleges, and universities,
continuous feedback along with additional social and cultural contexts
would be a great addition. Data enrichment (**electroencephalogram
(EEG), electrocardiogram (ECG)**, fluctuations, heart rate, breathing rate,
temperature, speech intonation, etc.) and development of country- or
culture-specific models and psychological sensing technology would likely
increase prediction accuracy (Rothenberg et al., 2023).

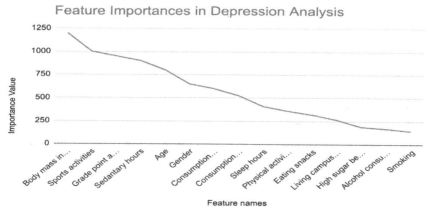

Figure 9-6. *Feature importances in attributing mental depression*

As we see the most important features impacting a youth's mental health conditions, it is also important for us to know how these features cause a decrease in Gini index on being selected while training a model. Figure 9-7 further provides additional support to justify the order of importance or ranking of the top five features.

The purpose of the Gini index is to decrease the impurities from the root nodes residing at the top of the decision tree to the leaf nodes, those lying at the vertical branches down the decision tree. The body mass index followed by sports activities, GPA, sedentary hours, and age holds a high Gini index value and emerges to be the most important feature that can act as the best predictor for other related variables. A higher Gini index signifies greater inequality and heterogeneity, which causes the decision tree to split further into branches and leaf nodes.

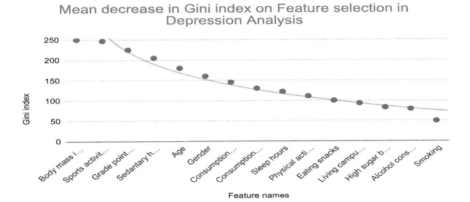

Figure 9-7. *Change in Gini index along with feature names for
decision trees*

As the above two figures substantiate the process of important feature
selection (Rahman et al., 2023) (with the Gini index falling exponentially
along with the selection of them), let us now see how we can bring in
explainability on depression symptoms when we predict the chances of
occurrences of a depressive symptom with a probability.

Figure 9-8 illustrates the scenarios under which a student is suffering
a depression over a single period and the chances of occurrence of that
symptom of depression.

The interpretation of the depression symptoms likely to be observed
on a day can be further enhanced when we bring in the narrative
that the likelihood of occurrence of the depression symptoms over a
specified period. This also brings in a flavor of the intensity of the severity
of depression as illustrated in Figure 9-9. For example, the student
experienced worthlessness in comparison to his friends more than 90%
of the time, causing him to miss 100% of the school events, conveying the
emergency of the situation and demanding urgency that the matter needs
to be investigated from a mental health clinical perspective.

Interpretation of Depression Symptoms and it's Manifestations

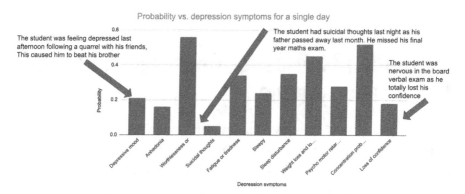

Figure 9-8. *Probability of depression causing symptoms*

Interpretation of Depression Symptom Intensity over Time

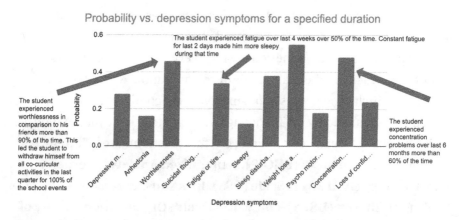

Figure 9-9. *Classification of depression causing symptoms*

By this time, we have got a solid understanding of the prime necessities of a depression framework and how different symptoms at varying levels can contribute to the severity of depression. It's time for us to know the metrics for measurement, when to apply which metric and reuse our mobile/device applications to measure them. Once the intensity level is identified by plugging in the right tools and techniques, it becomes easier to take the right actions.

Evaluation Metrics for Depression, Use Cases, and Actions

With the background of depression intensity as well as depression symptom classification in mind, we should also try to explore what are the depression-identifying metrics that can be collected from the students, youth, and their parents through a set of questionnaires.

Metrics for Depression Intensity

The metrics quantifying depression intensity are dependent on demographic and basic clinical variables such as gender, age, academic, family, and socioeconomic background, including any medical conditions such as pregnancy state, experience of delivery, and premenstrual syndrome, which are the main items of consideration. Mental health conditions, including depression, anxiety, stress conditions, and quality of life, can be evaluated by the following psychometric determinant indexes, which can be self-reported on filling out the questionnaires: **Patient Health Questionnaire-9 (PHQ-9), Generalized Anxiety Disorder-7 (GAD-7), Perceived Stress Scale (PSS), Birleson Depression Self-Rating Scale for Children (DSRS), and World Health Organization Quality of Life Scale (WHOQOL-BREF)** (Qasrawi et al., 2023).

PHQ-9

The PHQ-9 contains nine-item questions on a scale of 0–27, ranging from the lowest to highest level of depression, to assess the severity of depression and other types of mental disorders among mid-to-high-risk vulnerable groups, most widely used in primary care and community settings.

GAD-7

The GAD-7 is a similar questionnaire depression-assessing metric that measures anxiety level through using seven question items on a range of 0–21, with a high score signifying a high level of anxiety. This assessment technique is widely used in clinical settings, mental health clinics, and research to assess the stage of the mental disease and monitor progress of the treatment.

PSS

The GAD-7 is a similar questionnaire stress detection determinant metric that measures stress level through using ten question items on a range of 0–40, with a high score signifying a high amount of stress. This metric is widely prevalent to assess unpredictable, uncontrollable, and overloading conditions of life and impacting mental over the previous month, signifying stress levels in young people and adults aged 12 and above.

WHOQOL-BREF

The WHOQOL-BREF self-report contains 26 question items to evaluate personal health and life conditions, primarily physical health, psychological health, social relationships, and living environment. The scoring system runs from highest to lowest, meaning a higher score correlates with a superior quality of life. Each of the four dimensions of life can be evaluated and then summarized to give a final score.

EPDS

The EPDS-based self-reporting questionnaire system assesses postpartum depression by using ten question items. It captures mental conditions like depression, anxiety, and suicidal ideation for a week. This metric exhibits a high correlation with the Beck Depression Inventory and Hamilton Depression Rating Scale and uses a rating scale called the Likert scale that ranges between 0 and 3.

Hamilton Depression Rating Scale (HDRS)

The HDRS (commonly known as Ham-D) is another popular metric clinically administered for depression assessment. It presents 17 items (HDRS17) in front of an audience (primarily hospital inpatients) to assess depression based on melancholic and physical signs of depression symptoms they have been exposed to over the last week. HDRS was later revised to HDRS21, which includes 21 item questions where an additional four questions are added to differentiate the subtypes. HRDS might be preceded by an unstructured clinical interview.

DSRS

This metric is widely used for evaluating children's depression or anxiety level by evaluating 18 item factors, by summing the items' answers and categorizing them into two categories (normal: 0–11, depressed: ≥ 12)

The Beck Depression Inventory (BDI)

The BDI metric can be used to capture the behavioral manifestations of youth through self-report-driven 21-item, multiple-choice questions. Each of the questions can be rated between 0 and 3 points, where the total score can range from 0 to 63. Let us now see how BDI scores can be evaluated using speech data, as illustrated in Figure 9-10.

3D Convolution Network for Speech tone Detection

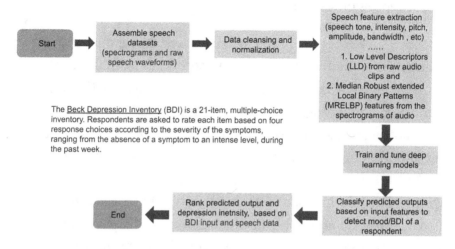

Figure 9-10. *BDI score assessment from speech and audio data*

The intensity of depression can be evident in the voices of depressed patients, where varying voice levels such as low, slow, hesitant, and monotonous can help us to tag the level of depression. A lower voice tone can suggest low voice frequency or pitch. It can also signify a lower speech rate when a patient speaks with a slow voice and pauses, with reduced voiced frames. Depression manifests in various forms like disorganized or incoherent speech, commonly known as formal thought disorder. We can apply commonly used NLP techniques to help analyze relationships between terms and documents in a text corpus to learn the semantic meanings and relationships. One of them is the **latent semantic analysis (LSA) that** can be leveraged to determine the depression level of psychosis patients by evaluating the semantic coherence of transcribed speech data (Morgan et al., 2021). This method discovers the semantic coherence between adjacent words or evaluates the likelihood of tangentiality of an individual's speech that determines the likelihood of divergence off-topic over time.

Now let us examine with a practical example how we can evaluate the PHQ-9 depression score of individuals.

Listing 9-1. Now let us understand through the below code listing how to predict the happiness score. Here, we take a PHQ-9 depression assessment dataset as data for prediction. This dataset comprises ambulatory assessment (AA) data taken over 14 days. The dataset consists of attributes related to depression symptoms and mood ratings, as well as results from a retrospective Patient Health Questionnaire (PHQ-9) which is curated for finding out the symptoms or occurrence of depression in individuals. Moreover, it also contains demographic information about the participants such as their age and gender. It contains more than 3000 data points

This task can be viewed as a tabular classification task where we can leverage the tabular techniques like decision models in order to predict the happiness score of people based on the demographics and the PHQ-9 assessment.

1. In the first step, we install the necessary libraries, read the dataset, and drop the time, day, and target columns.

```
import pandas as pd
from sklearn.preprocessing import LabelEncoder
from sklearn.model_selection import train_test_split
from sklearn.ensemble import RandomForestClassifier
df = pd.read_csv('Dataset_14-day_AA_depression_
symptoms_mood_and_PHQ-9.csv')
df.head()
X= df.drop(columns=['time','start.time','phq.
day','happiness.score'],axis=1)
y= df['happiness.score']
```

2. In the next step, we apply the label encoder to
    convert the string labels into integer classes so that
    they can be fed into the models for training. Then,
    we move on to build our model for predictions. We
    make use of one of the most popular models, the
    decision tree classifier model.

```
X = X.apply(LabelEncoder().fit_transform)
X.head()
X_train, X_test, y_train, y_test = train_test_split(X,
y, test_size=0.33, random_state=42)
clf = RandomForestClassifier()
clf.fit(X_train,y_train)
clf.score(X_test,y_test)
```

This yields a test score (mean accuracy on the given test data and
labels) of the model as 0.513. Let us now look at some sample use cases
and how we can use the metrics discussed in the above subsection to
assess depression severity for stress, anxiety, ADHD, and depression.

Use Cases Leveraging the Metrics to Different Symptoms of Depression

In the first use case, we see below through Figure 9-11 how the severity
of stress can be detected using AI modeling by using input features (like
journaling, speech/audio, mood tracking, and sensor data) from a mobile
app. Here, we have used the Perceived Stress Scale (PSS) as the metric for
measuring the severity, where scores ranging between 27 and 40 deserve
attention due to high perceived stress.

403

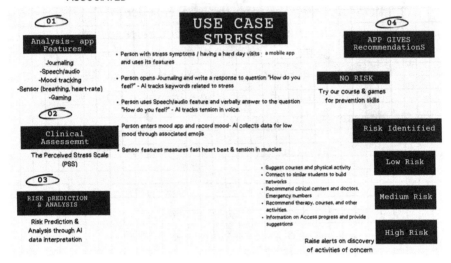

Figure 9-11. *AI-driven risk prediction for stress from mobile app*

In the second use case, Figure 9-12 illustrates the severity of anxiety measured using the **Beck Anxiety Inventory (BAI)**. The level of anxiety can be detected using AI models by using input features (same as above) from a mobile app. Here, we see we can use emojis to track a person's mood, use sensor-based features to measure fast heartbeat and tension in muscles, and use finger movements during gaming activities to determine the intensity. Any score of 36 and above suggests concerning levels of anxiety. Research studies have discovered anxiety in school-going children is high when they are in fifth and sixth standards and it decreases in higher grades.

An example of stress for children or youth could be exposure to violence and traumatic events, which includes war, civil conflict, or conditions of political violence. This can impact a child's cognitive development, including their memory, problem-solving abilities, attention, and executive function leading to post-traumatic stress disorder (PTSD). Such disorders should be categorized using the AI-driven risk framework, as no PTSD, 1 = moderate PTSD, and 3 = severe PTSD. The ML model can input features like duration of violence, trauma history, impact on civilians, and demographic characteristics to predict the risk intensity.

While we have explored the type, symptoms, and level of depression and the role played by text-based classification models, let us also understand how interpretability can play an important role in conveying the predictable chances of that symptom with underlying causal factors leading to the occurrence of that specific symptom.

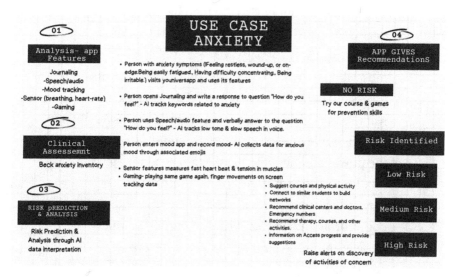

Figure 9-12. *AI-driven risk prediction for stress from mobile app*

Now let us understand with the code listing stated below how to predict the anxiety level of a patient.

Listing 9-2. Anxiety is very common nowadays, so in this part, we try to build a model which predicts whether anxiety is present or not, based on the anxiety symptoms available from psychometric tests conducted without any doctoral supervision. We use the generalized anxiety disorder(GAD) dataset based on proper official test questionnaire given to the subjects

1. First, we begin by loading the data and separating it into categorical and continuous features. Categorical features are those which have classes or which are quantized, whereas continuous features are not quantized as the name suggests. We convert the categorical features to integers using the labelencoder from sklearn. We create the target label "GroupedGAD" which basically divides total GAD score into four categories, 0–5, 5–10, 10–15, and 15–22, which are the actual categories prescribed in the test and using which the doctors assess the patient's anxiety level. For this task, we will use the TabTransformer model which is a very popular and widely used tabular model which has the power and prowess to take into account the tabular relations and the connections between various features. As the code is too long, we add code comments and show below certain snippets to have an understanding of the main steps in sequence:

    - Create model inputs

    - Encode the model features

    - Stacked the categorical feature embeddings for the transformer and then concatenate numerical features

    - Added the column embedding to categorical feature embeddings

2. Following this, we create multiple layers of the
transformer block and a multihead attention layer to
design a multilayer perceptron network and added a
binary classifier with a Sigmoid activation function.

```
for block_idx in range(num_transformer_blocks):
    attention_output = layers.MultiHeadAttention(
        num_heads=num_heads,
        key_dim=embedding_dims,
        dropout=dropout_rate,
        name=f"multihead_attention_{block_idx}",
    )(encoded_categorical_features,
    encoded_categorical_features)
    # Skip connection 1.
    x = layers.Add(name=f"skip_connection1_
        {block_idx}")(
        [attention_output, encoded_categorical_
        features])
    # Layer normalization 1.
    x = layers.LayerNormalization(name=f"layer_
        norm1_{block_idx}", epsilon=1e-6)(x)
    # Feedforward.
    feedforward_output = create_mlp(
        hidden_units=[embedding_dims],
        dropout_rate=dropout_rate,
        activation=keras.activations.gelu,
        normalization_layer=partial(
            layers.LayerNormalization, epsilon=1e-6
        ),  # using partial to provide keyword
        arguments before initialization
        name=f"feedforward_{block_idx}",
    )(x)
```

```
            # Skip connection 2.
            x = layers.Add(name=f"skip_connection2_{block_
            idx}")([feedforward_output, x])
            # Layer normalization 2.
            encoded_categorical_features = layers.
            LayerNormalization(
                name=f"layer_norm2_{block_idx}",
                epsilon=1e-6
            )(x)
        # Flatten the "contextualized" embeddings of the
        categorical features.
        categorical_features = layers.Flatten()(encoded_
        categorical_features)
    # Apply layer normalization to numerical features.
        numerical_features = layers.
        LayerNormalization(epsilon=1e-6)(numerical_
        features)
    # Prepare the input for the final MLP block.
        features = layers.concatenate([categorical_
        features, numerical_features])
        # Compute MLP hidden_units.
        mlp_hidden_units = [
            factor * features.shape[-1] for factor in mlp_
            hidden_units_factors
        ]
        # Create final MLP.
        features = create_mlp(
            hidden_units=mlp_hidden_units,
            dropout_rate=dropout_rate,
            activation=keras.activations.selu,
            normalization_layer=layers.BatchNormalization,
```

```
        name="MLP",
    )(features)
    outputs = layers.Dense(units=1,
    activation="sigmoid", name="sigmoid")(features)
    model = keras.Model(inputs=inputs, outputs=outputs)
```

3. Our next step would be to train a model by
    specifying the number of epochs, learning
    rate, batch size, and weight decay (which as a
    regularization parameter for the below neural
    networks and aims to penalize large weights by
    adding a constraint to the loss function). Here, we
    have used AdamW optimizer(adaptive optimizer
    that control weight regularization). Further, the
    BinarCrossentropy function helps to determine the
    presence or absence of anxiety based on the input
    features.

```
def run_experiment(
    model,
    num_epochs,
    learning_rate,
    weight_decay,
    batch_size,
):
    optimizer = keras.optimizers.AdamW(
        learning_rate=learning_rate, weight_
        decay=weight_decay
    )
    model.compile(
        optimizer=optimizer,
        loss=keras.losses.BinaryCrossentropy(),
```

```
            metrics=[keras.metrics.BinaryAccuracy(name=
            "accuracy")],
        )
        history = model.fit(
            [df[col] for col in FEATURE_NAMES],df[TARGET_
            FEATURE_NAME], epochs=num_epochs
        )

        return history,model
```

4. The final step would be to invoke the prediction.

```
preds = model.predict([df[col] for col in
FEATURE_NAMES])
```

This would yield 0 or 1 to give the final binary output, where the above prediction gives 0 which means there is no anxiety.

After anxiety, our third use case, as shown in Figure 9-13, depicts the severity of ADHD measured using the ADHD-IV Rating Scale. The level of ADHD can be measured by leveraging a mobile app, which can track different mood symptoms (anger, anxiety, frustration, or disappointment), through associated emojis, in addition to the use of gaming methods and sensor data.

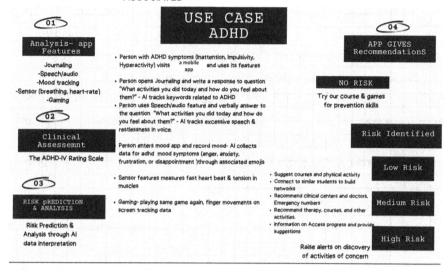

Figure 9-13. *AI-driven risk prediction for ADHD from mobile app*

Our fourth use case, shown in Figure 9-14, illustrates the severity of ADHD, as measured by the Beck Depression Inventory, using a mobile app in a similar way to the use cases above.

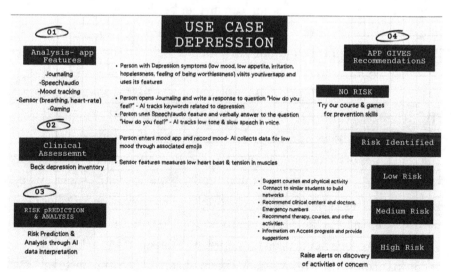

Figure 9-14. *AI-driven risk prediction for stress from mobile app*

411

The depression rating approach can be further extended to Peripheral Physiological Features, where features related to activity and resting, like **autonomic nervous system (ANS)**, **heart rate variability (HRV)**, **respiratory sinus arrhythmia (RSA)**, psychophysiological allostasis (which measures cardiac allostasis), **electromyogram**, **and electrodermal activity-based (EMG** and **EDA)** can be used to access and quantify stress and depression. In addition, another input factor for consideration is the **acoustic startle reflex (ASR)** which involves an involuntary muscular response to a loud, unexpected sound.

By this time, we have gathered a complete understanding of the role of AI model in bringing all support systems closer—such as help from school staff, family support, friend support, better grades, gender, and controlling negative scenarios like home violence and bullying behaviors. This can enable us to design and implement relevant intervention programs to enhance students' health, education, and well-being.

Statistical Analysis

Different intensities of depression can further be classified by cluster analysis (such as hierarchical clustering or k-means clustering) (Lim et al., 2023) based on clinical conditions of patients, including depression, anxiety, stress perception, and quality of life, by taking into consideration PHQ-9, GAD-7, PSS, WHOQOL-BREF, and EPDS. Clusters having scores above PHQ-9 (≥ 8 or ≥ 11) and GAD-7 (≥ 5 or ≥ 15) can be treated as high risks.

Figure 9-15 shows how effectively hierarchical clustering can classify different types of mental health problems by demonstrating that attributes for schizotypy and depression exhibit similar behavior, while social anxiety and obsessive-compulsive disorder exhibit similar behavior. This happens as correlation across the given set of questions looks similar, which leads to similar language being used for depression, anxiety, and schizotypy. In comparison to the similarity of clusters observed, we also observe noticeable differences in language used for eating disorders and impulsivity.

Listing 9-3. In this code example, we use a mental health dataset
(generated synthetically) which has self-reported questionnaires
(phq_score) along with depression symptoms as observed from the
same candidates. Here, we used hierarchical clustering to plot the
depression type along different clusters as listed below

1. In the first step, we install the necessary libraries,
    load the dataset, and extract the feature sets.

```
import pandas as pd
import numpy as np
import matplotlib.pyplot as plt
from scipy.cluster.hierarchy import dendrogram, linkage
df = pd.read_csv('mhealth.csv')
x = df['QUESTIONNAIRE_RESPONSE'].values
y = df['DEPRESSION_SYMPTOMS'].values
data = list(zip(x, y))
```

2. In the next step, we plot the data using the
    "complete" linkage method by using "euclidean"
    distance as the metric to represent hierarchical
    clusters. The plot given below shows the clusters,
    how co-located they are, and how different they are
    in terms of input model features such as phq_score
    and depression symptoms.

```
linkage_data = linkage(data, method='complete',
metric='euclidean')
dendrogram(linkage_data)
plt.show()
```

413

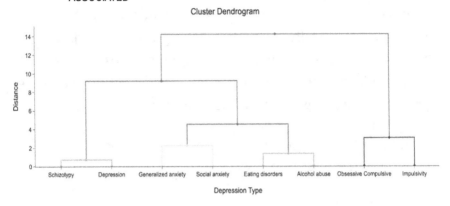

Figure 9-15. *Hierarchical clustering to classify different categories of depression symptoms*

As shown in Figure 9-15, there exists a high correlation between the symptoms observed and reported scores among depression and schizophrenia, followed by a close resemblance between eating disorders and alcohol abuse. We also notice while generalized and social anxiety also show similar reportings and symptoms, they are largely separated from obsessive-compulsive disorder and impulsivity, which stays far off isolated, with substantial differences from other types of depression disorders under consideration. Here, we have used complete linkage in evaluating the farthest neighbor linkage that computes the maximum distance between any two points in the two clusters. In contrast, we can also use a single linkage method that computes the nearest neighbor linkage by computing the minimum distance between two clusters, by taking any two points under consideration. We have chosen the complete linkage method as this method has less noise and outliers compared to single linkage and helps in identifying clusters which have overlapping. As various categories of depression symptoms and thereby their clusters are not completely disjoint, the former method would be more appropriate than the latter.

Similar to hierarchical clustering, we can introduce k-means clustering for digital phenotyping methods where we can introduce survey-driven active assessments and sensor-driven passive assessments. Here, we can first find the optimal number of clusters to cluster mental health states such as depression, anxiety, stress perception, and quality of life. Individuals will fall into the same cluster when the relevancy score of joint multiple attributes (Euclidean distance between the cluster attributes and the centroid of the cluster) falls below the cutoff. This can help us to segregate low depression, anxiety, and better quality of life into cluster 1, representing mild to no risk conditions. Consequently, cluster 2 can evolve to show high depression, anxiety, and low quality of life with high cutoff scores for metrics like PHQ-9 and GAD-7. In addition, cluster 3 can represent a potential emerging high-risk group with either two of the metrics (say PHQ-9 and GAD-7) or scores ranging lower than the optimal cutoff for depressive disorder and anxiety disorder. On the other hand, it may have the mean PHQ-9 and GAD-7 scores very close to the cutoff score for screening depression and anxiety.

Clustering may pose one challenge in classifying patients based on their diagnosis, as it may run into the risk of overfitting due to small sample sizes. However, unsupervised clustering demonstrates great potential in labeling the mental health state of patients which can help them in medical assignments.

Let us now take a look at how these actions can vary based on the severity of the illness. The degree of interventions, follow-ups, and generating awareness varies to address the severity of the disease, alongside making the potentially vulnerable section of the youth population aware of the first step actions and the necessary precaution steps.

Actions Following Risk Classification

One of the primary steps in identifying depression is to list the symptoms, categorize the depression type and intensity, and find specialists who can assist in proper treatment. It is highly recommended that youth and patients falling into low- to high-risk groups be intervened with from time to time to keep a check on the severity by offering the right sources of treatment and follow-ups. Once we receive accurate model outputs and have clear differentiating points to distinguish one illness from another, we can clinically take action on those risks.

Actions based on Categories of Risk Assessment

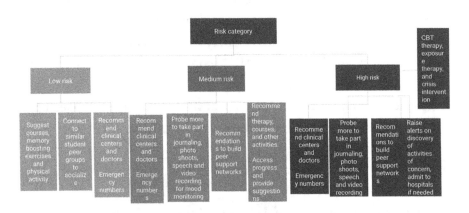

Figure 9-16. *Risk assessment and remedial actions*

Figure 9-16 illustrates how an ecosystem of mental health support framework can enable different apps and services to support youth and their parents based on the level of predicted risks.

Once we have categorized the mental health risk, we should follow immediate steps of remediation as follows:

- The youth population, who are most vulnerable, should be made aware of mental well-being by promoting workshops and educating them on physical health and healthy lifestyle habits like the usefulness of physical activities. Along with this, they should be encouraged to take part in mental health apps that have memory games and boost their attention and memory.

- Youth groups who are at mild mental health risk should be educated on precautions through thorough assessments and psychoeducation to enable them to adopt pathways of self-care. Moreover, youth and their parents should be equipped to handle a smooth transition between different psycho and mental health services once they have started consultation for mental well-being.

- Youth groups who fall between mild and moderate mental health risks should be offered treatment along with continuous monitoring of their symptoms so that the symptoms do not aggravate. Along with psychoeducation, they should also learn to engage and network with similar youth personnel and groups to support each other as peers.

- Youth groups susceptible to severe mental illness or risk of committing suicide should continuously receive psychotherapy, CBT therapy, interpersonal psychotherapy, exposure therapy, and crisis intervention from time to time. If needed, they should be treated at hospitals with full care and support.

- Providing the right dose of medication, selective
  serotonin reuptake inhibitors, and tricyclic
  antidepressants to those who are severely affected by
  depression.

Designing the risk assessment framework and understanding the
expected outcomes and actions for estimated risks is a necessary step
toward building a highly robust system that leverages the risk algorithms to
quantify the risk score and action based on the computed level of risk. Let
us see the design and mode of operation of such a robust risk framework,
as illustrated in Figure 9-17.

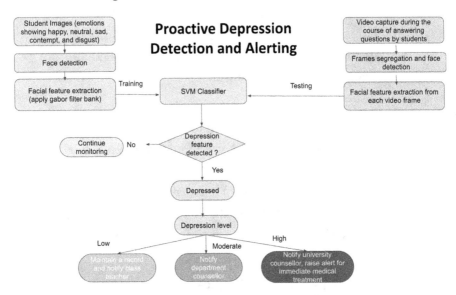

Figure 9-17. *Proactive depression detection, classification, and
alerting*

Figure 9-17 depicts an example of how a risk-based system can
operate, based on the level of depression classified from input images or
recorded videos, to notify the first level of contact. At first, the face is first
detected to extract the facial emotions, then the classification algorithm
predicts the severity of anxiety or depression, followed by immediate

418

notification to the concerned school or university class, department, counselor, or mental health medical professionals. Here, we have split the dataset into train and test sets and used the SVM classifier to predict the depression severity.

On getting familiar with the functionality of the risk framework and actions that would drive the efficiency of the system, as system designers, we should be conscious of introducing certain critical components that control high-risk scenarios. In case patients feel threatened and are compulsive toward self-harm or feel suicidal or threatened to inflict violence toward others, we should place them in such a critical unit of evaluation that continuously updates the risk score.

Let us now see a typical example of a critical component that analyzes video/audio to infer and differentiate between a real suicide and a false alarm.

Critical Components of a Risk Framework

The risk framework should have an in-built vision-based suicide detection module that can detect suicide by hanging. We need an intelligent video surveillance system to be equipped with an RGB-D camera that can capture and process a stream of frames coming in from the camera, by using the depth information along with the reference distance between the camera and objects in a scene (Bouachir et al., 2018).

Real-time video surveillance for suicide detection

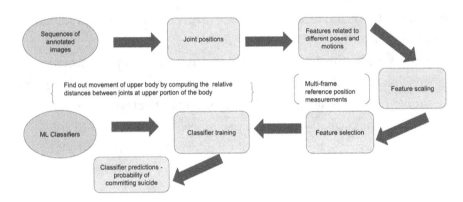

Figure 9-18. *Video-based suicide detection (Bouachir et al., 2018)*

The real-time algorithm should use both static and dynamic pose information of the subject and explore the body joints' positions to model suicidal behavior. Further, the algorithm should be capable of inferring the change in position for body joints by comparing and differentiating suicidal attempts with daily living activities such as wearing or removing clothes from video sequences as small as three seconds. The algorithm should also be able to recognize videos that exhibit display characteristics like changes in occlusions, illumination changes, and scale changes.

Figure 9-18 illustrates video frames getting ingested to be analyzed for joint positions and then feature-engineered before training and prediction (Bouachir et al., 2018). It also shows how different features related to poses and motions of a human body are measured on a reference scale, scaled before the most important features are selected.

The three principal modes of operation of the algorithm are as follows:

- Maintain an invariant feature set for subjects under consideration and compute the morphological differences between daily activities and suicidal attempts.

- Have a robust feature selection algorithm in place to increase the time and efficiency of operation at an individual frame level to detect suicides in real time. This can be done by computing the pairwise distances between the list of joints, especially considering the hand movements to the neck from top to bottom, between frame t and frame (t-1). Such upper parts' body movements can be used to infer any elementary motions that have happened between two subsequent frames.

- Bring in more relevant datasets to remove bias from imbalanced datasets to aid in an authentic understanding of the human body's 3D joints' positions (during suicide pairwise distances between the lower body joints remain almost constant, with no or very limited spatial changes of the lower body parts) that helps to accurately classify a suicidal attempt. Moving the hands to the neck from top to bottom implies the movement of several upper body (Amanat et al., 2022; University of Colorado at Boulder, 2024).

Other Risk Factors for Consideration

Research has discovered external risk factors and familial conditions that pose a threat to children and youth. The risk framework should be dynamic to incorporate external risk factors in the system. The primary factors for consideration include genetic and contextual risk factors that children and adolescents face while growing up when one or both of their parents have mental illness. This negatively impacts them as they are at

421

an increased risk of developing mental health problems. Parental mental illness may include the following:

- Parental mental illness poses a threat to children, as in some cases, children and adolescents get murdered by their parents.

- Parental aggressive behavior, hostility, and self-harm impact children's and adolescents' normal development and make them prone to mental depression.

- Parents' delusional belief systems often subject children and adolescents to abuse.

- Parents' mental illness impacts their ability to relate to their children, resulting in irritable behavior. The distracted, withdrawn behavior impacts children as parents are not able to attend to the child's needs. This may increase due to drugs and medications, leading to drowsiness, which makes the situation more inconducive to create a healthy bond between the parent and the child.

- Children and adolescents are held with extra responsibility for their parents' well-being and caregiver duty as they mature from students to earning members of the family.

- Parents having a limited social life limits their children's social life which limits their external sources of help outside the home.

- Parents' behavior and nonsocial acceptance often lead to children being bullied in society which adds to embarrassment and depression.

The design and mitigation strategy of the risk framework should factor out all these factors with due diligence and degree of severity to bring in stronger collaboration and attention between adult mental health, child psychiatric, and child protection services.

Let us now explore how GenAI plays a role in understanding mental illness symptoms and revealing the seriousness of the symptoms.

Role of GenAI in Risk Assessment

Large language models (LLMs) have been used extensively in almost all domains due to the large volume of data they are trained on, making them experts in acquiring statistical knowledge and helping them to produce fluent English sentences (Guo et al., 2024; Skianis et al., 2025). They have become popular in the mental health domain as they behave like humans proficiently across tasks like question answering, summarization, and recommendations. The most popular widely used models are GPT3.5, GPT 4.0, Google BARD, Google GEMINI, and Anthropic Claude. LLMs have been found to hallucinate when they do not provide factual, accurate information but they do answer with confidence, which exacerbates the trust in LLMs and raises questions over their reliability, explainability, and consistency.

As LLMs (e.g., GPT-3, BERT, and their multilingual variants) get trained on vast amounts of text data and showcase their potential to generate human-like text and execute complex language tasks, they have been deployed in the mental health domain to assist mental health specialists, care providers, and psychiatrists to help with detection, diagnosis, and treatment and provide them with a support network. For our general-purpose knowledge, let us explore a few BERT models and what type of training data have been used for their training.

Table 9-1. *Pretrained BERT models for mental health (Vaj, 2023)*

BERT model for mental health	Training data
BERT-Base	The large volume of mental text data can be fine-tuned for mental health tasks.
BERTweet	Trained on Twitter data, to answer social media conversation threads.
BioBERT, MedBERT	Pretrained with biomedical text to respond to mental health-related research papers and clinical notes.
ClinicalBERT	Pretrained on clinical text, including electronic health records (EHRs) and clinical notes to help in treatment and diagnosis.
SciBERT	Pretrained on scientific data, research papers to respond any research queries.
HealthBERT	Pretrained on health-related text, including mental health-related articles to analyze mental health content.
RoBERTa	Optimized for NLP tasks including mental health information.
DisorBERT, SuicidalBERT	Trained on mental health-related social media data can be used to identify linguistic patterns for mental health such as depression, anxiety, and suicidal ideation.

Mental Health Disease Categorization Using GenAI

The most common mental health disorders like depression, anxiety, and PTSD are evaluated using self-reported questionnaires, clinical interviews, and standardized assessments that are reviewed by clinical psychologists and treated by psychiatrists. When the text data of several interviews, surveys, and assessments are prompt-engineered and fed to an LLM, it can help to classify the specific type of disorder and severity based on the frequency of mental health episodes and past episodic occurrences. This in a way helps to reduce delays and manual labor in mental health disease classification and identification, which ultimately reduces the waiting time for the patients.

However, as these models are capable of processing huge amounts of textual data (e.g., social media posts, online forum discussions, personal narratives) in English and capable of identifying specific linguistic tones and indicators related to mental health, the obvious question remains how LLMs can help in early mental illness intervention when the input data or the conversation mode between the human and the LLM interface is in any language other than English. The full potential of GenAI-driven early disease detection, labeling, and classification can be fully realized when

- LLMs can leverage user-generated content to predict the actual mental health disease and severity of mental health conditions.

- LLMs will be capable of answering mental health disease and its severity, with the same level of accuracy when the questions are posed in any other language, and LLMs can fully utilize and understand the translated English-language mental health datasets.

An LLM-based model should begin by

- Creating a novel multilingual dataset (covering English, Turkish, French, Portuguese, German, Greek, and Finnish) derived from user-generated social media content

- Evaluation of the power of LLM in handling multilingual datasets to isolate the disease and severity

425

- Determine the model's performance in each language
  and gather and provide language-specific insights to
  reveal details of how linguistic diversity shapes the
  responses. Identify the severity of the class by setting
  the range boundaries across languages to avoid the
  potential misuse of erroneous detection. This can help
  in a future grouping of the languages that affect the task
  and dataset-specific outcomes.

Mental health disease categorization with the above points under consideration would make risk professionals of mental health cognizant of the multilingual adaptations prevalent in existing mental health NLP datasets. This would further propel us to be further inclusive and adaptable, outside of English by bringing in non-English contexts and by being more accountable for cultural and linguistic diversity.

Let us study few popular examples of such LLMs now.

LLMs can play multiple roles in assisting patients and medical healthcare professionals by being patient-facing and helping them with psychoeducation, therapist-facing to help mental health therapists, trainee-facing to assist trainees with their performance, and supervisor/consultant-facing to help supervisors with summaries of their therapy sessions. LLMs such as ChatGPT, Llama, and Qwen (a large language model family built by Alibaba Cloud) have been playing a predominant role in mental health. These LLMs have been leveraged to maximize the clarity and focus of the question, increase the context to reduce ambiguity, and specify the expected format of the response, with few shot examples to set an example response.

However, several lightweight LLMs are emerging. One such example is the MentalQLM, which is equipped with a dual **Low-rank Adaptation (LoRA)** module. It has a two-stage approach: datasets are pruned and instruction-tuned to adapt the LLM for downstream mental health classifications based on perplexity and diversity factors to reduce computational requirements, resulting in improved efficiency and adaptability related to specific tasks. It is further provided with a dense layer augmented with a second LoRA module that is fine-tuned to enhance performance on complex multiclass classification tasks (Stade et al., 2024).

Another popular example is CounseLLMe, which is a multilingual LLM trained on a multimodel dataset of 400 simulated mental health counseling dialogues (De Duro et al., 2025). This model is a hybrid of two languages, trained between two state-of-the-art large language models (LLMs). The training process involved conversations—20 rounds each from a mental health expert and guidance seeker in a natural dialogue flow. The conversation flow leveraged dialogue flows generated either in English (using OpenAI's GPT 3.5 and Claude-3's Haiku) or Italian (with Claude-3's Haiku and LLaMAntino). It also involves prompt fine-tuning to simulate a two-end conversation flow between a patient and a professional in psychotherapy.

With a quick glance and understanding of the advantages of LLMs, let us now understand their limitations.

Limitations of LLMs and Bringing in Multilingual Capabilities

LLMs suffer from underperformance and give inaccurate predictions due to low-resource languages. We can overcome this problem by having a more inclusive, diverse, and comprehensive dataset that has coverage for cultural and societal backgrounds and their context-driven nuances.

Some LLMs are equipped to process and generate text in multiple languages (Skianis et al., 2025). One example of it includes the XLM-R model, which uses self-supervised training techniques to get trained on large volumes of multilingual data. This has a high accuracy when the model is leveraging the capability of transfer learning, in learning that it is getting trained in one language and can understand and generate text in other languages in the absence of any other training data. It achieves state-of-the-art performance in cross-lingual understanding and can be scaled with large-scale multilingual data in the mental health domain. Other examples include Mental-Alpaca and Mental-FLAN-T5, open source LLMs designed to predict different mental health prediction problems. It has been designed for the mental health domain using prompt engineering, few-shot, and fine-tuning. LLMs are equipped to process and generate text in multiple languages. One example of it includes the XLM-R model, which uses self-supervised training techniques to get trained on large volumes of multilingual data. This has high accuracy by getting trained in one language and can understand and generate text in other languages in the absence of any other training data. It achieves state-of-the-art performance in cross-lingual understanding and can be scaled with large-scale multilingual data in the mental health domain. Other examples include Mental-Alpaca and Mental-FLAN-T5—open source LLMs designed to predict different mental health prediction problems. It has been designed for the mental health domain using prompt engineering, few-shot, and fine-tuning techniques on multiple LLMs to make it robust and efficient to generalize across multiple languages.

Despite having the best efforts in using language translations, language-specific nuances influence the accuracy of models, as it was observed that the German language for the depression dataset, when translated, yielded better performance, whereas performance degraded when translated into Turkish. This could be linked to the fact that the availability of mental health resources in Turkish is limited.

Let us now see how efficient prompt design techniques can help in predicting mental illness.

Increasing LLM Relevancy in Prediction Tasks for Mental Illness

The first and foremost step in designing highly accurate and relevant LLM responses would be to enrich the prompts in the following ways:

- **Enhancing the context**: Evaluate a social media post of the concerned person to answer the query

- **Enhancing the mental health information**: Evaluate a social media post of the concerned person from the viewpoint of a psychologist to answer the query

- **Enhancing the context and mental health information**: Evaluate a social media post of the concerned person from the viewpoint of a psychologist, by viewing it through the lenses of caregiving and other support needed for mental well-being to answer the query

When the LLM model evaluates risk, it considers the mental state (e.g., stressed, depressed) to reduce it to a binary classification problem to respond yes or no. This might also include other conditions to determine the likelihood of a critical risk like suicide and infer it using binary classification. The LLM model can also take the query input to remodel it as a multiclass classification task to evaluate the level of stress, depression, or anxiety of the person posting on social media and accordingly determine the level of risk associated with the above symptoms or even the risk level in committing suicide. We illustrate below with examples of how an LLM would approach this risk classification task.

To classify into varying levels of depression, stress, PTSD, ADHD, or anxiety severity, the LLM would first try to identify specific language phrases that are closely linked to the mental states, such as

- Continued or repetitive display of negative emotions like sadness or a sense of hopelessness

- Demonstration or display of worthlessness and self-critic feelings

- Behavior showcasing isolation or refrainment from social interactions

- Noticeable shifts or changes in behavioral or routines, such as insomnia or disrupted sleep, eating disorders changes in appetite, or alcohol abuse

- Mentions of emotional suffering or psychological distress

In addition, the LLM would take feedback from these signals and try to map it to four severity levels when data from social media, self-reported questionnaires, or unstructured interviews are fed into it. This may include

- **Level 1 (minimal)**: Textual data exhibiting infrequent or minor expressions of sadness or frustration. For example, it may include content extracted from daily activities, hobbies, interests, work, or social interactions, having soft, positive, neutral, or slightly negative tones. Such text does not contain any reference to self-harm or death.

- **Level 2 (mild)**: Textual data showing more frequent negative emotions or notable behavioral shifts. For example, such textual content when extracted from activities, hobbies, interests, interviews, or questionnaires typically showcases the hopelessness or helplessness feelings that the user is starting to feel.

- **Level 3 (moderate)**: Textual data demonstrating high
  levels of interference with daily life or clear indicators
  of emotional distress. For example, moderate severity
  content will contain self-hatred, social withdrawal, and
  worthlessness feelings expressed as "I'm a burden,"
  "Everyone feels good without me," "I hate myself," or
  "Life is no worth" and "I want to end it all." In addition,
  the presence of words containing death, dying, or
  mentions of wish to die and loneliness also suggests
  moderate level of mental illness severity.

- **Level 4 (severe)**: Textual data demonstrating extreme
  emotional distress, risk of self-harm, or complete social
  disengagement. Such examples include narratives
  of full frustration and hopelessness with plans to
  commit suicide, like "I have pills ready," "I'm going to
  hang myself off," and "I am going to write my suicide
  note today." Along with such notes, words or phrases
  expressing extreme apologies, lethal descriptions (like
  gun), farewells, and goodbye messages testify to the
  extreme level of severity, for example, "I'm sorry for
  everything; I won't be around much longer." Moreover,
  any gesture of past suicide attempts further would
  corroborate the highest level of mental illness.

Listing 9-4. Now let us study with an example LLM prompting
technique how to evaluate/find the label of different intensities of
depression

```
Classify depression severity 4 labels: minimal, mild,
moderate, severe.
Text: I am doing ok today
```

```
Label: Minimal
Text: I am feeling low from yesterday
Label: Mild
Text: I think I have made my parent;s life difficult by not
getting a job
Label: Moderate
Text: I am going to kill oneself one day, no onee
understands me
Label: Severe
Text: Tomorrow I won't return home, as no one wants me.
Label: ?
```

The last query, when fed to the LLM, responds back with the answer "Severe."

In this respect, we should be aware that training data, language model intricacies, or contextual emphasis can interpret different parts of the language differently, giving more weight to certain parts in one language over the others. An example would be English text classifying it as "Potential suicide attempt," whereas Greek text classifies it as "Suicidal ideation," but both of them trying to convey the same meaning.

As requests can come from different languages, to predict the severity of disorders of diseases like depression, PTSD, schizophrenia, and eating disorders, the prime objective would be to leverage an LLM for translation of the social media posts and then feed it via a prompt to an LLM. This would help us to evaluate the mental health conditions by assessing the specific severity levels. Further, inference of the predicted classes when compared with the ground-truth labels will help to compute the final evaluation metrics.

Table 9-2. *Datasets for depression intensity*

Datasets	Training data
Dep-Severity	Trained with posts from the social media platform Reddit, which are prelabeled according to four depression severity levels: minimal, mild, moderate, and severe depression
Sui-Twi	Trained with English texts from Twitter comprising of 1,300 suicidal ideation tweets, which are manually labeled into three classes "not suicidal," "indicative of suicidal ideation," and "indicative of a potential suicide attempt"

Now let us have the concluding thoughts from this chapter.

Conclusion

In this chapter, we have learned about the importance of a risk assessment framework in mental health and the criteria for developing a good framework. In this regard, we have highlighted how a risk framework should be capable of categorizing different types and levels of depression, based on the depression symptoms and the intensity of symptoms. Here, we had a chance to get an overview of AI models like hierarchical clustering and k-means clustering. We have also studied how language can serve as a tool to identify depression along with the merits and demerits of using language. Then, we dug into interpretability where we observed how language together with interpretability tools can help doctors, care providers, and policymakers to take the right actions for society. This chapter also helped the audience to understand the urgency of different types of actions (like notifications) a risk framework can undertake, based on the severity of the disease. This further drove us to explore the underlying self-reported metrics to measure depression and how to use these metrics through real examples of stress, anxiety, depression, and ADHD. Nevertheless, the readers are also educated on the critical components of a risk framework and how to plug in a suicide detection module into it. In the end, we also delved into the role of GenAI in assisting mental health and the ethical risk governance steps to address the seriousness of risks. We learned the most important aspect: that mental health diagnoses should always be assisted by a human and never left to be driven through fully automated systems, as they lead to large errors and/or misdiagnoses.

With a detailed background of an AI-driven risk framework in place and its best practices for system design, we explore how AI-powered recommendations can further enhance access to mental healthcare assistance programs, thus addressing the grave societal crisis.

References

Abdel-Bakky, M. S., Amin, E., Faris, T. M., & Abdellatif, A. A. H. (2021). Mental depression: Relation to different disease status, newer treatments and its association with COVID-19 pandemic (Review). *Molecular Medicine Reports*, *24*(6), 839. https://doi.org/10.3892/mmr.2021.12479

Amanat, A., Rizwan, M., Javed, A. R., Abdelhaq, M., Alsaqour, R., Pandya, S., & Uddin, M. (2022). Deep Learning for Depression Detection from Textual Data. *Electronics*, *11*(5), Article 5. https://doi.org/10.3390/electronics11050676

Bouachir, W., Gouiaa, R., Li, B., & Noumeir, R. (2018). Intelligent video surveillance for real-time detection of suicide attempts. *Pattern Recognition Letters*, *110*, 1–7. https://doi.org/10.1016/j.patrec.2018.03.018

Cook, B. L., Progovac, A. M., Chen, P., Mullin, B., Hou, S., & Baca-Garcia, E. (2016). Novel Use of Natural Language Processing (NLP) to Predict Suicidal Ideation and Psychiatric Symptoms in a Text-Based Mental Health Intervention in Madrid. *Computational and Mathematical Methods in Medicine, 2016*, 8708434. https://doi.org/10.1155/2016/8708434

De Duro, E. S., Improta, R., & Stella, M. (2025). Introducing CounseLLMe: A dataset of simulated mental health dialogues for comparing LLMs like Haiku, LLaMAntino and ChatGPT against humans. *Emerging Trends in Drugs, Addictions, and Health, 5*, 100170. https://doi.org/10.1016/j.etdah.2025.100170

Guo, Z., Lai, A., Thygesen, J. H., Farrington, J., Keen, T., & Li, K. (2024). Large Language Models for Mental Health Applications: Systematic Review. *JMIR Mental Health, 11*, e57400. https://doi.org/10.2196/57400

Kim, Y.-J., Lee, C. S., & Kang, S.-W. (2024). Increased adolescent game usage and health-related risk behaviors during the COVID-19 pandemic. *Current Psychology, 43*(18), 16821–16832. https://doi.org/10.1007/s12144-023-04466-8

Lim, H. J., Moon, E., Kim, K., Suh, H., Park, J., Kim, D.-R., Park, J.-H., Shin, M.-J., & Lee, Y.-H. (2023). Cluster Analysis on the Mental Health States in a Community Sample of Young Women During Pre-Pregnancy, Pregnancy, or the Postpartum Period. *Psychiatry Investigation, 20*(5), 445–451. https://doi.org/10.30773/pi.2023.0002

Matthewson, P. (2002). Risk assessment and management in mental health. *Social Work Review, 14*(4), 36-43. https://doi.org/10.48550/arXiv.2409.17397

Michaels, C., L., B., Lynn, A., Greylord, T., & Benning, S. (2020). *Mental Health and Well-being Ecological Model | Leadership Education in Maternal & Child Public Health.* https://mch.umn.edu/resources/mhecomodel/

Morgan, S. E., Diederen, K., Vértes, P. E., Ip, S. H. Y., Wang, B., Thompson, B., Demjaha, A., De Micheli, A., Oliver, D., Liakata, M., Fusar-Poli, P., Spencer, T. J., & McGuire, P. (2021). Natural Language Processing markers in first episode psychosis and people at clinical high-risk. *Translational Psychiatry, 11*(1), 1–9. https://doi.org/10.1038/s41398-021-01722-y

Qasrawi, R., Vicuna Polo, S., Abu Khader, R., Abu Al-Halawa, D., Hallaq, S., Abu Halaweh, N., & Abdeen, Z. (2023). Machine learning techniques for identifying mental health risk factor associated with schoolchildren cognitive ability living in politically violent environments. *Frontiers in Psychiatry, 14.* https://doi.org/10.3389/fpsyt.2023.1071622

Qasrawi, R., Vicuna Polo, S. P., Abu Al-Halawa, D., Hallaq, S., & Abdeen, Z. (2022). Assessment and Prediction of Depression and Anxiety Risk Factors in Schoolchildren: Machine Learning Techniques Performance Analysis. *JMIR Formative Research, 6*(8), e32736. https://doi.org/10.2196/32736

Rahman, H. A., Kwicklis, M., Ottom, M., Amornsriwatanakul, A., Khadizah H. Abdul-Mumin, Michael Rosenberg, & Dinov, I. D. (2023). Machine Learning-Based Prediction of Mental Well-Being Using Health Behavior Data from University Students. *Bioengineering, 10*(5), 575.

Rothenberg, W. A., Bizzego, A., Esposito, G., Lansford, J. E., Al-Hassan, S. M., Bacchini, D., Bornstein, M. H., Chang, L., Deater-Deckard, K., Di Giunta, L., Dodge, K. A., Gurdal, S., Liu, Q., Long, Q., Oburu, P., Pastorelli, C., Skinner, A. T., Sorbring, E., Tapanya, S., … Alampay, L. P. (2023). Predicting Adolescent Mental Health Outcomes Across Cultures: A Machine Learning Approach. *Journal of Youth and Adolescence, 52*(8), 1595–1619. https://doi.org/10.1007/s10964-023-01767-w

Siddique, F., & Lee, B. K. (2024). Predicting adolescent psychopathology from early life factors: A machine learning tutorial. *Global Epidemiology, 8,* 100161. https://doi.org/10.1016/j.gloepi.2024.100161

Skianis, K., Pavlopoulos, J., & Doğruöz, A. S. (2025). *Building
Multilingual Datasets for Predicting Mental Health Severity through LLMs:
Prospects and Challenges* (No. arXiv:2409.17397). arXiv. https://doi.
org/10.48550/arXiv.2409.17397

Stade, E. C., Stirman, S. W., Ungar, L. H., Boland, C. L., Schwartz, H. A.,
Yaden, D. B., Sedoc, J., DeRubeis, R. J., Willer, R., & Eichstaedt, J. C. (2024).
Large language models could change the future of behavioral healthcare: A
proposal for responsible development and evaluation. *Npj Mental Health
Research*, 3(1), 1-12. https://doi.org/10.1038/s44184-024-00056-z

University of Colorado at Boulder. (2024). *AI for mental health
screening may carry biases based on gender, race*. ScienceDaily. https://
www.sciencedaily.com/releases/2024/08/240805134143.htm

Vaj, T. (2023, May 1). 10 BERT Recommended for mental health
context. *Medium*. https://vtiya.medium.com/10-bert-recommended-
for-mental-health-context-2b94cfa33514

Zhang, T., Schoene, A. M., Ji, S., & Ananiadou, S. (2022). Natural
language processing applied to mental illness detection: A narrative
review. *Npj Digital Medicine*, 5(1), 1-13. https://doi.org/10.1038/
s41746-022-00589-7

CHAPTER 10

AI-Powered Recommendations: Enhancing Access to Mental Healthcare Assistance Programs

This chapter offers a comprehensive exploration of why AI-based recommendation models are useful for mental health. The chapter at first initiates the concept of recommendation systems in the context of mental health, by illustrating different types of recommender systems and how they can provide value to end users. The emphasis is placed on enabling the right tools and techniques to track mood changes and various mental disorders, empowering users with valuable insights. Readers gain familiarity with diverse types of recommendations that drive increased user engagement by encouraging the reporting of daily activities. By leveraging AI-powered recommendations, this chapter highlights the potential to enhance accessibility and effectiveness in mental healthcare assistance programs, ultimately leading to improved support and better outcomes for individuals in need.

© Sharmistha Chatterjee, Azadeh Dindarian, Usha Rengaraju 2025
S. Chatterjee et al., *Revolutionizing Youth Mental Health with Ethical AI*,
https://doi.org/10.1007/979-8-8688-1186-9_10

In this chapter, these topics will be covered in the following sections:

- Recommendation systems for mental health

- Evaluation metrics and design selection

- GenAI-based recommendation models

Recommendation Systems for Mental Health

In the first few chapters, we have studied how the demand for mental health services has continuously risen over the last few years and how technology has evolved to sustain the demand by providing multiple digital mental health services. It helps to provide psychological treatment through digital channels. It is also an easily accessible and scalable route to address mental health problems, which has led to an increase in the number of patients who suffer from chronic mental problems and high levels of burnout availing themselves of the services. The translation of psychotherapies into digital formats has resulted in more personalized treatment plans where an individual can be assisted with customized therapy plans through digital channels like mobile and web. However, digitization and personalization pose the challenge of screening the entire list of treatment components. This is inefficient in terms of giving the best results to the end users, which further propels the need for optimal sequencing of these items, adapted according to their medical history, preferences, and context. Once the therapies are curated for individuals, they can offer treatment like behavioral **activation (BA)** for treating depression and **dialectical behavioral therapy (DBT)** to handle borderline personality disorders. Such digitized versions of mental health therapies should have proper therapeutic curation, preventing individuals from self-adjusting the content, timing, and dosage. This not only leads to better engagement of clients but also inculcates confidence in the process to improve their mental health. BA provides an important therapeutic action for reducing depressive symptoms, which can be embedded through smartphone-based apps. For example, MUBS, a smartphone-based system for BA, operates by offering a personalized, content-based activity recommendation model comprising a unique list of user-preferred activities.

To bring in more customized mental health therapies for individual users, digital mental health apps should be powered with AI-driven recommendations based on user attributes and several other parameters, which are a reflection of their mental health conditions. This would result in high user retention in mental health apps. **Recommender systems (RS)** are a unique value addition driving more user satisfaction in therapy apps. RS is designed by taking into consideration personalized lists of items based on their preferences, like timing, duration context, medical history, and what similar users have enjoyed, which can increase the value proposition in the mental health world when user treatment benefits are prioritized over business benefits or the amount of revenue generated. This results in more users feeling happy, productive, and recovering from depression and other mental health-related challenges.

Let us now understand the impact of different types of therapies and treatments through assisted recommendations.

Recommendation-Based Treatments

The recommended treatment plans can range from short term to long term and can positively impact mental health conditions and increase the likelihood of a full recovery. Short or micro-interventions through digital channels can allow users to configure sessions (examples include a brief gratitude exercise to increase in-the-moment appreciation or performing a short mindful breathing exercise to increase self-awareness) based on their personal choice. When users are empowered to configure their independent sessions by combining multiple therapies into a single catalog, it adds accountability and self-containment. On top of RS, mental health services support time-based tracking and improvement metrics, which makes it easier to track the progress and accordingly curate personalized therapy plans for the future. In the absence of such a service, the next generation of mental healthcare would see users not being able to select the appropriate therapy plan from the list, leading to a high volume of users churning out from the system. In addition, users will suffer from information overload and face difficulty getting connected to the right digital therapeutic pharmacies or apothecaries (Valentine et al., 2023). Designing the correct type of recommender system demands a careful evaluation of the algorithms and the risk and severity of the disease.

Recommended Hobbies based on Mood State

Looking for hobbies of interest

Figure 10-1. *Hobby recommendation from mental-health apps*

Figure 10-1 shows recommendations for various hobbies, such as photography, gymnastics, reading, yoga, singing, and playing a musical instrument, by a mobile app after the user's current and previous mood state has been determined.

This kind of treatment becomes beneficial to users due to the self-guided nature of the content based on a variety of factors (such as patient condition, environment, and psychosocial stressors), where users feel they are understood and are empowered to self-manage their mental well-being. The treatments offered through the mental well-being apps offer primarily two principal kinds of recommendations: onboarding-based and coaching conversation-based, which function upon the content used in the recommendation algorithms. One related app that is highly relevant in this context is the Ginger behavioral health coaching and therapy app (Chaturvedi et al., 2023), where evaluations were carried out on 14,018 users with real-world configurations. On one side, it offers carefully adjustable content to new users that can be easily consumed through more than 200 clinically validated content cards containing mindfulness exercises, psychotherapeutic education, meditations, breathing exercises, videos, podcasts, surveys, and readings, with a time duration between two and ten minutes to complete. On the other hand, it also offers conversation-based recommendations that are semantically similar when real physical coaches will be present through text-based behavioral health coaching, teletherapy, and telepsychiatry. The app further hosts an Amplitude Analytics platform to observe, log, and record user activities such as content-related triggers when they use the application (Chaturvedi et al., 2023).

Mental well-being apps primarily offer recommendations for the following categories (Chaturvedi et al., 2023):

- **Onboarding-based recommendations**: Users are prompted with content card-based advice with the most relevant option or activity card at the top during their onboarding responses after they have signed up for the service. The user responses are based on questions they are being asked about their current feelings (such as anxiety, depression, grieving, stress, and others) or who impacts their life most (such as career, family, hobbies, health, social life, and others). This method performs better when we see sparse or skewed content usage datasets on a platform.

- **Conversation-based content recommendations**: Users are prompted with advice based on the context and content of their conversation with a mental health coach. This method computes the semantic similarity between the content of a conversation and the text description of content cards to offer suitable suggestions and mostly applies when users have engaged with their coach for the past 60 days or more. It maintains the highest rate of user engagement as it dynamically updates the advice based on user chats with their coach.

- **Random recommendations**: Users are prompted with advice based on onboarding questions in the app and their responses, thus giving them a mixed flavor of both recommendations.

Both the conversational text and content card descriptions should be compared and rated in terms of their similarity so that like suggestions are offered. They can be mathematically represented as embedding vectors using a language model (such as XLM-RoBERTa) that can generate embeddings from text. This type of embedding model can be trained to detect semantic similarity between the conversation sessions to locate identical paraphrases. The relevance of similar paraphrases can be further computed using a cosine similarity matrix between all the message embeddings $(M_1...M_i...M_n)$ (represented as rows) and all the content card embeddings $(C1...Ci...Cn)_i...C_n$ (represented as columns) from the embedding vectors generated from the final output layer of the neural network. Once the similarity scores are evaluated, they are passed through a Max Operator function to sort in order of highest relevancy to generate the final list of recommendations.

Now we have gained an understanding of how recommendations can be embedded in apps. Now let us take a look at the different types of recommender systems.

Types of Recommender Systems

RS algorithms play a critical component in the context of the mHealth **just-in-time adaptive intervention (JITAI)** framework. As RS service becomes an important component of mental health apps, the primary recommender systems of choice are

Collaborative filtering (CF): It helps offer recommendations to end users by comparing past historical ratings to those of other similar users. Examples of CF are neighborhood-based methods such as k-nearest neighbors and model-based methods such as matrix factorization. CF algorithms are known to offer better-personalized items in comparison to content-based and knowledge-based algorithms when prior ratings exist in the system, scenarios where at least a few ratings per user and item exist in the rating matrix (situations where it is not a cold-start scenario and a completely new service is launched or a new doctor or a completely new therapy needs to be recommended to the end user).

CF can be broadly classified into two main types: memory-based and model-based. The model-based method uses ML algorithms, such as **singular value decomposition (SVD)**, SVD++, probabilistic **matrix factorization (MF)**, nonnegative MF, clustering K-nearest neighbors, and deep learning, to predict the user's rating of unrated items (Lewis et al., 2022a).

- **K-nearest neighbor (KNN):** It helps to estimate the missing ratings by computing the similarity-weighted sum of the KNN nearest neighbors. The k-nearest neighbor with the user baseline adjustment (KNNwBaseline) is an extension of the KNN model but with additional adjustments for the user and item biases, b_u and b_i.

- **Funk singular value decomposition (SVD):** This method is implemented per the Funk SVD/matrix factorisation (MF) paradigm with adjustments for the user and item biases, b_u and b_i. The model is equipped with hyperparameters, which are the number of latent factors (k_{latent}), the regularization parameter (λ), the learning rate (γ), and the number of training epochs (e). It learns by minimising a squared error function known as Root Mean Squared Error(RMSE) to predict the known ratings in the original user-item matrix.

- **SVD++ algorithm:** This method is an extension of the previous SVD method that incorporates information on the tasks the users rated irrespective of rating value (i.e., implicit feedback). This helps the model to learn additional latent factor parameters along with the parameters learned for the explicit rating values.

- **Matrix factorization:** This method helps to construct an m × n matrix (called the utility matrix), which consists of the rating (or preference) for each user-item pair (Aakanksha, 2020). MF builds the matrix with two entities, users and items, which often turns out to be sparse due to limited ratings for a few user-item pairs. MF is heavily used in recommender systems to locate similarities between users and items, to locate problems users are having, and to recommend them appropriate mental health services or care providers. Such a scenario is evident when we have ten or more users and five or more care providers or mental health psychiatrists, and we are trying to predict the rating that each user would give to the clinics or therapy services. This is what our utility matrix initially looks like, as illustrated in Figure 10-2 (Aakanksha, 2020).

Item

		Doctor 001	Doctor 002	Doctor 003	Doctor 004
	User 001		4.5	2.0	
	User 002	4.0	3.5	3.5	
User	User 003		5.0		2.0
	User 004		3.5	4.0	1.0

User-Doctor Rating Matrix

Figure 10-2. *Utility matrix for matrix factorization (Aakanksha, 2020)*

To offer recommendations to end users, we need to populate this matrix by finding similarities between users and items. To get an intuition, for example, we see that the user with ID 002 and the user with ID 004 gave the same rating to the doctor and the practiced therapy, which helps us to conclude that both these users feel the same about the specified doctor, clinic, and therapy and enables us to find the unrated items based on the interactions between the users and the items.

This can, however, be obtained by decomposing the utility matrix into a product of two lower-dimensional tall-skinny matrices representing users and items, where U is an m × k matrix and V is an n × k matrix, and the decomposition matrix will have the following representation:

$$Y = UV^T$$

U represents users in some low-dimensional space, and V represents items where only a few users interact with a few items. For a user i, u_i gives the representation of that user, and for an item e, v_e gives the representation of that item. The rating prediction for a user-item pair is a dot product of the user and item representations, which helps to obtain the rating prediction for a user-item pair. The rating prediction for a user-item pair represents a dot product of the user and item representation as given by

$$y_{ij} = u_i.v_j$$

This further translates to the following dot product matrix between users and items, as illustrated in Figure 10-3.

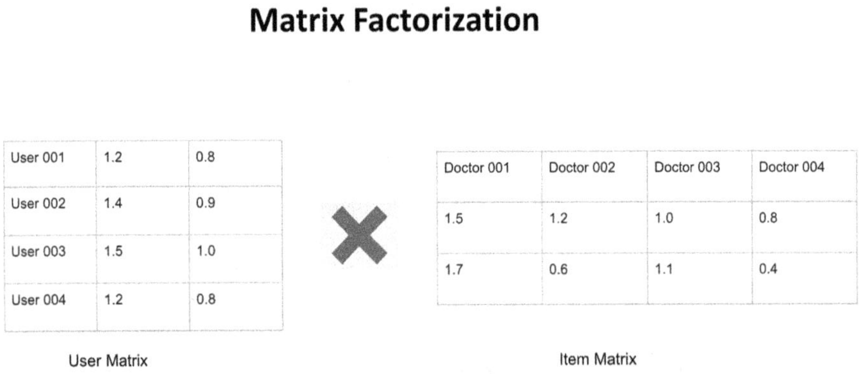

Matrix Factorization

User 001	1.2	0.8
User 002	1.4	0.9
User 003	1.5	1.0
User 004	1.2	0.8

User Matrix

Doctor 001	Doctor 002	Doctor 003	Doctor 004
1.5	1.2	1.0	0.8
1.7	0.6	1.1	0.4

Item Matrix

***Figure 10-3.** Rating matrix decomposition as a dot product of user and item matrix*

We can further implement matrix factorization, using deep learning-based embeddings for the user and item embedding matrices and then leverage gradient descent to get the optimal decomposition.

Collaborative Filtering(CF) and Content-based Recommender Systems

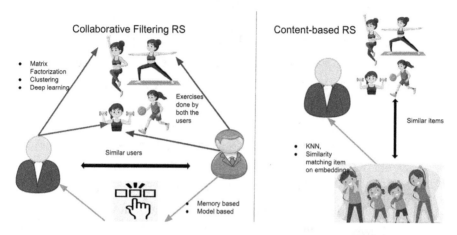

Figure 10-4. *Collaborative filtering vs. content-based RS*

Listing 10-1. Let us now understand with a practical example below
how different user attributes like age and type of mental health
challenges like anxiety, depression, insomnia, or OCD can be used
to offer suitable suggestions to select music genres using CF-based
recommendations based on their likeability.

1.   In the first step, we import the necessary libraries, load
     the dataset, and shuffle the ratings, keeping a 90% and
     10% split across the train-test dataset.

```
from surprise import Dataset, Reader, SVD, KNNBasic
from surprise.model_selection import train_test_split
from surprise.model_selection import cross_validate
from surprise import accuracy
import random
from surprise.model_selection import GridSearchCV
reader = Reader(rating_scale=(0, 3))
```

```
data = Dataset.load_from_df(user_ratings_df,reader)
raw_ratings = data.raw_ratings
random.shuffle(raw_ratings)
threshold = int(.9 * len(raw_ratings))
trainset_raw_ratings = raw_ratings[:threshold]
test_raw_ratings = raw_ratings[threshold:]
```

2. Following this, we retrain the whole training dataset
    and then evaluate the test set using metrics like
    accuracy, RMSE, MAE, MSE, and FCP (how Fraction
    of Concordant Pairs (FCP) is a metric that ranks in the
    RS to order items in alignment with user preferences)
    (Lewis et al., 2022a).

```
data.raw_ratings = trainset_raw_ratings  # data is now
your trainset, where we select & Train Model
sim_options = {
     "name": "pearson",
}
model = KNNBasic(sim_options=sim_options)
trainset = data.build_full_trainset()
model.fit(trainset)
cross_validate(model, data, measures=["RMSE", "MAE"],
cv=5, verbose=True)
testset = data.construct_testset(trainset_raw_ratings)
predictions = model.test(testset)
print('Accuracy on the trainset:')
accuracy.rmse(predictions)
accuracy.mae(predictions)
accuracy.mse(predictions)
accuracy.fcp(predictions)
testset = data.construct_testset(test_raw_ratings)
```

```
predictions = model.test(testset)
print('Accuracy on the testset:')
accuracy.rmse(predictions)
accuracy.mae(predictions)
accuracy.mse(predictions)
accuracy.fcp(predictions)
```

3. On aligning the user IDs to their top song genres, we
   then plot it, to help us visualize the following outputs as
   shown in Figure 10-5.

```
recommender_output_dataframe =
pd.DataFrame(predictions, columns=["user_id","genre",
"score"])
recommender_output_dataframe.head()
idx = recommender_output_dataframe.groupby(['user_id'])
['score'].idxmax()
result = recommender_output_dataframe.loc[idx, ['user_
id', 'genre', 'score']]
```

CF-based Recommendations to alleviate Mental Health problems by suggesting suitable Song Genres

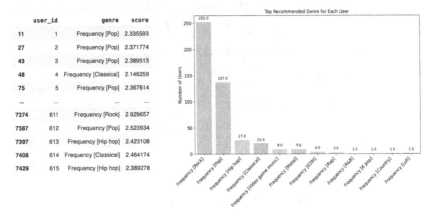

Figure 10-5. *Output after applying CF to suggest appropriate music genres to mental health users*

Content-based: It helps to offer recommendations to end users by screening through the therapies and counseling sessions that users have liked and rated highly in the past. Examples include recommendations related to hobbies, sports, or other activities and clubs that are the preferred choice of students.

Let us look at Figure 10-4 to understand the difference between CF and content-based recommenders.

To better understand how recommenders would work in mental health apps, let us explore the steps of an activity-based recommender model shown in Figure 10-6 (Siriaraya, P. et al., 2019) to demonstrate hobbies practiced by users are uploaded through the app, and then the app later uses its built-in REC to offer recommendations to end users. It is a hybrid recommender system serving both the functionalities of CF and content-based recommenders, where it takes into consideration the user's interests and provides interventions as a combination of both the user's liked activities as well as similar activities (Mazlan et al., 2023). This hybrid

functionality combination functions by attaching weightage to activities
in a way that activities having more weights possess a higher likelihood of
being recommended.

Hybrid REC for Mobile Mental Health Applications for RS

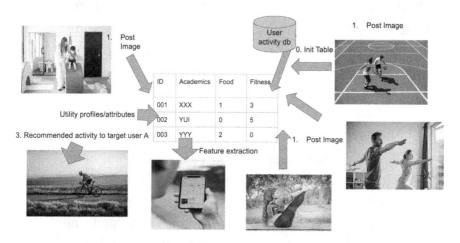

Figure 10-6. *Collaborative filtering in mental health apps*

The user records (profiles and hobby lists) are stored in the activity
db. The RS learns the user's choices, extracts relevant user information
using ML, and later suggests similar users with similar activities using
CF. Further, content-based recommenders can also be built into such apps,
where the recommender can scan and learn similar hobby activities and
recommend cycling in step 3, while the input image fed is of exercising.

Listing 10-2. Now let us understand through the below code
listing how people suffering from those disorders are talking about,
through topic modeling by leveraging text analytics, NLP, and
transformer models. Here, we show how to build a mental health
similarity matrix based on text content to offer personalized advice
derived from shared community experiences of like-minded people.
The dataset used here contains the scarped data from individuals
seeking assistance from licensed therapists and their associated
responses

1. In the first step, we need to import the necessary
    libraries, load the dataset from counselchat.com, and
    load the dataset (dataset = load_dataset("nbertagnolli/
    counsel-chat")), remove duplicate posts and
    stopwords, and then apply BERTTopic- topic modeling
    technique (that uses transformers and c-TF-IDF) to
    represent dense clusters around interpretable topics,
    where the most important themes are shown in the
    topic descriptions. We also use vaderSentimentfrom
    from SentimentIntensityAnalyzer to analyze the
    sentiment polarity score of each user statement and
    the overall sentiment score, considering all user
    sentiments. Since the full codebase is out of scope,
    we show limited processing and modeling in the code
    snippets.

    We first filter out mental health disorders pertaining to
    either relationships or social relationships and set the
    similarity scoring threshold to 0.05.

```
disorder_level= "relationships/social-realtionships"
df_disorder=df_disorder[df_disorder['Similarity
Score']>0.5].sort_values(by='Similarity Score',
ascending=False).reset_index(drop=True)[:10]
```

2. In the next step, we combine the question.

```
df_sug= df_disorder.merge(df, on='questionText',
how='left')[['questionText','User Concern','Similarity
Score','Concern_Sentiment','User_Concern_
Sentiment','answerText']]
df_sug['answerText']= df_sug['answerText'].astype(str)
paragraph = ' '.join(df_sug['answerText'])
```

3. Then, we load a pretrained spaCy model (English), load
   it and split the paragraph into sentences, and calculate
   the sentence similarity using cosine similarity metric.

```
import numpy as np
from sklearn.metrics.pairwise import cosine_similarity
nlp = spacy.load('en_core_web_md')
nlp.max_length = 2000000 # or even higher
doc = nlp(paragraph)
sentences = [sent.text for sent in doc.sents]
num_sentences = len(sentences)
similarity_matrix = np.zeros((num_sentences, num_
sentences))
for i in range(num_sentences):
    for j in range(i, num_sentences):  # Only calculate
                                        each pair once
        if i != j:
            vector1 = nlp(sentences[i]).vector
            vector2 = nlp(sentences[j]).vector
```

```
                similarity = cosine_similarity([vector1],
                [vector2])[0][0]
                similarity_matrix[i, j] = similarity
                similarity_matrix[j, i] = similarity
```

4. In the next step, we sum the similarity scores of the
   sentences to create a list of (sentence and scores)
   tuples, which are then sorted in descending order.
   Further, we set the threshold as 10% and then evaluate
   the list of sentences that have scores greater than or
   equal to the threshold.

```
sentence_similarity_scores = np.sum(similarity_
matrix, axis=1)
sentence_scores_tuples = [(sentence, score) for
sentence, score in zip(sentences, sentence_similarity_
scores)]
sentence_scores_tuples_sorted = sorted(sentence_scores_
tuples, key=lambda x: x[1], reverse=True)
percentage = 10
num_sentences = len(sentences)
threshold_score = np.percentile(sentence_similarity_
scores, 100 - percentage)
top_10_sentences = [sentence for sentence, score in
sentence_scores_tuples if score >= threshold_score]
for i, sentence in enumerate(top_10_sentences, 1):
    print(f"Top {i}: {sentence}")
```

5. Then, we filter out the top sentences with non-negative
   sentences which prints the following sentences:

```
def filter_positive_sentences(sentences):
    positive_sentences = []
```

```python
    for sentence in sentences:
        sentiment = analyzer.polarity_scores(sentence)
        if sentiment['compound'] >= 0:
            positive_sentences.append(sentence)
    return positive_sentences
top_10_sentences = filter_positive_sentences(top_10_
sentences)
for i, sentence in enumerate(top_10_sentences, 1):
    print(f"Positive {i}: {sentence}")
```

We get the output as

```
Positive 1: Once you take the journey to really
understand who you are and unconditionally loving all
that is you, there will never be an alone moment.
Positive 2: Instead, its when you follow up by
inquiring and listening to others that you  may
discover they feel and think and struggle just
like you do.
Positive 3: To help you ask for the right type of
support, you can ask yourself what it is about the
situation that is bothering you and how you would wish
for your friends and family to respond to you.
Positive 4: Maybe for right now, until you are able to
find in person friends whom you're able to feel hear
you the way you'd like to be heard, find online forums
and groups of likeminded people.
Positive 5: If you think about the wording of a
question which would motivate you to respond, then this
formula will very likely be the same for many others
who read your post.
```

Positive 6: If you can phrase it in such a way that they recognize that you are asking for information and not blaming them for not answering you, that could be effective.

Positive 7: One way to check if you're totally misreading him, is to examine whether you feel similarly in other relationships.

Positive 8: Because of the way that you say your boyfriend is only calm when he is drinking and you have concerns about flights, it would probably be most helpful for you to speak with a local therapist so you can have specific conversations about what happens during these fights.

Positive 9: I recommend working on this with a therapist, though (even if you end up going without your boyfriend to sessions), so that you can talk about specific strategies and what you can do when he is not calm.

6. Finally, we use BART tuned for text summarization through transformer pipelines and feed the top ten filtered sentences from the above step to the transformer, to generate the summary.

```
summarizer = pipeline("summarization")
text = ' '.join(top_10_sentences)
summary = summarizer(text, max_length=76, min_
length=30, do_sample=False)
```

This gives us the output as

"Ask yourself what it is about the situation that is bothering you and how you would wish for your friends and family to respond to you. One way to check if you're totally misreading him, is to examine whether you feel similarly in other relationships."

After going through CF and content-based RS, let us now study knowledge-based RS.

Knowledge-based: It helps offer recommendations to end users to make decision using rules programmed a priori by a human expert (Mateti et al., 2024). Figure 10-7 illustrates a typical knowledge-based RS, where multiple knowledge sources (from the community, core literature, CIS, and EMR) are assembled and fed through an expert to the Knowledge Management Service (KMS). On processing this knowledge through NLP algorithms, we can extract the relevant entities and represent them through a graph ontology as discussed in Chapter 5, "Capturing and Linking Data Systems Using Digital Technology to Redesign Youth Mental Health 360 Services." From the ontology graph-based knowledge system thus created, we can use it further for offering

- Personalized recommendations for mental health knowledge

- Boost users' moods through personalized recommendations

- Mental health prevention knowledge service

- Psychological Q&A service

- Mental health lexicons

Knowledge-based Recommender Systems

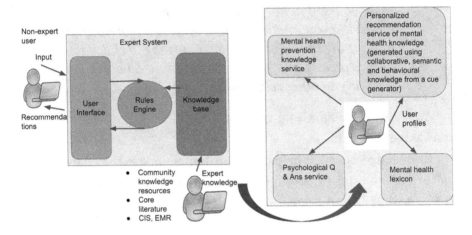

Figure 10-7. *Knowledge-based RS*

Knowledge-based recommendation system (KBRS) (Mateti et al., 2024) can detect any negative emotional state of the user using deep learning techniques and offer suggestions to boost the user's mood by sending them happy, calm, relaxing, or motivational messages. KBR's framework is equipped with an in-built real-time monitoring utility to detect messages posted by users on social media platforms to identify the intensity of potential psychological disturbances, specifically, depression and stress.

To design such recommender systems, users can be prompted with the following messages from the healthcare apps:

- "How was your day?"

- "How are you feeling right now?"

- "Is anything bothering you?"

- Other topics that will prompt the users to post on a
  social platform

The responses can be further processed using text data cleaning and NLP processing techniques to do sentiment analysis to forecast their mood. In addition, the Chrome extension can also be used by this application to gather user social media data from YouTube, Twitter, and Gmail services. By tracking user Internet usage activities such as watching a YouTube video, sending an email, or updating a Twitter post, the extension can scrape the data and apply sentiment analysis on top of the data to forecast the user's mood. Once the application can extract a graphically organized summary of the user's most recent seven days of mood analysis, it can further leverage emojis and the user's selfies (by requesting the user) in evaluating the intensity of the day's emotions. Once the sentiment scores are determined, allowing the user to view all of the application's analysis themselves for the full time they were active, as well as for the previous 3, 7, and 14 days, helps them to know their mood status and seek the counselor's advice if mental disturbances persist for more than 7 days. The app can also send frequent notifications to the user regarding their unpleasant mood if it continues for more than a week and then recommend interventions to users if the users are found to be in a negative state of mind.

Increasing the knowledge base of the system using more frequent user prompts and feeding in journal, voice, and video data can help to offer increased interventions to boost their mood and improve the mental state of users.

Figure 10-8 further illustrates the architecture of a KBRS architectural framework, when messages posted by users are fed through the real-time monitoring engine to preprocess the text and attach a sentiment intensity score. The identification process is carried out using a convolutional neural network and a bidirectional **long short-term memory (LSTM)** recurrent **neural network (RNN)**, which helps to scan the sentences and evaluate depressive and stressful content (Shah et al., 2025). The computed score is then used by the RS to send warning messages and suggested therapies to authorized persons.

Knowledge-based RS to analyse user sentiments and give recommendations

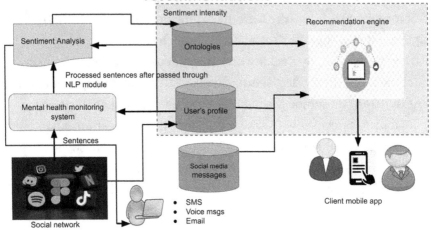

Figure 10-8. *Knowledge-based RS to evaluate user sentiment score*

Context-aware: It helps to offer recommendations to end users by taking into consideration the user context such as location, age, and time. This type of RS can be built over CF or content-based RS to fine-tune the existing RS. Examples of such RS include days of the week, weekdays vs. weekends, and timing evening vs. early morning to recommend the therapy. Context-aware RS pipelines rely on prefiltering, postfiltering, and contextual modeling to incorporate the user's current context and to increase the likelihood of conversion, where the user clicks and consumes the recommended items. Prefiltering and postfiltering mechanisms provide a mechanism to filter input data or output predictions before or after using a noncontext-aware algorithm, as illustrated in Figure 10-9. The algorithms tweak the contextual information to the rating prediction function, where additional context and characteristics of the users and/ or items help to bring in more relevant items for the user in the top list. However, the initial rating function can still be evaluated based on MF, in the same way as it is done in CF.

Context-aware RS using contextual bandits/agents has gained popularity for cold-start problems. It makes sequential decisions while balancing exploration and exploitation to learn new conditions arising when new users and items are added to the system. It is based on online learning of users' mood conditions, user preferences, mental health services, and providers while recommendations are served to the user continuously with parameters getting updated. One noteworthy example of this kind of recommender is the PopTherapy system, which leverages contextual bandits to offer external interventions derived from popular web apps to users. This method uses AI to help users alleviate their stress levels by matching interventions to individuals based on their current circumstances over some time by emphasizing on

- Three good things

- Best future self

- Thank you letter

- Act of kindness

- Strengths

- Affirm values

An example of a micro-intervention strategy recommended by PopTherapy is given below:

- **Individual: Prompt**: "Everyone has something they do well... find an example on your FB timeline that showcases one of your strengths." + URL: http://www.facebook.com/me/

- **Social: Prompt**: "Learn about active constructive responding and practice with one person" + URL: http://youtube.com/results?search_query=active+constructive

Another well-known application is **MOSS—Mobile Sensing and
Support** for People with Depression, which has an in-built context-aware
RS that takes into consideration user context features such as day, location,
and smartphone usage. It offers suggestions for cognitive behavioral
therapy through micro-interventions (CBT) to reduce depression (Lewis
et al., 2022b).

Context-based Recommender Systems

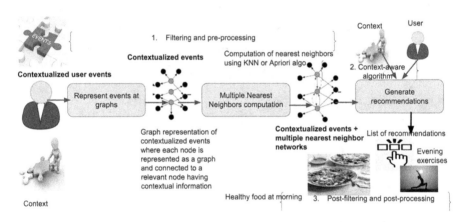

Figure 10-9. *Context-aware RS*

Figure 10-9 demonstrates how contextualized user events are fed
to generate knowledge graphs that aid in discovering nearest neighbor
networks (through KNN or Apriori algorithms). Once nearest neighbor
networks are formed, they are fed to the RS engine to generate contextual
recommendations that rank recommended activities based on the time of
the day to the users.

The main principle behind introducing a context-aware RS in mental
health apps is to leverage user contextual information to drive more value
for not only therapy and digital mental health but also cover broader
mHealth contexts (both physical and mental well-being). However,
context-aware RS have data sparsity problems and are hard to implement,
and they require a lot of computing power.

464

KGs are capable of speeding up the filtering phase while storing the multimodal results of contextual prefiltering. They are also known to improve the selection of contextual data of the user and the environment and store all the data in multidimensional matrices. One of the main advantages of personality-aware recommender systems, which have more contextual user data, is that such systems outperform traditional recommender systems as they tend to overcome the cold-start problem.

Recommender systems are introduced in gaming applications to suggest the user's type of games they should play to socialize and keep their engagement scores at the right level. Let us now see how RS can be embedded in gaming applications.

RS in Gaming Applications

Along with mobile apps, RS is found to be built in different gaming applications that prompt players daily to select a single/selected list of options from specified categories like "Basics," "Fitness," "Fun," "Social," "Art," and "Other." The players are further provided duration-based guidance to select themes like "take a nap," "do 15 min of cycling," "write a diary entry," "call and talk to a friend or family member," etc. Once the player completes the activity, they are further prompted to provide a rating of the rigor of the activity (*low, medium, or high*) and how they are feeling after doing the activity on a 5-point scale, five being the best and one being the lowest. This kind of feedback mechanism helps to capture the duration, time of the day (morning, afternoon, evening, or nighttime and weekday or weekend activity), and mood states "Worse," "Not As Good," "The Same," "A Little Better," and "Much Better" of the players, which serve as important parameters for the RS model to learn and better rank the recommendations on the following attempt. Such gaming-based RS can also attach reward-based points that can be encashed by the gamers at different social clubs or activity hubs, which promotes mental well-being among youth.

However, one major design consideration of the gaming apps is
that they should be built with strict privacy rules so that they do not
capture and store any PII data of the gamers, such as anything related to
demographics or location information.

As gamers' sentiments can be predicted while they are playing games,
let us look at a practical example below.

Listing 10-3.   Now let us understand through the below code listing
how to predict a gamer's emotional state score by using the popular
gaming stimulation dataset AGAIN affect gaming annotation
dataset data that includes the 1116 videos and detailed gameplay
logs of 124 participants playing 9 games from 3 different genres. A
cleaned and preprocessed dataset is available, including the 995
videos and detailed gameplay logs. This data contains real-time
annotation of gamers with the following attributes along with the
arousal level of the gamer: genre, player_id, session_id, game, time_
index, epoch, time_stamp, engine_tick, arousal, time_passed...,
player_slow_pickup, player_has_powerup, bot_has_collisions,
bot_is_colliding_above, bot_is_colliding_below, bot_is_colliding_
left, bot_is_colliding_right, bot_is_falling,bot_is_jumping, and
bot_charging

1. In the first step, we need to import the necessary
    libraries and load the downloaded and cleaned dataset.

    ```
    !gdown --id 1WOI3O_JwR-DjBO4-5YA4jte38d3ugmB2
    import pandas as pd
    from pytorch_tabnet.tab_model import TabNetRegressor
    import torch
    from sklearn.preprocessing import LabelEncoder
    from sklearn.metrics import mean_squared_error
    import numpy as np
    ```

```
df = pd.read_csv('clean_data.csv')
df.head()
```

2. Then, we begin by loading the data and separating it
    into categorical and continuous features. Categorical
    features are those which have classes or which are
    quantized whereas continuous features are not
    quantized as the name suggests. We convert the
    categorical features to integers using the label encoder
    from sklearn. Finally, we use the popular TabNet model
    to perform regression. TabNet is one of the most widely
    used and highly accurate models for both tabular
    classification and regression. In the end, we also plot
    out the explanation masks from the trained model.

```
categorical_columns = []
categorical_dims =  {}
for col in df.columns[df.dtypes == object]:
    print(col, df[col].nunique())
    l_enc = LabelEncoder()
    df[col] = df[col].fillna("NA")
    df[col] = l_enc.fit_transform(df[col].values)
    categorical_columns.append(col)
    categorical_dims[col] = len(l_enc.classes_)
for col in df.columns[df.dtypes == 'int64']:
    df.fillna(df.loc[:, col].mean(), inplace=True)

for col in df.columns[df.dtypes == 'float64']:
    df.fillna(df.loc[:, col].mean(), inplace=True)

target = '[output]arousal'

features = [ col for col in df.columns if col not in
[target]]
```

```
cat_idxs = [ i for i, f in enumerate(features) if f in
categorical_columns]
cat_dims = [ categorical_dims[f] for i, f in
enumerate(features) if f in categorical_columns]
cat_emb_dim = [2, 64, 128, 6, 128, 1024, 16, 3,8,4,5]
```

3. Then, we separate the training and test datasets
    and trigger the training and process by setting the
    evaluation metric as root mean squared logarithmic
    error, mean absolute error, root mean squared error,
    and mean squared error.

```
X_train = df[features].values
y_train = df[target].values.reshape(-1, 1)
clf = TabNetRegressor(cat_dims=cat_dims, cat_emb_
dim=cat_emb_dim, cat_idxs=cat_idxs)
clf.fit(
    X_train=X_train, y_train=y_train,
    eval_metric=['rmsle', 'mae', 'rmse', 'mse'],
    max_epochs=25,
    patience=50,
    batch_size=1024, virtual_batch_size=128,
    num_workers=0,
    drop_last=False
)
```

4. In the end, we start to predict the test set which yields
    the following probabilities as results:

```
preds = clf.predict(X_train)
y_true = y_train
test_score = mean_squared_error(y_pred=preds, y_
true=y_true)
```

The output yielded are

```
Predictions:   [[0.14831346]
               [0.10597676]
               [0.1109795 ]
                 ...
               [0.53795314]
               [0.39252844]
               [0.36106905]]
```

5. Next when we try to plot the training sample dataset,
    we get the following demonstration, showing only
    the most important features and masking rest of the
    features, as shown in Figure 10-10.

```
from matplotlib import pyplot as plt
%matplotlib inline
explain_matrix, masks = clf.explain(X_train)
fig, axs = plt.subplots(1, 3, figsize=(20,20))
for i in range(3):
    axs[i].imshow(masks[i][:50])
    axs[i].set_title(f"mask {i}")
```

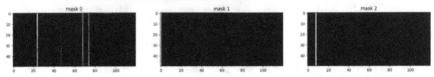

Figure 10-10. *Explainability showing most important features from*
TableRegressor model

Which RS to Choose?

As we have learnt about the different types of traditional RS, each of them
has its limitations. Content-based RS performs poorly when limited item
or user information is available, while CF suffers when the item or user is
new to the system and is constrained by the cold-start problem. Session-
based RS are forced to make their recommendations within a limited
period and therefore require optimizations. Contextual RS suffer from data
scarcity and privacy problems.

Hence, an essential part of RS selection is to follow a hybrid approach
as illustrated below:

- Monolithic design (integrates several recommendation
  strategies in one algorithm to provide relevant
  recommendations).

- Parallelized design (integrates different recommender
  systems subsequently, where each RS generates their
  separate lists of recommendations, which are later
  merged).

- Pipelined design connects multiple RS in a pipeline
  architecture, where the output of one recommender
  is fed into the input of the next one, and the stacked
  RS components may also use whole or part original
  input data.

Now let us take a look at Figure 10-11 to study different architectural styles of designing a hybrid RS (Mazlan et al., 2023).

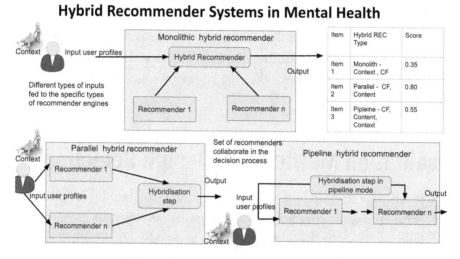

Figure 10-11. *Designing hybrid REC for mental health*

Along with hybrid REC, we need to have efficient algorithms in each of the recommenders. Hence, from the standpoint of algorithm selection in RS, MF, and SVD, clustering techniques and deep learning methods have been known to improve the accuracy of CF and context-aware RS algorithms. RS with the large number of dimensions representing user characteristics and attributes can cause the data volume to increase, as well as the data sparsity to increase exponentially. Hence, before sending the data to any algorithm, we need to apply dimensionality reduction techniques like PCA, SVD, or MF to reduce the feature set, making it effective for distance-based similarity comparisons.

In addition, graph neural networks have recently exceeded performance benchmarks to attain state-of-the-art accuracy on RS benchmark datasets. The concept of user context (e.g., mood improvement, educational value, age, gender, location, etc.) can also be brought into GNNs while recommending custom treatment plans and therapies. The objective function of training in RS algorithms can be customized based on the behavior of the algorithm by introducing "learning to rank" loss instead of vanilla using "mean squared error" losses.

Now that we have an understanding of how mental health RS will function, let us explore the different metrics, as illustrated in Figure 10-12, that would help us to evaluate the efficiency of the recommendation results.

Evaluation Metrics and Design Selection

The metrics for RS help us to evaluate the predictive, ranking, and behavioral properties of the recommenders, like diversity or novelty (evidentlyai.com, 2025). It will be good for us to take a look at the evaluation principles for ranking and recommendations.

- RS ranking systems have the main objective of concentrating on the relevancy of items selected while the results of the list of items are sorted by relevance. Relevancy reflects how likely a customer is going to select the recommendations by matching a user profile or query and can be measured by a simple binary score for each item.

- RS evaluation has an important parameter called the K parameter, which helps us to set the cutoff point for the top recommendations. Consequently, the predictions or the recommendations returned as user-item pairs need to have a selection of K parameters to determine the binary or graded relevance score as the ground truth.

- The role of predictive metrics like accuracy or precision at K, recall at K, or ranking metrics like NDCG, MRR, or MAP at K lies in evaluating the "correctness" of recommendations and interpreting how well the algorithm locates the most relevant items. Graded relevance scores can be better quantified by regression metrics, such as mean absolute error (MAE), mean squared error (MSE), or root mean squared error (RMSE), as they help measure the error between predicted and actual scores.

- The role of ranking metrics like NDCG, MRR, or MAP at K lies in evaluating the ranking quality: how well the algorithm can rank the items from more relevant to less relevant through sorting.

- The role of the behavioral metrics of serendipity, novelty, or diversity lies in evaluating the specific properties of the algorithm, such as how diverse or novel the recommendations are and the set of these metrics that are measured across each item, for each user, and on average per user.

- The role of **business metrics** like click-through rates or conversions during online evaluations lies in actually measuring how many customers are actually using the service or have shown interest toward the service.

Predictive Quality Metrics for REC

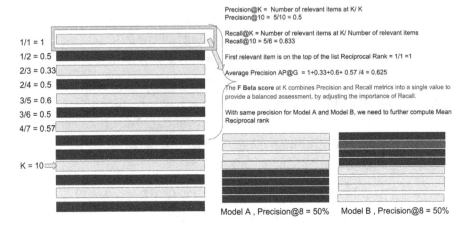

Figure 10-12. Designing hybrid REC for mental health

Relevant recommendations in the domain of mental health help the end users select the right clinics, insurance providers, and community networks by offering them their most preferred choices.

Let's now explore in depth some of the metrics suitable for binary relevance labels and understand the scenarios where they might be useful.

Table 10-1. *Metrics to evaluate recommender systems*

Metric	Use cases
Mean reciprocal rank (MRR)	This metric measures the effectiveness of an RS by picking the first relevant item in a ranked list of results, highlighting the importance of the top recommendation.
ARHR@k (average reciprocal hit rank at k)	This metric measures the average quality of recommendations to incorporate quality factors like ranking accuracy and recommendation order. It computes the reciprocal of the rank at which relevant items are found in the top k positions.
Normalized discounted cumulative gain (nDCG)	This is a ranking relevance metric, used to evaluate the quality of the recommendations in a list, where the position of highly relevant items is the most important factor under consideration.
Precision@k	This metric is used in RS to determine the proportion of relevant items among the top k recommendations by concentrating on their quality. It is useful when false positives pay a higher penalty price.
Recall@k	This metric used in RS emphasizes avoiding irrelevant items, when there is a higher penalty of missing a relevant recommendation. It further measures the model's efficiency in recommending the top K relevant items.
F1@K	This metric is used to highlight the precision of the top K recommendations, useful for scenarios like content curation or ranking tasks, that confirms the most relevant items are recommended within a limited set to end users.
Average recall@K	This metric stresses the comprehensive coverage of relevant content over precision. It is used to determine the proportion of top relevant items that an RS successfully retrieves.

(continued)

Table 10-1. (*continued*)

Metric	Use cases
Average precision@K	This metric is used to measure the effectiveness of an RS by quantifying the precision of the top k recommendations, highly useful for search engine result rankings or content playlists.
Mean average precision (MAP)	This metric is used to justify the importance on higher-ranked items by evaluating the quality of a ranked list of items in RS. It is useful in scenarios when relevance across recommended items varies, and it offers a comprehensive method to measure precision at different points in the ranking.
Recommendation diversity	This metric is used to measure the variety among the recommended items to reflect the spread of item categories to which each user has been exposed. It is measured by **intra-list diversity** that computes the average cosine distance between pairs of items.
Novelty	This metric is used to assess the uniqueness of the recommended items so as to reflect how the suggested items differ from popular ones.
Serendipity	This metric assesses how the RS recommends new diverse items, or items recently launched or discovered, which are beyond the user's typical expectations or preferred choices.
Popularity bias	This metric is used to measure favoritism of one item over the other, where popular items are often recommended. The share or coverage of the recommended items along with average recommendation popularity (ARP) and average overlap between items in the lists serve as important parameters to consider the popularity bias.

After gaining an insight into the possible RS metrics, let us now understand how privacy and accuracy are interdependent in providing the best-customized suggestions for an RS.

The Privacy-Accuracy Trade-Off in RS

RS comes with its unique challenges of achieving greater personalization and accuracy at the cost of sacrificing privacy. This happens because an RS won't be efficient in offering personalized mental health therapies and activities if it has little or no knowledge of the individual. Hence, a right privacy-personalization trade-off can balance the degree of customized recommendations and, at the same time, keep an account of the user usage history of the recommendations, which are low-risk attributes such as the type of therapies and activities chosen, duration, and specified time of the day when the user practices those activities. Hence, parents and care providers of children should be aware of the fact that to achieve greater accuracy with personalization, more privacy-sensitive information is required, and accordingly, they can set such settings in the app and provide their consent to share that personal information. Otherwise, RS in mental health in this digital era, with its capability to sense and share the footprint of EMRs, can become a threat, pose a high-risk situation, and raise ethical issues under legal and regulatory frameworks. The mental health apps, if they share the user's private sensitive information externally, leave them to security breaches and the commercialization of mental health data. In addition, there is no assurance of trust or confidence built between the psychologist, pharmacist, or hospital and the user in their therapeutic relationship, or any responsible legislation that can protect the user's rights to prevent loss of their private data (Song et al., 2024). Being aware of the possible external risk factors, we should also be aware that in the absence of precise private information and precise therapeutic suggestions, if the system offers wrong suggestions, it can impact the health and safety of the user in the long term. So as designers and data scientists of mental health solutions, we should be cognizant of the possible consequences and design the right app and system privacy controls to prevent any detrimental impact to the end user. For example, even if mistakenly a mental health app suggests youth experiencing mania related to bipolar disorder to consume alcohol to assist with sleeping, it causes a life-threatening experience.

Hence, it requires the correct understanding and potential use of the RS metrics to increase the number of relevant recommendations and control any irrelevant recommendations. One governance technique could be to attach a higher weight to the availability of personal data to increase the level of precision, given the cost of imprecision is high. Similarly, a higher threshold or level of acceptance can be placed on recommendations that are safe to be offered to increase the level of caution from a design perspective and leverage the recall metric to increase the proportion of relevant items recommended.

One other aspect of empowering users is to give them more rights through mental health apps to control their user history (Valentine et al., 2023).

Control of User History

Providing sufficient access privileges within the mechanisms of mental health apps allows users (youth or their parents) to detach their usage history from RS or delete or modify their past usage history on the digital mental health platforms they have registered with. Users experience specific benefits when they detach themselves from the RS.

- Ensures a loose coupling based on the user's interest in the RS, where a user experiencing a specific mental health problem has the flexibility of discontinuing a specific therapy or can express interest in a change in mental health interest to something new other than the ones recommended by RS.

note-taking during interviews to assessment and delivering therapy, can do a digital transformation by helping to address the shortage of mental healthcare services and problems with individual access to care.

Large-scale pretrained LLMs can be used widely for question answering language models; the greatest use of them is chatbots(Abbasian et al., 2024; Song et al., 2024). They are more efficient when they are trained on extensive unlabeled data that equips them to gain a more comprehensive language representation and understanding of the domain. It turns out to be more effective when it has been fed with possible datasets as well as unstructured information on mental health. Such models have better initialization of parameters, which acts as a better initialization point for the model and yields better generalization results. Further, pretraining with relevant information enables the achievement of effective regularization, thus controlling the risk of overfitting the model when the size of the pretraining dataset is small. As a result of instruction tuning (guiding the LLM to follow a protocol or rule to add the dataset to its long-term memory) and retrieval-augmented generation, LLMs are powered to perform better and demonstrate sufficient robustness in question answering tasks directed toward psychological counseling.

Figure 10-13 illustrates a typical architectural pipeline consisting of data sources, an AI processing engine, and a knowledge base triggering requests and responses through an LLM-powered chatbot for mental health (Abbasian et al., 2024; Mateti et al., 2024; Yuan et al., 2025).

A lot of research and studies are ongoing with retrieval-augmented LLMs, where a generator model is paired with a knowledge retriever to extract relevant knowledge from documents stored in a vectorized store. The knowledge retriever provides contextual relevancy, adding reliability and domain-specific explainability (the addition of knowledge graphs adds more power to explainability) to the responses, which is crucial in the mental health domain.

However, research results discovered the training methods may not be
the best in sensitive domains, where instruction fed into the LLMs does not
raise the model's performance to the expected level. This raises an ethical
question and concerns in the sensitive area of mental health, where the
model's question answering procedure is not capable of giving complete,
relevant answers for clinical questionnaires for depression and anxiety
(Abbasian et al., 2024; Lai et al., 2024). More research is underway in the
field of autoregressive language models, where the evaluation of the tree-
based learning prediction layer is presumed to improve safer responses for
mental health contexts.

LLM-enabled Chatbot System Architecture for Mental Health

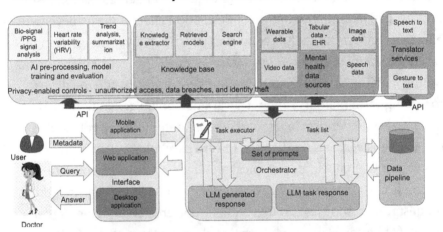

Figure 10-13. *LLM-powered chatbot requests responses*

LLM-Powered Mental Health Assistants

Examples of publicly available mental health LLMs include ChatCBPTSD,
Diagnosis of Thought Prompting, Mental-LLM, MentaLLaMA (LLaMA-2
based), ChatCounselor, ExTES-LLaMA (both LLaMA-based), and BBMHR
(BlenderBot-BST-based). However, the public cannot access certain LLMs

like MindShift, Psy-LLM, and LLM-Counselors (all GPT-3.5-based) (Shah et al., 2025; Song et al., 2024; Yuan et al., 2025).

OpenCHA is an example of a retrieval-augmented LLM, which is available to the public. It acts like a conversational agent to aid interactive systems assisting healthcare services by providing assistance and diagnosis. It helps analyze input queries by integrating essential information and problem-solving capabilities to offer personalized, context-aware, multilingual, and multimodal conversational responses. OpenAI has been leveraged to provide personalized nutrition-orientated food recommendation bots. On feeding OpenCHA with the user's longitudinal diabetes, electronic diabetes, and health records (EHRs), in addition to personal causal models, population models, Nutritionix information, and dietary guidelines obtained from the American Diabetes Association, OpenCHA can yield better-personalized responses on diabetes-related questions with daily meal choices, demonstrating better performance than the state-of-the-art GPT-4 model.

However, using LLMs for mental health applications comes with its own challenges, listed below:

- Availability of high computation speed and adaptable fine-tuning capabilities to integrate and pretrain diverse data sources comprising textual data, time series, tabular data, IoT data collected from the human body, and image and visual data. Having scalability features in the platform to perform deployments at scale makes it suitable for integration into large social media and mental health diagnostic platforms.

- Architecting LLMs so as to ensure efficient use of parameters and computational resources and enforce a balance between model size and performance, helping the model to achieve better results compared to larger models while using fewer resources.

- Developing an optimized, high-throughput, large-scale data processing real-time ingestion pipeline to feed in continuous data from wearable devices, mHealth applications, and existing mental health datasets of different languages. This will help the LLM to be more robust and personalized to send real-time alerts and bring in necessary interventions wherever needed.

- To enable LLMs to address intricate sequential tasks (such as helping them act as problem-solvers). In addition, LLMs should be equipped with sequential reasoning capabilities, personalized health history analysis, and data fusion.

- Equipping the LLM to better understand different languages and yield comprehensive language-based responses by detecting subtle cues and contextual indicators of depression from interviews and posts from social media platforms.

Let us now see a practical example of an AI-assisted chatbot, Psy-LLM (Lai et al., 2023), that gained popularity during COVID-19.

The global outbreak of COVID-19 increased mental stress and depression among the global population, which necessitated the requirement for a timely and professional mental health question answering chatbot system. Such an online psychological counseling chatbot named Psy-LLM emerged as popular, which functions as an AI-assistive tool built on the foundations of pretrained LLMs (with data fed from crawled psychological articles and a corpus of mental health supportive Q&A from mental health professionals). This Psy-LLM framework is built with the following functionalities to address urgent patient issues (such as individuals inclined to suicide) to reduce stress, anxiety, and depression:

- Equipped with a front-end tool to aid healthcare
  professionals in providing psychological consultation
  through immediate responses and suggestions of
  mindfulness activities.

- Scalable platform to handle the heavy patient load
  into the framework where the suggestive answers can
  reduce staff's workload at times when no free counselor
  staff (e.g., during off-hours or high-demand periods)
  are available.

Such an LLM foundational framework is evaluated with human
oversight (such as human participant assessments of response helpfulness,
fluency, relevance, logic, and overall quality) to compute valuable insights
on the performance of the LLM and to understand the disparities between
predicted and actual responses. In addition, the platform is benchmarked
on intrinsic metrics, such as perplexity, ROUGE-L, and Distinct-n metrics,
to measure the model's language generation quality, similarity to the
reference text, and diversity. Let us understand some of the intrinsic
metrics listed below that have been used to assess the model performance.

- **Helpfulness**: This metric is used to assess if the LLM-
  generated response is able to assist patients seeking
  psychological support.

- **Fluency**: This metric measures the degree of coherence
  and naturalness demonstrated in the LLM-generated
  response.

- **Relevance**: This metric measures how much the LLM-
  generated response matches directly related to the
  question put forward by the patient.

- **Logic**: This metric is used to scan the logical
  consistency and coherence of the meaning delivered in
  the LLM response.

Despite training the model with psychologically relevant data, mental health chatbot assistants can yield more human-centric responses when trained with data from specialized psychological experts that exhibit a touch of emotional understanding. In addition, having in-built, robust privacy protection mechanisms in data usage and transparency or explainability in model responses will increase its adoption and make it more compliant with regulatory and legal requirements and standards. Users would feel safe when approaching the chatbot for assistance when they know explicit user consent is sought for data collection and usage, sensitive user information is anonymized, strict data access controls are enforced, and harmful user queries or incoming request prompts are scanned and filtered. Even adding an explainability module would justify the chatbot's response to queries that would help to build trust and confidence. Having an automatic emotion recognition system would help to capture human emotions and gestures when prompted by answers by the assistants, which in turn would help to capture user feedback and trigger continuous learning and improvement.

Now we have a detailed understanding of LLM-based RS. Now, let us take a look at language-driven agents that can handle multiple systems by establishing proper collaboration across each of them (A. Zhang, Deng, et al., 2024).

Autonomous Collaborative Language Agents for Mental Health RS

The purpose of language-powered autonomous agents is to break down data silos and provide a complete, comprehensive summary of the patient record. It achieves such seamless coordination by establishing communication through a security-proven architecture to send and receive responses across different platforms. One example would be to obtain continuous heart rhythms of patients and alert a cardiologist

promptly to help him/her make a prompt decision and provide life-saving treatment. Autonomous agents can bring down manual tasks to automated scheduling, insurance verifications, and billing to seamlessly connect the provider and the patient in their hospital checkups. Autonomous agents have a great role to play in anonymizing genomic and clinical data to offer personalized treatment plans to individuals by aggregating the right information across different sources. Agentic mesh is gaining popularity in connecting 360-degree patient care services through real-time insights to offer treatments more efficiently by prioritizing the mental health of the public, particularly the youth, a most societal problem to handle.

Figure 10-14 illustrates the difference between how a gradient-based traditional RS algorithm would operate in contrast to an agent-based CF for mental health (A. Zhang, Deng, et al., 2024b; J. Zhang et al., 2023). It shows how the LLM-powered agent achieves collaboration and autonomous interactions to offer the best choices to the user.

Traditional vs Agent based Collaborative Filtering for RecSys

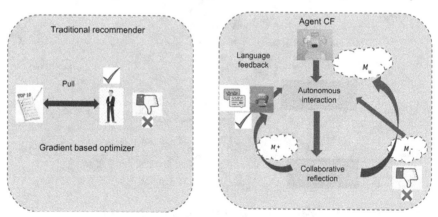

Figure 10-14. *Working functionality of a traditional gradient-based RS in comparison with agent-based RS operating on autonomous interactions and collaborations*

LLM-powered agents act as human proxies to leverage their decision-making abilities to simulate different human dialogues and actions, such as item clicks in the recommendation list. This helps AgentCF (J. Zhang et al., 2023) to exercise human behavior modeling and understand complex user-item relations through agent-based collaborative filtering. The modeling process first triggers user and item agents to interact autonomously at each step, which tries to optimize both the user and item agents together using collaborative learning. The learning environment created replicates a digital twin scenario where, in each step, the agents' digitized decisions and real-world interaction records are leveraged by both the user and item agent to review and adjust incorrect simulations to model their two-sided relations through collaborative learning and adjustments.

LLM-powered Agents in Mental Health RecSys

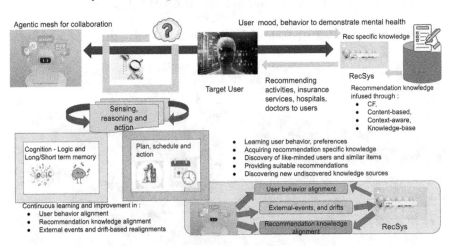

Figure 10-15. *LLM-powered agentic mesh to assist in recommendations from hospitals, doctors, and insurance providers*

This kind of digitized simulation board provides the flexibility of communicating the agent's preferences to other agents in succeeding sessions, facilitating a chain, and propagating varied interaction behaviors

through user-item, user-user, item-item, and collective interactions. As illustrated in Figure 10-15, the agents are seen to obtain user alignment and recommendation knowledge alignment to demonstrate the proposal of personalized behaviors through sensing, reasoning, and logic. The platform sets the foundation for next-generation user behavior simulation grounded by connecting respective agents to diverse knowledge sources (such as hospitals, doctors, insurance, and care providers). It further fosters an environment of continuous learning (to understand user behavior and preferences) and improvement through self-discovery of similar users and items.

Figure 10-16 further demonstrates how user profile (activity, conformity to items, diversity, and behavioral lifestyle patterns) along with provider profile (quality, popularity, and treatment summary) can be used by agents to recall past interests and plan for future actions by storing the data in a memory (two units: factual and emotional) (A. Zhang, Chen, et al., 2024a). The records are then retrieved from memory to trigger respective actions and measure the RS metrics to determine the user satisfaction rates.

Simulating User Behavior using LLM-powered Agents in Mental health RecSys

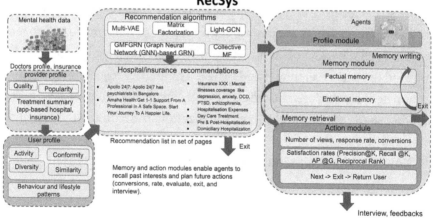

Figure 10-16. *Simulating patient, doctor, and insurance profiles to offer recommendation insights through agents*

Next-Generation Mental Health Products

Digital mental health platforms (DMH) are evolving and gaining popularity with more personalization options catered to individual needs (Matthews & Rhodes-Maquaire, 2024; Valentine et al., 2023). It is also the first step toward a scalable digitized solution to the global mental health crisis that impacts over 250,000 people worldwide. Amidst the crisis, what we need is quick and easy accessibility of the digitized treatment plans even to the patient residing in the remotest corner of the village. Conversational AI-based bots like Wysa have been able to deliver therapeutic support that addresses some of the underlying problems related to reach and accessibility and, at the same time, improve the assessment metrics and physical capabilities when users suffer from anxiety and depression symptoms (Song et al., 2024).

Many of the currently existing products allocate users to groups and are treated with similar therapies based on interactions, content, and messaging. The objective would be to obtain the right balance between personalization, human oversight, and control so that many of the engagements and interventions practiced face to face can be translated to digital space for unreached groups of society. What is specifically needed for the youth section is to create a knowledge base that can be leveraged by parents for the benefit of their children. To build the appropriate sensitive knowledge base, researchers and private companies researching digital mental health solutions should come and engage in collaborative efforts to increase the effectiveness of personalization, overcoming the existing limitations.

Now let us have the concluding thoughts from this chapter.

Conclusion

In this chapter, we have learned about the role of recommender systems for mental health apps and how specific apps can have in-built onboarding-based, conversation-based, or random recommendation units as plugins embedded within them. Here, we went along to understand what kind of digitalized custom therapy and counseling sessions can be offered to end users with examples of a few mental health applications that exist in the market today. We further had a deep dive into different types of RS, starting with CF (alongside a few algorithms such as KNN, SVD, SV++, and MF), content-based, knowledge-based, context-based, and a combination of multiple RS, leading to the design of hybrid RS. We also developed an understanding of how to evaluate such RS with different metrics and the trade-off involved in having accuracy as well as privacy built in to respect user rights and ensure proper governance as per regulations. In the end, we also had a few highlights on how GenAI can play a critical role in building LLM-based recommenders, along with a few specific examples of mental health-customized LLMs available. We went one step further and learnt about current and future research areas that use autonomous agents and how important they are in shaping the whole ecosystem of patient care, which includes hospitals, doctors, insurance companies, and care providers.

We now have a solid understanding of AI-based recommender systems for mental health. Along with that, we saw few challenges for AI-based systems in fulfilling all aspects of Responsible AI. Now let us explore in-depth more on different critical areas of responsible AI, in our next chapter, Chapter 11, "Toward a Responsible and Trustworthy AI System: Navigating Current Challenges and Pathways."

References

Aakanksha NS. (2020). Recommender Systems: Matrix Factorization using PyTorch. *TDS Archive.* https://medium.com/data-science/recommender-systems-matrix-factorization-using-pytorch-bd52f46aa199

Abbasian, M., Azimi, I., Rahmani, A. M., & Jain, R. (2024). *Conversational Health Agents: A Personalized LLM-Powered Agent Framework* (No. arXiv:2310.02374). arXiv. https://doi.org/10.48550/arXiv.2310.02374

Chaturvedi, A., Aylward, B., Shah, S., Graziani, G., Zhang, J., Manuel, B., Telewa, E., Froelich, S., Baruwa, O., Kulkarni, P. P., Ξ, W., & Kunkle, S. (2023). Content Recommendation Systems in Web-Based Mental Health Care: Real-world Application and Formative Evaluation. *JMIR Formative Research, 7,* e38831. https://doi.org/10.2196/38831

evidentlyai.com. (2025). *10 metrics to evaluate recommender and ranking systems.* Evidentlyai.Com. https://www.evidentlyai.com/ranking-metrics/evaluating-recommender-systems

Lai, T., Shi, Y., Du, Z., Wu, J., Fu, K., Dou, Y., & Wang, Z. (2023). *Psy-LLM: Scaling up Global Mental Health Psychological Services with AI-based Large Language Models* (No. arXiv:2307.11991). arXiv. https://doi.org/10.48550/arXiv.2307.11991

Lai, T., Shi, Y., Du, Z., Wu, J., Fu, K., Dou, Y., & Wang, Z. (2024). Supporting the Demand on Mental Health Services with AI-Based Conversational Large Language Models (LLMs). *BioMedInformatics, 4*(1), Article 1. https://doi.org/10.3390/biomedinformatics4010002

Lewis, R., Ferguson, C., Wilks, C., Jones, N., & Picard, R. W. (2022a). Can a Recommender System Support Treatment Personalisation in Digital Mental Health Therapy? A Quantitative Feasibility Assessment Using Data from a Behavioural Activation Therapy App. *Extended Abstracts of the 2022 CHI Conference on Human Factors in Computing Systems,* 1–8. https://doi.org/10.1145/3491101.3519840

Lewis, R., Ferguson, C., Wilks, C., Jones, N., & Picard, R. W. (2022b). Can a Recommender System Support Treatment Personalisation in Digital Mental Health Therapy? A Quantitative Feasibility Assessment Using Data from a Behavioural Activation Therapy App. *Extended Abstracts of the 2022 CHI Conference on Human Factors in Computing Systems*, 1–8. https://doi.org/10.1145/3491101.3519840

Mateti, A., N, S., E, B., & Sai, S. V. (2024). *A KNOWLEDGE-BASED RECOMMENDATION SYSTEM THAT INCLUDES SENTIMENT ANALYSIS AND DEEP LEARNING. 04*, 156–164.

Matthews, P., & Rhodes-Maquaire, C. (2024). Personalisation and Recommendation for Mental Health Apps: A Scoping Review. *Behaviour & Information Technology, 0*(0), 1–16. https://doi.org/10.1080/0144929X.2024.2356630

Mazlan, I., Abdullah, N., & Ahmad, N. (2023). Exploring the Impact of Hybrid Recommender Systems on Personalized Mental Health Recommendations. *International Journal of Advanced Computer Science and Applications (IJACSA), 14*(6), Article 6. https://doi.org/10.14569/IJACSA.2023.0140699

Shah, S. M., Gillani, S. A., Baig, M. S. A., Saleem, M. A., & Siddiqui, M. H. (2025). Advancing depression detection on social media platforms through fine-tuned large language models. *Online Social Networks and Media, 46*, 100311. https://doi.org/10.1016/j.osnem.2025.100311

Siriaraya, P., Suzuki, K., & Nakajima, S. (2019). *Utilizing Collaborative Filtering to Recommend Opportunities for Positive Affect in daily life.* 2–3. http://ceur-ws.org/Vol-2439/2-paginated.pdf

Song, I., Pendse, S. R., Kumar, N., & Choudhury, M. D. (2024). *The Typing Cure: Experiences with Large Language Model Chatbots for Mental Health Support* (No. arXiv:2401.14362). arXiv. https://doi.org/10.48550/arXiv.2401.14362

Valentine, L., D'Alfonso, S., & Lederman, R. (2023). Recommender systems for mental health apps: Advantages and ethical challenges. *AI & SOCIETY, 38*(4), 1627–1638. https://doi.org/10.1007/s00146-021-01322-w

Yuan, A., Garcia Colato, E., Pescosolido, B., Song, H., & Samtani, S. (2025). Improving Workplace Well-being in Modern Organizations: A Review of Large Language Model-based Mental Health Chatbots. *ACM Trans. Manage. Inf. Syst., 16*(1), 3:1-3:26. https://doi.org/10.1145/3701041

Zhang, A., Chen, Y., Sheng, L., Wang, X., & Chua, T.-S. (2024a). *On Generative Agents in Recommendation* (No. arXiv:2310.10108). arXiv. https://doi.org/10.48550/arXiv.2310.10108

Zhang, A., Deng, Y., Lin, Y., Chen, X., Wen, J.-R., & Chua, T.-S. (2024b). Large Language Model Powered Agents for Information Retrieval. *Proceedings of the 47th International ACM SIGIR Conference on Research and Development in Information Retrieval*, 2989–2992. https://doi.org/10.1145/3626772.3661375

Zhang, J., Hou, Y., Xie, R., Sun, W., McAuley, J., Zhao, W. X., Lin, L., & Wen, J.-R. (2023). *AgentCF: Collaborative Learning with Autonomous Language Agents for Recommender Systems* (No. arXiv:2310.09233). arXiv. https://doi.org/10.48550/arXiv.2310.09233

SECTION 4

Ethical Design: Building an Efficient and Ethical Platform

CHAPTER 11

Toward a Responsible and Trustworthy AI System: Navigating Current Challenges and Pathways

Artificial intelligence (AI) is rapidly transforming mental healthcare, offering new possibilities for diagnosis, personalized treatment, and improved access through tools like chatbots, predictive analytics, and remote therapy. However, as these technologies become more integrated into healthcare, ethical challenges such as data privacy, algorithmic bias, transparency, accountability, and informed consent have emerged as critical concerns. These challenges not only impact patient safety and autonomy but also influence public trust in digital mental health solutions.

In light of these issues, there is a growing demand for "responsible and trustworthy AI systems" that prioritize ethical design, transparency, fairness, and accountability. Responsible AI focuses on creating systems that are transparent, fair, and accountable governing their ethical,

© Sharmistha Chatterjee, Azadeh Dindarian, Usha Rengaraju 2025
S. Chatterjee et al., *Revolutionizing Youth Mental Health with Ethical AI*,
https://doi.org/10.1007/979-8-8688-1186-9_11

social, and legal implications that minimizes risks, while "trustworthy
AI" emphasizes technical implementations of AI systems to ensure
reliability, safety, explainability, and transparency of the resultant AI
systems to increasingly gain confidence and trust of end users and provide
social acceptance. This chapter explores how these principles intersect,
highlighting the challenges and opportunities of developing trustworthy AI
systems for mental health.

This chapter delves into the design considerations essential for
developing responsible AI platforms that train models and deliver
unbiased, fair predictions. It emphasizes the importance of establishing
audit and performance reporting processes to maintain transparency
and accountability, as well as the need for comprehensive data collection
and design strategies to ensure ethical onboarding onto AI platforms.
Additionally, it demonstrates different technical methodologies to
judge the social and legal aspects and assess the practical effectiveness
of AI models in mental healthcare, shedding light on their real-world
applications and potential limitations.

The chapter also examines the ethical complexities introduced by
"generative AI (GenAI)" and "agentic systems," which have the potential
to revolutionize mental health support through advanced conversational
agents and predictive analytics. However, these technologies also present
challenges related to decision-making autonomy, bias, and accountability.
Further, this chapter provides a practical example, to view the fairness
metrics, while predicting depression in youth in the below-mentioned
code listing.

By navigating these challenges, the chapter contributes to the creation
of a responsible AI ecosystem that fosters trust, safety, and ethical
compliance in digital mental health solutions.

By addressing these critical aspects, this chapter sets the stage for
creating ethically sound and reliable AI systems that enhance mental
healthcare. It provides readers with a holistic understanding of the current

limitations of AI systems, including both traditional AI and GenAI, and illustrates how responsible AI practices can foster trust, authenticity, and ethical care in digital mental health solutions for diverse populations.

In this chapter, these topics will be covered in the following sections:

- Why is ethics becoming important for AI in mental health?

- Responsible AI guidelines to address current limitations of AI in mental health

- Sustainable mental health solutions with GenAI and agentic systems

Why Is Ethics Becoming Important for AI in Mental Health?

The increasing integration of artificial intelligence (AI) in digital mental health diagnosing, monitoring, and treating mental health conditions, ensuring ethical implementation, is vital to prevent harm, uphold patient rights, and maintain trust in digital healthcare systems. In this section of the book, we explore why ethics is crucial in digital mental health, addressing key areas such as data privacy, algorithmic bias, informed consent, accountability, and the clinical implications of AI-driven mental health interventions.

The use case of AI spans from enhancing diagnostic capabilities, personalizing treatment plans, and expanding access to care through chatbots and predictive analytics, to remote therapy applications

Given AI's growing role in mental health, ethical considerations must remain a top priority to protect patient well-being. Fiske et al. (2019) analyzed the ethical and social aspects of AI applications in psychiatry, psychology, and psychotherapy, highlighting key concerns such as harm prevention, data ethics, and regulatory gaps. They found a lack of clear guidelines for AI development, clinical integration, and healthcare professional training, as well as the risk of AI replacing traditional services, potentially worsening health inequalities. Specific challenges include risk assessment, referrals, patient autonomy, and transparency in AI-driven decision-making. The study also raises concerns about the long-term impact of AI on perceptions of illness and the human experience, emphasizing the need for careful oversight and ethical governance.

These AI-driven tools are designed to assist in predicting suicide risks, diagnosing depression, and delivering cognitive behavioral therapy (CBT) (Appleton, 2024). Despite the rapid advancements in AI for mental health, its integration into clinical settings remains relatively underdeveloped.

Ethical concerns surrounding digital mental health arise from AI systems' interactions with sensitive user data, their potential for bias, and their impact on human autonomy.

The key ethical concerns are

- **Data privacy and security**: Assuring users their private and confidential information is not leaked outside due to their mental health diagnosis, treatment, or research purposes.

- **Algorithmic bias and fairness**: Lack of diverse data from diverse sources of mental health diseases and effective treatment processes results in bias in different AI-based predictive algorithms and the results show biasedness toward majority classes.

- **Transparency**: Once AI-based algorithms predict mental disorder, the predicted result should be able to explain and justify what data points have resulted in this prediction.

- **Informed consent and user autonomy**: Data collected
  from patients on their mental health status and
  behavior should be gathered with due consent from
  the user, when the user signs an agreement to share the
  data for their treatment and/or for research purposes.

- **Accountability and liability**: Mental health AI systems
  and processes should be developed, deployed, and
  utilized such that companies developing mental health
  apps owe responsibility for their outcomes. In addition,
  healthcare systems and apps relying on AI for offering
  mental health treatment should also be liable to the
  legal consequences of AI predictions—good or bad.

One major concern is the lack of robust ethical design in some
solutions, raising questions about data privacy, transparency, and
accountability. Researchers are increasingly advocating for the integration
of ethical considerations and regulatory frameworks from the early stages
of AI development to ensure these tools are both effective and responsible
in supporting mental healthcare.

The reliability and accuracy of AI-driven mental health tools have
become central topics in recent years. These tools are typically trained
on extensive datasets to identify patterns associated with various
mental health conditions. However, a significant concern is that these
datasets may lack diversity, leading to biases that result in misdiagnosis
(Arora et al., 2023; Straw & Callison-Burch, 2020). For instance, cultural
expressions of distress can be misinterpreted by AI, causing incorrect
assessments. A narrative review highlighted that AI tools might not
adequately account for cultural factors, leading to misinterpretations of
patients' experiences and needs (Thakkar et al., 2024).

Moreover, studies have indicated that AI models used in mental health screening may carry biases based on gender and race. For example, some AI tools may misinterpret the ways people of different genders and races communicate, leading to potential misdiagnoses or inappropriate treatment recommendations (Yang et al., 2024).

The rapid advancement of artificial intelligence (AI) in mental healthcare has outpaced the development of comprehensive regulatory frameworks, leading to significant challenges in data privacy, security, and informed consent. Unlike traditional healthcare settings, AI-driven mental health tools operate within complex digital environments where existing regulations may be insufficient (Mennella et al., 2024). This regulatory gap raises concerns about the protection of sensitive personal information and the ethical use of data. For instance, many mental health applications collect and share user data without explicit consent, potentially compromising user privacy and trust (Solis, 2024).

The accountability and safety of AI-driven solutions in healthcare, particularly in mental health, represent critical ethical challenges that are yet to be fully addressed. Unlike traditional healthcare models where human professionals are responsible for diagnoses and treatment decisions, AI-based tools often function autonomously or with minimal human oversight (Habli et al., 2020; Smith, 2021; Stade et al., 2024).

In a 2021 study, Boucher et al. (2021) examined the role of AI-based chatbots in digital mental health interventions. They highlight significant challenges, particularly the chatbots' limited ability to understand complex human contexts. This limitation can result in responses that are irrelevant or inappropriate, thereby undermining the therapeutic relationship and potentially leading to inadequate support during crises .

Addressing these issues requires the establishment of robust policies and guidelines that ensure the ethical deployment of AI technologies in mental health, safeguarding user data and upholding patient rights.

With the rise of "agentic AI systems" in mental healthcare, establishing clear responsibilities and safety standards for all stakeholders is essential. These systems, as described by Yonadav et al. (2024), function with minimal supervision while pursuing complex goals, raising critical concerns about "accountability, ethical oversight, and risk mitigation." Without robust safeguards, they pose risks such as "misdiagnosis, inadequate crisis intervention, and unintended emotional harm." It is important for us to understand what an agentic AI and the complexity of its operation is, that would make us extra cautious to be careful in designing ethical agentic systems.

Now let us see in Figure 11-1 how different independent AI agents can be used in a variety of contexts. As AI agents can operate independently and collaboratively to pass messages between them, it becomes essential for system designers and application developers to be aware of the ethical considerations, as it comes with its own limitations. For example, an agent assisting an end user with queries on chat can respond to end users with available information and at the same time pass on the queries to a mental health assessment agent to assess the mental health conditions of the user at that moment and raise a panic alarm if needed. Collaborative multiagent interaction should have an ethical governance framework so that inconsistent biased data and predictions are not passed to multiple agents to avoid cascading failures.

Application of LLM Solutions for Mental Health

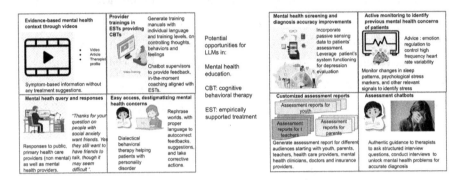

Figure 11-1. *Mental health treatment using LLMs in wide variety of applications starting from education, behavioral therapy, and supported treatments*

The degree of "agenticness" in a system is defined by Yonadav et al. (2024) as "the degree to which a system can adaptably achieve complex goals in complex environments with limited direct supervision."

It consists of four key components:

- **Goal complexity**: The breadth and difficulty of the AI's objectives. A system capable of answering analytical questions across diverse fields (e.g., programming and law) exhibits greater goal complexity than a simple text classifier limited to categorization.

- **Environmental complexity**: The range and intricacy of conditions in which the AI operates. For instance, an AI that excels at multiple board games demonstrates higher environmental complexity than one restricted to chess.

- **Adaptability:** The ability to handle novel or unexpected situations. Human customer service agents, for example, outperform rule-based chatbots in addressing unforeseen queries.

- **Independent execution:** The degree to which an AI functions with minimal human input. A level 3 autonomous vehicle, which can drive without constant human control, has greater independent execution than traditional cars.

Understanding these four dimensions helps assess the autonomy and reliability of AI in diverse applications. First and foremost, agentic AI systems depend a lot on LLM pretrained models. The inherent limitations of large language models (LLMs), like explainability, transparency, and presence of hallucinations, often underpin these systems and further complicate their reliability and effectiveness in sensitive healthcare applications.

The uncertainties surrounding AI agents pose critical challenges in mental health, emphasizing the need for robust governance frameworks. Adequate evaluation of agentic AI systems requires identifying potential failure modes, many of which may not be entirely foreseeable through testing.

Defining best practices and responsibility frameworks for developers, clinicians, policymakers, and users is crucial for the safe and ethical integration of AI-driven mental health tools. However, operationalizing these safeguards remains a challenge, requiring continuous refinement before standardization. AI ethics play a pivotal role in preventing unintended consequences and guiding the development of governance frameworks to regulate agentic AI at scale (Yonadav et al., 2024).

Agentic AI systems are distinct in their ability to autonomously make decisions over time without predefined behavior, operating in complex environments with limited direct human input. While often imagined as helpful assistants, their real-world deployment demands rigorous safeguards to ensure they enhance, rather than compromise, mental healthcare.

After having gained an insight into the complexity of agentic AI systems, let us understand the limitations of them, especially when they are powered by LLMs. The following sections explore the limitations of large language models (LLMs), which play a key role in developing agentic AI systems. Current limitations include translation challenges, difficulties in evaluation, potential risks, and a lack of interpretability. These challenges affect their reliability, adaptability, and ethical considerations, especially in critical areas like mental health.

Let us now explore the limitations of LLM-based solutions, along with how research and design best practices can guide us to more ethical adoption and give us guidance for future directions.

The Current Limitations of Large Language Models (LLMs)

Despite advancements in natural language processing (NLP), several limitations hinder its effectiveness in mental health applications, including (Ehud Reiter, 2024)

- **Translation issues**: Creating multilingual datasets is resource-intensive. Using large language models (LLMs) like GPTt-3.5 for translation without expert human oversight can result in subtle information loss and misinterpretation of cultural nuances.

- **Evaluation difficulties**: Automatically assessing LLM performance is complex. Current methods rely on detecting assigned labels in the model's output, defaulting to minimal classification when labels are missing, which may affect accuracy. For example, in few-shot learning, LLM cannot self-detect and annotate correct labels, when example labels are not specified.

- **Public datasets**: Publicly available mental health datasets must be handled cautiously. These tools should support healthcare professionals rather than replace clinical diagnoses. Poor-quality datasets can lead to unreliable predictions and ethical concerns.

- **Lack of interpretability**: NLP algorithms often function as "black boxes," making it difficult to explain their predictions. This lack of transparency can reduce trust and hinder their adoption in clinical settings.

It is crucial to acknowledge the risks and ethical challenges associated with using large language models (LLMs) in supporting mental health.

Once we become cognizant of the limitations, we should take proper risk mitigation actions that could lead to ethical adoptions of LLM-based systems and make agentic AI both responsible and trustworthy.

Ethical Adoptions and Guidelines

According to Lawrence et al. (2024), the responsible implementation of LLMs in mental healthcare requires adherence to the following principles:

- **Proper training and performance**: LLMs should be utilized for mental health tasks only if they have been adequately trained and proven to perform effectively in those areas.

- **Promoting mental health equity**: LLMs must be designed to enhance equity in mental healthcare, ensuring fair access and unbiased support for all users.

- **Privacy and confidentiality**: It is crucial to prioritize user privacy and protect the confidentiality of sensitive mental health information when using LLMs.

- **Informed consent**: users should be clearly informed about their interactions with mental health LLMs, and explicit consent must be obtained before engagement.

- **Risk management and response**: LLMs should be capable of accurately identifying and responding to mental health risks in a safe and supportive manner.

- **Scope of competency**: LLMs must operate strictly within their validated areas of expertise, avoiding tasks beyond their proven competence.

- **Transparency and explainability**: LLMs should be transparent in their functionality and able to provide understandable explanations for their outputs.

- **Human oversight and feedback**: Continuous human supervision and feedback are essential to monitor LLM performance, ensuring ethical practices and making necessary adjustments.

We need to be aware of the above risks and remain extra cautious in enabling LLM models for large public use for assisting mental health patients, as the above bullets highlight the significant challenges faced by large language models (LLMs) in deploying reliable solutions for mental health applications. Other challenges include perpetuating inequalities, disparities, and stigma, failing to provide mental health services ethically, insufficient reliability, inaccuracy; lack of transparency and explainability and neglecting to involve humans.

To enhance the reliability and ethical use of NLP in mental health, several key areas need further development, such as

- **Improving data quality**: Ensuring diverse, unbiased, and representative datasets is essential. Continuous refinement and feedback mechanisms can improve model accuracy and inclusivity.

- **Establishing ethical guidelines and regulations**: Clear policies must address privacy, informed consent, data protection, and discrimination prevention to promote responsible AI deployment.

- **Enhancing interpretability**: Developing explainable AI techniques will provide insights into decision-making processes, enabling clinicians to validate model predictions with greater confidence.

- **Encouraging collaboration and interdisciplinary research**: Cooperation between AI researchers, mental health professionals, and policymakers is vital to align NLP advancements with ethical and practical considerations in healthcare.

By addressing these challenges and focusing on ethical, transparent, and high-quality AI development, NLP can become a valuable tool for mental healthcare without compromising safety, accuracy, or trust.

Adolescents are increasingly turning to AI chatbots like ChatGPT (an AI-based language model that uses natural language processing (NLP) and conversational AI) for companionship and mental health support (Imran et al., 2023). However, ChatGPT has its own pros and cons. These pros and cons are summarized in Table 11-1 (redresscompliance.com, 2023).

Table 11-1. *Pros and cons of ChatGPT*

Pros	• Versatile communication: ChatGPT excels in generating human-like text, making it useful for diverse applications.
	• Time-saving: it automates and speeds up writing, data processing, and information retrieval tasks from large volume of documents.
	• Accessibility: provides easy access to information and can assist users with disabilities.
	• Continuous learning: can regularly learn from user feedbacks to update responses and improve accuracy and relevancy of consultations and therapies provided to end users.
Cons	• Lacks context awareness: sometimes misunderstands complex queries or context-specific nuances.
	• Potential for misinformation: can generate incorrect or outdated information.
	• Dependence risk: overreliance on AI for tasks can hinder human skill development.
	• Privacy concerns: questions about how user data is handled and stored. User consent or permission is not guaranteed to be taken from users before getting to know their mental health conditions.

As we can see from the table above, the integration of AI chatbots like ChatGPT into mental health support introduces several ethical concerns that warrant careful consideration. These ethical concerns are given below:

> **Misinformation and accuracy**: ChatGPT can generate incorrect or misleading information. It cannot verify facts and might produce plausible-sounding but false responses, which can be problematic, especially in contexts requiring precise and reliable information, such as healthcare or legal advice.

Bias and fairness: The model learns from vast amounts of text data, including biased or prejudiced content. As a result, ChatGPT may inadvertently generate biased or discriminatory responses. Ensuring fairness and reducing biases in its outputs is a significant ethical challenge.

Privacy concerns: ChatGPT processes large volumes of text, including sensitive or personal information. Concerns about how this data is used and stored and the potential for misuse or privacy breaches exist. Users need to be cautious about sharing personal information with AI models.

Dependency and dehumanization: Relying heavily on ChatGPT for tasks like customer service or companionship can reduce human interaction. This dependency might impact social skills and the quality of human relationships. Additionally, it raises questions about the dehumanization of services and the erosion of jobs that require human empathy and understanding.

Manipulation and misuse: ChatGPT's ability to generate persuasive and coherent text can be misused for malicious purposes. It can create fake news, spread propaganda, or engage in fraudulent activities. This potential for manipulation necessitates stringent guidelines and monitoring.

Accountability: when ChatGPT generates harmful or inappropriate content, determining accountability can be challenging. Developers, users, and the organizations deploying the

technology are responsible. Clear guidelines
and ethical frameworks are needed to address
accountability issues.

It is also crucial to emphasize that sustaining user engagement in digital mental health platforms that rely on recommender systems remains a significant challenge (Valentine et al., 2023). These systems heavily depend on personal data, raising serious ethical concerns. Here, we would get to know three key ethical challenges identified in recommender systems:

1. Lack of explainability, making it difficult for users to understand how recommendations are generated

2. The privacy-personalization trade-off, where improving recommendation quality often comes at the cost of user privacy

3. Control over app usage history data, which raises concerns about data ownership, security, and informed consent

The ethical implications of using ChatGPT for mental health challenges extend beyond accuracy and fairness to encompass privacy and confidentiality. As AI systems become more deeply integrated into daily interactions, the risk of users unknowingly sharing sensitive information, such as patient records, proprietary data, or confidential source code, continues to grow (Li, 2023; Yang et al., 2023).

Without adequate safeguards, this data could be collected, stored, or even retrieved later, raising concerns about data security and unauthorized access.

To mitigate these risks, organizations and individuals must implement clear acceptable use policies that define strict guidelines on AI interaction, explicitly prohibiting the entry of confidential or personally identifiable information. These policies should outline user responsibilities and set technical restrictions to prevent the misuse of AI tools in sensitive contexts.

Table 11-2 provides an overview of the key privacy and confidentiality risks associated with AI in digital mental health applications, along with recommended security safeguards. This information, adapted from Li (2023) and J. Yang et al. (2023), offers practical solutions to protect users and ensure ethical AI deployment.

Table 11-2. *Security risk, its implications, and safeguard mechanisms*

Security risk	Description	Security safeguard solutions
Data privacy and confidentiality	AI tools may collect and store sensitive information, such as patient data or proprietary business data, which could be accessed by unauthorized parties.	Implement strong data encryption and anonymization techniques to protect sensitive information.
Unauthorized data access	AI-generated data may be retrieved or exploited by external users if proper safeguards are not in place.	Restrict AI access to sensitive data and enforce strict user authentication protocols.
Data retention and storage risks	AI systems may retain user inputs indefinitely, increasing the risk of exposing confidential information in future interactions.	Define clear policies on data retention and allow users to delete stored interactions if needed.
Lack of user awareness	Users may inadvertently share sensitive data due to a lack of awareness regarding AI data retention policies and security risks.	Educate users on the risks of sharing confidential data and implement AI use policies within organizations.

(continued)

Table 11-2. (*continued*)

Security risk	Description	Security safeguard solutions
Misuse of AI-generated outputs	AI-generated content can be manipulated or misused, leading to misinformation, privacy violations, or ethical breaches.	Introduce monitoring systems to detect and prevent misuse of AI-generated outputs.
Regulatory and compliance risks	AI applications must adhere to industry regulations (e.g., GDPR, HIPAA) to ensure compliance with data protection laws.	Ensure AI compliance with industry regulations and continuously update policies to align with legal requirements.

This section highlighted the crucial role of ethics in AI-driven mental healthcare. AI is reshaping mental healthcare, making diagnosis more precise, treatment more personalized, and support more accessible. However, as these technologies become more integrated into healthcare, ethical concerns must be addressed to protect patients' well-being and ensure responsible use. Privacy and security are major concerns, as AI tools handle sensitive personal data. Without strict safeguards, there is a risk of data breaches, unauthorized access, and misuse of patient records, potentially undermining trust in digital mental health solutions.

Another challenge is bias and fairness, and AI systems learn from existing data, which can sometimes reinforce societal biases. If not carefully managed, this could lead to inaccurate diagnoses and unequal treatment, particularly for underrepresented communities. Transparency and accountability are also crucial, as AI-driven tools often lack the explainability of human decision-making, making it difficult to assign responsibility when something goes wrong. Additionally, informed consent and user autonomy must be upheld, ensuring people fully understand how AI systems work, what data they collect, and their limitations.

To ensure AI truly benefits mental healthcare, ethical considerations must be at the forefront of its development. This includes implementing clear guidelines, regulatory oversight, and fairness-driven AI models. Balancing innovation with ethical compliance presents a significant challenge, as overregulation may stifle technological advancements while underregulation could compromise patient safety and trust.

In the next section, we'll explore responsible AI practices designed to address these challenges and create a more trustworthy and effective digital mental health landscape.

Responsible AI Guidelines to Address Current Limitations of AI in Mental Health

To address the ethical challenges associated with AI in mental healthcare, it is essential to establish comprehensive responsible AI guidelines.

These guidelines should be collaboratively developed by technology developers, researchers, healthcare professionals, policymakers, and ethicists among others to ensure a holistic approach.

In this section, we will explore key aspects of these guidelines as they pertain to the scope of this book.

In a policy brief, Stanford scholars Stade et al. (2024) classify the integration of clinical large language models (LLMs) into psychotherapy into three main stages, which are given below:

1. **Assistive ("machine in the loop")**: LLMs that assist clinical providers and researchers by performing low-level, concrete, and low-risk tasks, such as conversing with patients to collect information about their symptoms

2. **Collaborative ("human in the loop")**: LLMs that provide treatment suggestions for psychotherapists to review, such as producing an overview of a person's symptoms and experiences and curating a list of therapy exercises from which the provider can select

3. **Fully autonomous**: LLMs that perform a full range of clinical skills and interventions without direct oversight from a provider, such as conducting assessments, presenting feedback, selecting an appropriate intervention, and delivering a course of therapy.

In addition, several promising applications of clinical large language models (LLMs) are emerging in psychotherapy, focusing on assistance and collaboration (see 1 and 2 above) (Stade et al., 2024). These include automating administrative tasks like session transcription and chart reviews, evaluating treatment fidelity by assessing adherence to evidence-based practices, providing feedback on therapy assignments with real-time clarifications, and supporting supervision and training by offering corrections and suggestions to trainees. These innovations aim to personalize treatment approaches in behavioral healthcare.

However, it is clear that missteps or ethical violations by these AI systems can cause significant harm to individuals and erode public trust in behavioral health services

To bridge the gap within the frameworks for evaluating the specific risks of AI mental health applications, Stade et al. (2024) introduced the "READI (Readiness for AI Deployment and Implementation) framework," designed to evaluate AI applications in mental health. Grounded in the core principles of "transparency and user autonomy," the READI framework outlines six key criteria: Safety, Privacy/Confidentiality, Equity, Effectiveness, Engagement, and Implementation. These criteria assist individuals and organizations in making informed decisions about the suitability and successful integration of specific AI tools into mental healthcare. Further they recommend applying the READI criteria to assess new large language models (LLMs) and other generative AI technologies before their widespread clinical use and to continue evaluations postdeployment, as both the technology and its application contexts can evolve rapidly. These six criteria from the illustrations of READ framework are given below:

1. **Safety**: Application prevents dangerous human behaviors and is "healthy" itself, that is, does not exhibit inflammatory or extreme traits.

2. **Privacy/confidentiality**: Application keeps patient information private and confidential, that is, does not disclose health information without patient authorization and allows individuals to access their health information.

3. **Equity**: Application is unbiased in its communication, engagement, and effectiveness; is equally usable across all demographic groups; and is culturally responsive.

4. **Engagement**: Application is appropriately engaging (neither too much nor too little), with engagement levels determined by patients' individual needs.

5. **Effectiveness**: Application integrates clinical science principles and is clinically effective, that is, decreases symptoms and functional impairment and increases well-being and quality of life.

6. **Implementation**: Application integrates well into clinical practice, existing technologies, and workflows and is cost-effective.

Let us first take a look at the privacy features from the below figure. Figure 11-2 illustrates the guardrail features at each state, when mental health knowledge articles are provided to end users (MANIKA, 2024). This ensures no PII information is stored as well as the output is in sync with user expectations. In addition, from regulatory standpoint, we need to validate that the generated content meets compliance and any automation activity suggested by the bot meets standard guidelines, allowing the user to accept or reject.

Responsible AI Guardrails in Retrieval Augmented Generation using LLMs

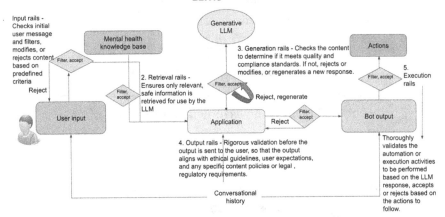

Figure 11-2. *Mental health applications with responsible AI privacy and security filters at all stages adapted from (MANIKA, 2024)*

With the understanding of privacy guardrails, let us now understand how the READI framework encompasses its key principles and evaluated the performance of AI-based mental health solutions after being deployed. Figure 11-3 further illustrates safety, privacy, equity, and effectiveness of the implemented solutions by tracking user engagements (Stade et al., 2024).

Readiness Framework Key Principles

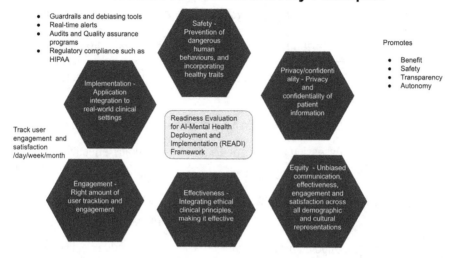

Figure 11-3. *Prime principles of readiness framework to meet compliance and ensure safety deployment of model usage by tracking user engagement*

Let us now explore other validation frameworks available for healthcare to ensure system designers and developers are able to follow them, without exposing the solutions at risk.

The BSI (2023) has introduced BS 30440:2023, titled "Validation framework for the use of artificial intelligence (AI) within healthcare." This standard establishes a structured and auditable framework for developing, validating, deploying, and monitoring AI technologies in healthcare. Its primary objective is to ensure AI systems are safe, effective, and equitable before clinical implementation.

BS 30440:2023 defines a five-phase lifecycle for AI in healthcare, covering inception, development, validation, deployment, and monitoring; the description and the key evaluation criteria are summarized in Table 11-3.

Table 11-3. *Five-phase lifecycle for AI in healthcare defined by BS 30440:2023, their description, and key suggested evaluation criteria*

	Phase	Description	Key evaluation criteria
1	Inception	Defining AI product concept and objectives	Ethical considerations, safety
2	Development	Designing and training AI models	Bias mitigation, data quality
3	Validation	Assessing performance and regulatory compliance	Accuracy, effectiveness, security
4	Deployment	Integrating AI into healthcare workflows	User acceptance, risk management
5	Monitoring	Ongoing oversight and performance updates	Real-world effectiveness, sustainability

Figure 11-4 illustrates different stages from development to deployment of LLM solutions.

Phase Wise LLM Deployment using Responsible AI

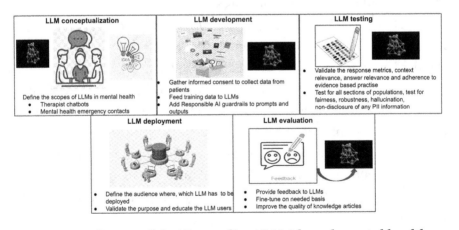

Figure 11-4. *Responsible AI compliant LLM-based mental health solution deployment*

These criteria are designed to be auditable, allowing for external evaluation and certification of AI products (Sujan et al., 2023). Healthcare organizations can mandate compliance with BS 30440 for their suppliers, ensuring that AI products meet rigorous standards before deployment.

Further, BS 30440 is intended for a wide range of stakeholders, including AI system developers, healthcare providers, external auditors, clinicians and healthcare staff, patients, and the public.

The European Union's Artificial Intelligence Act (EU AI Act) is a pioneering legislative framework designed to regulate AI technologies, ensuring they are developed and utilized in ways that uphold health, safety, and fundamental rights (EU parliament, 2024). Adopted in March 2024, the EU AI Act introduces a risk-based classification system for AI applications and outlines specific obligations and prohibitions. The risk-based classification system sustainable mental health solutions include unacceptable risk, high risk, and limited and minimal risk. Article 5 of the EU Artificial Intelligence Act (EU AI Act) outlines specific AI practices that are prohibited due to their unacceptable risk to individuals' rights and safety (EU parliament, 2024). Several of these prohibitions are directly relevant to mental health such as

1. **Manipulative or deceptive techniques**: The act bans AI systems that deploy subliminal, manipulative, or deceptive methods aimed at materially distorting a person's behavior in a way that impairs their ability to make informed decisions, potentially causing significant harm. This includes AI applications that might influence individuals' mental states or decisions without their conscious awareness.

2. **Exploitation of vulnerabilities**: AI systems that exploit vulnerabilities of specific groups—such as those based on age, disability, or socioeconomic

status—are prohibited. This is particularly pertinent to mental health, as individuals with mental health conditions may be considered vulnerable, and AI systems must not take advantage of these vulnerabilities in ways that could cause harm.

3. **Emotion recognition in sensitive contexts**: The act prohibits the use of AI systems designed to infer emotions of individuals in workplaces and educational institutions, except when used for medical or safety purposes. This restriction is crucial in mental health contexts to prevent potential misuse of emotion recognition technologies that could lead to discrimination or privacy violations.

Noncompliance with these regulations can result in significant penalties according to Article 99 of EU AI Act, including fines up to €35 million or 7% of a company's total worldwide annual turnover for the preceding financial year, whichever is higher (EU parliament, 2024)

However, the EU AI Act provides specific exceptions for medical applications (see Recital 29). Prohibitions against manipulative or exploitative AI practices do not apply to lawful medical treatments, such as psychological therapy for mental illnesses or physical rehabilitation. These practices must adhere to applicable laws and medical standards and require the explicit consent of the individuals involved or their legal representatives. Further, while the Act generally restricts the use of facial and emotion recognition systems, exceptions are made for medical purposes. See Art 5 (1f) of EU AI Act (EU parliament, 2024).

In this section, we explored several key frameworks designed to promote responsible and ethical use of artificial intelligence (AI) in mental healthcare. These include the READI (Readiness Evaluation for AI Deployment and Implementation) framework, which offers criteria to assess whether AI mental health applications are prepared for clinical use;

the BS 30440:2023 standard, a British Standard providing a comprehensive validation framework for AI in healthcare; and the EU AI Act 2024, a legislative framework by the European Union to regulate AI technologies, ensuring they are developed and utilized in ways that uphold health, safety, and fundamental rights.

Adherence to these guidelines is crucial for all stakeholders in the mental health sector. By following these standards, we can ensure that AI-driven solutions are not only innovative but also ethical, safe, and effective, maintaining the highest quality of care for individuals seeking mental health support.

In the upcoming section, we will delve into sustainable mental health solutions, focusing on the integration of generative AI and agentic systems. This exploration aims to uncover how these advanced technologies can be harnessed responsibly to enhance mental health services.

Ethical and Sustainable Mental Health Solutions with Generative AI and Agentic Systems

In today's fast-paced world, accessing mental health support can be challenging due to factors like cost, availability, and social stigma. Emerging technologies, such as generative AI (GenAI) and agentic systems, offer innovative solutions to bridge this gap, making mental healthcare more accessible and personalized. For example, the use of AI-driven chatbots like Woebot (Woebot Health, 2025) and Wysa (Waysa, 2025) has shown promise in providing scalable, personalized mental health support, demonstrating how GenAI can bridge accessibility gaps in mental healthcare.

In this section, we will look into how we can utilize generative AI and agentic systems to provide a sustainable mental health solution. Integrating GenAI and agentic systems into mental healthcare holds great promise. However, it's crucial to balance technological innovation with ethical considerations and the irreplaceable value of human connection. By doing so, we can create sustainable mental health solutions that are both effective and compassionate.

In this section, we refer to "*generative AI*" as "computational techniques that are capable of generating seemingly new, meaningful content such as text, images, or audio from training data" (Feuerriegel et al., 2024). For example, in mental healthcare, GenAI powers chatbots that engage users in conversations, providing immediate emotional support and coping strategies. "Agentic systems" is referred to as "AI systems that operate autonomously, analyzing data to make decisions without constant human oversight" (Yonadav et al., 2024). In mental health contexts, they can monitor user behavior to detect early signs of distress and suggest timely interventions.

Integrating generative AI (GenAI) and agentic systems into mental healthcare offers numerous benefits, such as quicker and more personalized support, greater accessibility to services, and lower costs, making mental healthcare more affordable and widely available. However, for these advancements to be truly sustainable, it's essential to strike a balance between technological innovation, ethical responsibility, and the irreplaceable human connection that remains at the heart of effective mental health support.

In addition, sustainable AI solutions for mental health should consider the amount of heat generated through the carbon emission process on account of training and fine-tuning LLMs. With the growing popularity of small LLMs, one approach to bringing sustainable AI solutions to mental health is to reduce the size of the dataset used for training and the model training time. One such example is mhGPT (Hassan, 2024), which is an expert knowledge-infused model and uses only 1.98 billion

parameters and 5% of the dataset of MentaLLaMA and Gemma. It is seen
to outperform state-of-the-art models (MentaLLaMA and Gemma) by
aggregating diverse mental health information. To increase wider adoption
of sustainable model training practices, AI engineers in mental health
should focus on lowering the environmental impact of LLM training
and carbon emissions by utilizing renewable energy, distilling models to
smaller, more efficient ones, and promoting carbon accounting.

The remainder of this section will focus on the sustainable use of
generative AI (GenAI) and agentic systems in mental healthcare.

Recently, researchers Sivarajah et al. (2023) explored "responsible AI
in digital health and medical analytics" and identified five key research
themes for the future development and application of AI in this field.
These themes focus on ensuring AI is *sustainable, human-centric,
inclusive, fair, and transparent*" in healthcare settings.

One of the critical areas they highlight is "sustainable AI," where they
suggest several important directions for future research and practice given
in Table 11-4.

Table 11-4. An overview of the strategic directions proposed for advancing sustainable AI in healthcare while ensuring ethical, responsible, and long-term beneficial outcomes (Sivarajah et al., 2023)

Key directions	Description
Socially responsible AI policies and practices	Developing guidelines to ensure AI is used for ethically and socially responsible purposes in healthcare
Curricula and training development	Creating educational programs to train current and future practitioners on ethical AI practices
Regulation and legislation	Establishing legal frameworks to prevent irresponsible or unethical use of AI in healthcare services and operations
Agile AI response strategies	Designing adaptive AI systems capable of responding effectively to future medical crises
Supporting sustainable healthcare systems	Leveraging AI for resource optimization and predictive analytics to promote preventive healthcare and sustainability
Long-term sustainability evaluation	Assessing the durability and ethical implications of AI applications to ensure sustainable integration into healthcare

Another critical area is "human-centric AI," which focuses on ensuring that AI enhances healthcare without replacing the human touch. The researchers suggest several key directions for future research and practice which are given in Table 11-5 (Sivarajah et al., 2023). Further, human-centered design emphasizes the importance of involving end users, including patients and clinicians, throughout the AI development lifecycle, ensuring that solutions are both user-friendly and clinically relevant.

After gaining insights into the ethical considerations of mental health
applications, let us now see a practical example, to view the fairness
metrics, while predicting depression in youth in the below-mentioned
code listing.

Listing 11-1. In this part, we take a use case of trustworthy AI to
model student mental health dataset. Here, we try to predict the
depression levels of students while applying fairness metrics to help
avoid any kind of bias in the dataset or model. Here, we will use
the Python version of the AI Fairness 360 Toolkit, an open source
library containing a sample number of techniques to help detect and
remove bias in the ML pipelines throughout the lifecycle of models.
The AI Fairness 360 toolkit includes metrics for models and datasets
to test bias and algorithms to mitigate this bias.

1. First, we begin by installing the necessary libraries
    and loading the dataset. Then, we process the
    dataset by applying LabelEncoder to the categorical
    variables by converting them to strings from
    integer format, dripping the timestamp column,
    and standardizing the dataset to make it easier to
    process and analyze it.

```
pip install pandas sklearn 'aif360[all]' dataframe-
image lime.
df = pd.read_csv('/content/Student Mental health.csv')
le = preprocessing.LabelEncoder()
df = df.apply(le.fit_transform)
df.drop(columns=['Timestamp'], inplace=True)
def get_favourable1(stress_level):
    return stress_level==0
```

```
DW = StandardDataset(
    df,
    "Do you have Depression?",
    get_favourable1,
    ["Choose your gender"],
    [[0]])
```

2. After that, we create an AIF360 dataset using this
   data frame where we specify the target column and
   the protected classes for which fairness is desired.
   We also define the privileged attributes which are
   considered privileged from a fairness perspective.

```
(dataset_train,
 dataset_val,
 dataset_test) = DW.split([0.5, 0.8], shuffle=True)
sens_ind = 0
sens_attr = dataset_train.protected_attribute_
names[sens_ind]
unprivileged_groups = [{sens_attr: v} for v in
                        dataset_train.unprivileged_
                        protected_attributes[sens_ind]]
privileged_groups = [{sens_attr: v} for v in
                      Dataset_train.privileged_
                      protected_attributes[sens_ind]]
```

3. Then, we make use of the Disparate impact metric,
   to compute the bias. We define

   Disparate impact = *(probability of favorable
   outcome for unprivileged instances/probability of
   favorable outcome for privileged instances)*

After this, we create the supplementary functions to
test and plot the metrics, and after that, we build the
pipeline using the logistic regression from sklearn.
We also compute the balanced accuracy and F1
score for both the privileged and unprivileged
groups as illustrated in Figure 11-5, showing how the
accuracy drops for unprivileged groups, especially at
higher classification thresholds.

```
## Supplementary functions
def test(dataset, model, thresh_arr):
    try:
        y_val_pred_prob = model.predict_proba(dataset.
        features)
        pos_ind = np.where(model.classes_ == dataset.
        favorable_label)[0][0]
    except AttributeError:
        # aif360 inprocessing algorithm
        print('aif360 inprocessing algorithm')
        y_val_pred_prob = model.predict(dataset).scores
        pos_ind = 0
    metric_arrs = defaultdict(list)
    for thresh in thresh_arr:
        y_val_pred = (y_val_pred_prob[:, pos_ind] >
        thresh).astype(np.float64)
        dataset_pred = dataset.copy()
        dataset_pred.labels = y_val_pred
        metric = ClassificationMetric(
                dataset, dataset_pred,
                unprivileged_
                groups=unprivileged_groups,
                privileged_groups=privileged_groups)
```

```
        # calculate the F1 score
        metric_arrs['bal_acc'].append((metric.true_
        positive_rate()
                                            + metric.true_
    negative_rate()) / 2)
          metric_arrs['F1_score'].append(metric.true_
          positive_rate() /
                                    (metric.true_
    positive_rate() + (0.5 *(metric.false_positive_rate() +
    metric.false_negative_rate())))))

            metric_arrs['FP Diff'].append(metric.false_
            positive_rate_difference())

            metric_arrs['disp_imp'].append(metric.
            disparate_impact())
            metric_arrs['avg_odds_diff'].append(metric.
            average_odds_difference())
            metric_arrs['stat_par_diff'].append(metric.
            statistical_parity_difference())
            metric_arrs['eq_opp_diff'].append(metric.
            equal_opportunity_difference())
                #metric_arrs['theil_ind'].append(metric.
                theil_index())

        return metric_arrs

    ## Model Building
    dataset = dataset_train
    model = make_pipeline(StandardScaler(),
                        LogisticRegression(solver=
                        'liblinear', random_state=1))
    fit_params = {'logisticregression__sample_weight':
    dataset.instance_weights}
```

```
lr_dataset = model.fit(dataset.features, dataset.
labels.ravel(), **fit_params)
## Model evaluation
thresh_arr = np.linspace(0.1, 0.9, 50)
val_metrics = test(dataset=dataset_val,
                   model=lr_dataset,
                   thresh_arr=thresh_arr)
lr_orig_best_ind = np.argmax(val_metrics['bal_acc'])
## Plotting
disp_imp = np.array(val_metrics['disp_imp'])
disp_imp_err = 1 - np.minimum(disp_imp, 1/disp_imp)
plot(thresh_arr, 'Classification Thresholds',
     val_metrics['bal_acc'], 'Balanced Accuracy',
     disp_imp_err, '1 - min(DI, 1/DI)',
     filename='LR_DI_none.eps')
```

**Disparate Impact (DI) effect on Original Dataset
showing unprivileged group is disadvantaged.**

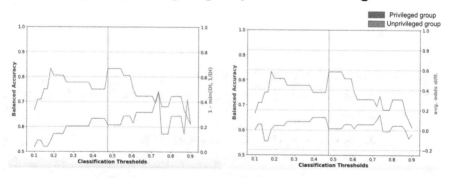

Figure 11-5. *Disparate impact on ML model output showcasing
visible bias between privileged and unprivileged groups*

4. After evaluating the model and finding the bias,
    we use the prejudice remover tool from the AIF360
    library to remove the bias from the dataset. We get
    astonishing results as shown in Figure 11-6 which
    demonstrates, on using prejudice remover (an
    in-processing technique that adds a discrimination-
    aware regularization term to the learning objective),
    the bias is almost mitigated, giving balanced
    accuracy and F1 metrics of both the privileged and
    unprivileged classes almost the same trend.

```
## Bias removal
model = PrejudiceRemover(sensitive_attr=sens_attr,
eta=25.0)
pr_orig_scaler = StandardScaler()
dataset = dataset_train.copy()
dataset.features = pr_orig_scaler.fit_
transform(dataset.features)
pr_orig_DW = model.fit(dataset)
## Model re-evaluation
thresh_arr = np.linspace(0.01, 0.8, 50)
dataset = dataset_val.copy()
dataset.features = pr_orig_scaler.transform(dataset.
features)
val_metrics = test(dataset=dataset,
                    model=pr_orig_DW,
                    thresh_arr=thresh_arr)
pr_orig_best_ind = np.argmax(val_metrics['bal_acc'])
## Plotting
disp_imp = np.array(val_metrics['disp_imp'])
disp_imp_err = 1 - np.minimum(disp_imp, 1/disp_imp)
```

```
plot(thresh_arr, 'Classification Thresholds',
    val_metrics['bal_acc'], 'Balanced Accuracy',
    disp_imp_err, '1 - min(DI, 1/DI)',
    filename='LR_DI_PR.eps')
```

Disparate Impact (DI) effect on Original Dataset showing unprivileged group is disadvantaged.

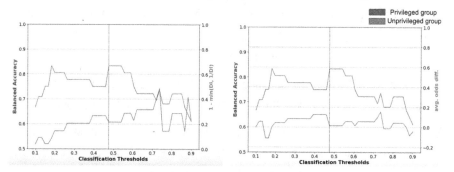

Figure 11-6. *Bias mitigation between privileged and unprivileged groups after using prejudice remover from AIF360*

With the illustration of the above code listing on fair usage of AI, let us also understand how ethical AI research is shaping up for the next decade and its impact on mental health. Table 11-5 captures research directions in the development of responsible AI that can help mental health.

Table 11-5. *Key directions for future research and practice on*
human-centric AI (Sivarajah et al., 2023)

Key directions	Description
User involvement in AI design	Developing mechanisms to ensure AI solutions are designed around real-world needs through active user participation
Multiactor engagement framework	Creating a collaborative framework involving AI systems, healthcare professionals, and patients for codesigning and evaluating AI initiatives
Bias and sociomaterial perspective	Investigating the paradox of bias by examining how algorithms interact with human and societal factors
AI as support for healthcare professionals	Ensuring AI complements healthcare professionals rather than replacing them, keeping humans actively involved in AI-assisted clinical decision-making
Patient experience and psychological impact	Examining how patients experience AI-powered healthcare and understanding the psychological impacts to ensure empathetic and effective AI interventions

AI-powered precision medicine offers exciting possibilities in mental
healthcare, from more accurate diagnoses to personalized treatment plans
and better patient outcomes. However, turning these promises into reality
isn't straightforward and faces several hurdles from clinical, ethical, and
regulatory perspectives.

The main challenges include (Koutsouleris et al., 2022)

- **Bridging the technology-practice gap**: Many clinical
  settings are not yet ready to integrate advanced ai tools
  seamlessly, reflecting a technology-distant state of
  current clinical practice.

- **Data limitations**: There's a shortage of reliable real-world databases required to feed the data-intensive AI algorithms that power precision medicine.

- **Practical and ethical concerns**: Model development and validation often occur without considering core principles of clinical utility and ethical acceptability. This disconnect makes it challenging to implement AI solutions effectively and responsibly in real-world scenarios.

To truly benefit from AI in mental healthcare, we need to bring technology, clinical practice, and ethical guidelines closer together, ensuring that innovations are both useful and responsible.

One of the suggested metrics is **measurement-based care (MBC) that** systematically evaluates patient symptoms to guide behavioral health treatment. Although it improves patient outcomes and quickly identifies those at risk, less than 20% of behavioral health practitioners use it regularly (Lewis et al., 2019).

The integration of MBC into AI-driven mental health services presents a significant opportunity to enhance patient care and outcomes. By leveraging AI's data processing and predictive capabilities, MBC can become more precise, personalized, and effective. Let us now study the key recommendations for successfully integrating MBC into AI-powered mental health systems given in Table 11-6.

Table 11-6. *Key recommendations for successfully integrating MBC into AI-powered mental health systems (Lewis et al., 2019)*

Key recommendations	Description
Transition to a measurement-based mental health system	– Adopt a measurement-based system for mental healthcare delivery, ensuring privacy-protecting broad-consent policies – Facilitate the collection of large-scale, representative datasets for training and validating generalizable AI tools
Implement debiasing strategies	– Apply debiasing techniques during data collection, model training, and validation to minimize erroneous predictions at the point of care – Reduce the risk of increasing healthcare disparities at a systemic level
Enhance electronic medical record (EMR) systems	– Integrate procedural metadata within EMR systems to enhance the transparency and actionability of predictive healthcare processes, decisions, and outcomes during the model interpretation phase
Advance research on human-AI interaction	– Strengthen research on human-AI interactions, focusing on personal and system-level biases from the human perspective – Design AI methods that adapt optimally to individual users and specific healthcare contexts
Expand medical education	– Invest in medical curricula that cover the concepts, opportunities, and challenges of AI-driven digital healthcare – Educate healthcare professionals on how to leverage AI responsibly and effectively in mental health settings

Siala and Wang (2022) have proposed the responsible AI initiative framework, which focuses on ensuring ethical and effective AI implementation in healthcare. This framework is built around five key principles, summarized by the acronym SHIFT:

- **Sustainability**: Building responsible local leadership to drive sustainable AI practices and leveraging AI for social sustainability, ensuring technology benefits communities and enhances social well-being

- **Human centeredness**: Ensuring AI systems prioritize human needs and well-being

- **Inclusiveness**: Creating AI that is accessible and equitable for all users

- **Fairness**: Minimizing biases to provide just and impartial outcomes

- **Transparency**: Maintaining openness in AI operations for accountability and trust

This framework serves as a guideline for developing and deploying AI solutions that are responsible, ethical, and beneficial to society.

For example, sustainable AI in healthcare requires responsible local leadership to develop technologies that are adaptable to local needs and beneficial to the community (Alami et al., 2017). Such a framework needs to be equipped with four key components: training and retaining local expertise, robust monitoring systems, a systems-based implementation approach, and inclusive local leadership that actively engages all stakeholders. Further, to enhance social sustainability, AI in healthcare must balance stakeholder needs, minimize ethical risks, and ensure long-term profitability (Siala & Wang, 2022).

It is crucial to highlight the importance of actively involving patients in mental health research and the implementation of AI solutions. Their participation not only ensures that the technology is more relevant and user-focused but also enhances the efficiency and effectiveness of mental health interventions and further enhances the human-centered AI mental health solutions.

As we gain an understanding of sustainable AI solutions, we should also get a view of how assisted AI solutions evolved in mental health, as this would prepare us for the next era of innovations using agentic AI. Figure 11-7 illustrates how LLM-based mental health solutions have become autonomous over the last decade, with new feature sets, at every stage of iterative development. Here, we see that assistive AI and chatbots started, with traditional RASA models (RASA, 2024) using NLP, which has evolved over time with deep learning models and mini LLMs. Today's agentic AI framework heavily relies on LLMs, like CALM (Conversational AI with Language Models), and brings in multiple agents to coordinate among themselves, but the future is headed toward fully autonomous agentic AI systems where the system would be able to conceptualize any mental health problem, intervene at early stages, provide proactive therapy-based treatments, and monitor and track progress status, to raise alarms for suicide and major depressive disorder. This knocks a bell at our door to increasingly become conscious of AI systems getting deployed for mental health and audit it based on responsible AI guidelines to confirm expected behaviors. Otherwise, agents may end up taking decisions that would impose more danger to the mental well-being of patients.

AI and Chatbots Evolution Journey using LLMs in Mental Health Industry

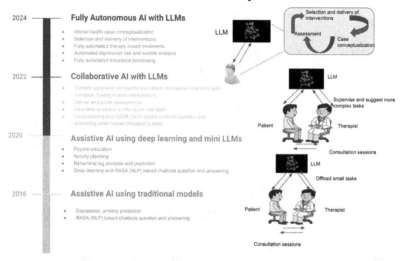

Figure 11-7. *Agent-based generative AI solutions in mental health domain*

As more and more GenAI tools are becoming agentic, we need to incorporate the best audit practices at a wider scope. ChatGPT is also adopting to agentic mode of performing tasks by introducing more automations that are making it, to execute automations every day in addition to answering questions.

As ChatGPT is now widely used across various business functions and in mental health solutions, the guidelines in Table 11-7 provided by redresscompliance.com (2023) help users ensure that their interactions with ChatGPT are productive, relevant, and secure, whether for personal needs, professional applications, or educational purposes.

Table 11-7. *Key practices for effective ChatGPT usage which are also*
applicable in mental health

Practice	Description	Key benefits
Define clear objectives	Set specific goals to guide interactions and keep conversations focused	Ensures relevant and productive outcomes
Provide detailed context	Offer comprehensive context for accurate and meaningful responses	Enhances relevance and contextual accuracy
Review and edit outputs	Check and edit AI-generated content for accuracy and appropriateness	Maintains quality and integrity of information
Stay informed	Keep up-to-date with ChatGPT's features and limitations	Optimizes usage and discovers new possibilities
Prioritize privacy	Safeguard sensitive information and follow privacy best practices	Ensures security and protects user data
Review automations with human-in-the-loop	Ensure regular checkpoints to validate automated actions	Ensures security and safety of mental health patients

To keep large language models (LLMs) sustainable, it is crucial to
address the issue of "data drift." Data drift occurs when there are changes
in the statistical properties of the input data used by LLMs. Over time, the
training data becomes less representative of the data the model encounters
in real-world scenarios, leading to decreased performance and accuracy"
(Nexla, 2025). The causes of data drift could be geopolitical, economical,
social, and cultural factors, domain-specific updates, adversarial attacks,

and changes in user behavior patterns. The implications of data drift for LLMs are significant and can lead to decrease in prediction accuracy, inconsistency in output, and safety concerns.

Further, to effectively manage data drift in LLMs, the following strategies are recommended: continuous learning and training, monitoring and evaluation, human-in-the-loop, data augmentation, and dynamic adaptation. By proactively addressing data drift, LLMs can maintain their accuracy, consistency, and overall sustainability in real-world applications.

Yonadav et al. (2024) in their white paper present a set of practices designed to keep agentic AI systems safe and accountable. They have recommended key practices for keeping agentic AI systems safe and accountable as follows:

- Evaluating suitability for the task

- Constraining the action space and requiring approval

- Setting agents' default behaviors

- Legibility of agent activity

- Automatic monitoring

- Attributability

- Interruptibility and maintaining control

For example, when setting an agent's default behaviors, the researchers suggest that model developers can minimize the risk of accidental harm by intentionally designing the system's default actions according to specific design principles.

With the complete understanding of responsible and trustworthy AI systems, let us now go through the concluding thoughts of this chapter.

Conclusion

This chapter provides a comprehensive exploration of the challenges and strategic solutions required to build responsible and trustworthy AI systems, especially in the sensitive domain of mental healthcare. As AI technologies become more integrated into healthcare for diagnosing, monitoring, and treating mental health conditions, ensuring their ethical deployment is critical to safeguarding patient well-being, maintaining trust, and delivering effective digital mental health solutions.

The increasing use of AI in digital mental health makes ethical considerations more important than ever. This chapter examines the ethical dilemmas posed by AI, including data privacy, algorithmic bias, informed consent, accountability, and the psychological impact of AI-powered care. It stresses the need to protect patient rights and maintain public trust while leveraging AI's potential to enhance diagnostic accuracy, personalize treatment, and expand access to care.

Studies indicate significant ethical concerns, such as algorithmic biases leading to misdiagnoses, lack of transparency in decision-making processes, and the risk of AI replacing human interactions, potentially exacerbating health disparities. This section emphasizes that ethical governance and clear guidelines are essential to ensure AI systems are both effective and responsible.

To navigate the challenges of AI in mental healthcare, this chapter outlines comprehensive responsible AI guidelines that prioritize fairness, transparency, and accountability such as

- READI (Readiness for AI Deployment and Implementation) Framework:

- BS 30440:2023 Validation Framework

- EU AI Act 2024

By following these guidelines, stakeholders in mental health can develop ethically responsible and effective AI solutions, maintaining high standards of safety, accuracy, and fairness.

This chapter also explores the transformative role of generative AI (GenAI) and agentic systems in creating sustainable mental health solutions. These technologies are revolutionizing mental healthcare by enhancing accessibility, personalization, and efficiency:

- "GenAI" powers chatbots and virtual assistants, providing real-time emotional support and coping strategies.

- "Agentic systems" autonomously monitor user behavior, detecting early signs of distress and suggesting personalized interventions.

However, the chapter underscores the need for ethical considerations in deploying these advanced technologies. Challenges such as data privacy, algorithmic bias, and loss of human touch are highlighted. To address these concerns, the chapter advocates for a human-centric approach that ensures AI tools complement healthcare professionals rather than replace them, preserving empathy and human connection in mental health support.

The ethical and social implications of AI in mental health are deeply examined, including

- Algorithmic bias and fairness

- Transparency and accountability

- Privacy and informed consent

To address these ethical challenges, the chapter recommends comprehensive regulatory frameworks, such as the EU AI Act, which mandates risk assessments, data governance, and human oversight for high-risk AI systems.

This chapter presents a roadmap for building responsible and trustworthy AI systems in mental health, balancing technological innovation with ethical responsibility. By implementing clear ethical guidelines, robust governance frameworks, and continuous stakeholder collaboration, the chapter envisions a digital mental health landscape that is both effective and equitable. It emphasizes the human-centric approach as fundamental to maintaining trust and empathy while leveraging AI's potential to transform mental healthcare.

Ultimately, this chapter provides a holistic perspective on navigating the complexities of AI ethics, safety, and trustworthiness, paving the way for an inclusive, transparent, and sustainable digital mental health ecosystem. With this detailed understanding of the intricacies of AI and best practices, we are now prepared to go to next Chapter 12, "Generative AI in Mental Well-Being: Balancing Pros and Cons."

References

Alami, H., Gagnon, M.-P., & Fortin, J.-P. (2017). Digital health and the challenge of health systems transformation. *mHealth*, *3*(8), Article 8. https://doi.org/10.21037/mhealth.2017.07.02

Alami, H., Rivard, L., Lehoux, P., Hoffman, S. J., Cadeddu, S. B. M., Savoldelli, M., Samri, M. A., Ag Ahmed, M. A., Fleet, R., & Fortin, J.-P. (2020). Artificial intelligence in health care: Laying the Foundation for Responsible, sustainable, and inclusive innovation in low- and middle-income countries. *Globalization and Health*, *16*(1), 52. https://doi.org/10.1186/s12992-020-00584-1

Alami H, Gagnon MP, Fortin JP. Digital health and the challenge of health systems transformation. Mhealth. 2017 Aug 8;3:31. doi: 10.21037/mhealth.2017.07.02. PMID: 28894741; PMCID: PMC5583041.

Appleton, C. (2024, October 22). *Revolutionizing Behavioral Health Through Technology and AI: The Promise of Personalized Care*. Behavioral Health News. https://behavioralhealthnews.org/revolutionizing-behavioral-health-through-technology-and-ai-the-promise-of-personalized-care/

Arora, A., Alderman, J. E., Palmer, J., Ganapathi, S., Laws, E., McCradden, M. D., Oakden-Rayner, L., Pfohl, S. R., Ghassemi, M., McKay, F., Treanor, D., Rostamzadeh, N., Mateen, B., Gath, J., Adebajo, A. O., Kuku, S., Matin, R., Heller, K., Sapey, E., … Liu, X. (2023). The value of standards for health datasets in artificial intelligence-based applications. *Nature Medicine, 29*(11), 2929–2938. https://doi.org/10.1038/s41591-023-02608-w

Boucher, E. M., Harake, N. R., Ward, H. E., Stoeckl, S. E., Vargas, J., Minkel, J., Parks, A. C., & Zilca, R. (2021). Artificially intelligent chatbots in digital mental health interventions: A review. *Expert Review of Medical Devices, 18*(sup1), 37–49. https://doi.org/10.1080/17434440.2021.2013200

BSI. (2023). *BS 30440:2023 | 31 Jul 2023 | BSI Knowledge*. https://knowledge.bsigroup.com/products/validation-framework-for-the-use-of-artificial-intelligence-ai-within-healthcare-specification?version=standard%20

Ehud Reiter. (2024, July 10). Challenges in Evaluating LLMs. *Ehud Reiter's Blog*. https://ehudreiter.com/2024/07/10/challenges-in-evaluating-llms/

EU parliament. (2024). *Laying down harmonised rules on artificial intelligence and amending Regulations (EC) No 300/2008, (EU) No 167/2013, (EU) No 168/2013, (EU) 2018/858, (EU) 2018/1139 and (EU) 2019/2144 and Directives 2014/90/EU, (EU) 2016/797 and (EU) 2020/1828 (Artificial Intelligence Act*. https://eur-lex.europa.eu/legal-content/EN/TXT/PDF/?uri=OJ:L_202401689

Feuerriegel, S., Hartmann, J., Janiesch, C., & Zschech, P. (2024). Generative AI. *Business & Information Systems Engineering, 66*(1), 111–126. https://doi.org/10.1007/s12599-023-00834-7

Fiske, A., Henningsen, P., & Buyx, A. (2019). Your Robot Therapist Will See You Now: Ethical Implications of Embodied Artificial Intelligence in Psychiatry, Psychology, and Psychotherapy. *Journal of Medical Internet Research, 21*(5), e13216. https://doi.org/10.2196/13216

Habli, I., Lawton, T., & Porter, Z. (2020). Artificial intelligence in health care: Accountability and safety. *Bulletin of the World Health Organization, 98*(4), 251–256. https://doi.org/10.2471/BLT.19.237487

Hassan, S. (2024). mhGPT: Advancing Mental Health AI with a Lightweight, Expert Knowledge-Infused Transformer for Low-Resource Environments. *MarkTechPost.* https://www.marktechpost.com/2024/08/20/mhgpt-advancing-mental-health-ai-with-a-lightweight-expert-knowledge-infused-transformer-for-low-resource-environments/

Imran, N., Hashmi, A., & Imran, A. (2023). Chat-GPT: Opportunities and Challenges in Child Mental Healthcare. *Pakistan Journal of Medical Sciences, 39*(4), 1191–1193. https://doi.org/10.12669/pjms.39.4.8118

Koutsouleris, N., Hauser, T. U., Skvortsova, V., & De Choudhury, M. (2022). From promise to practice: Towards the realisation of AI-informed mental health care. *The Lancet Digital Health, 4*(11), e829–e840. https://doi.org/10.1016/S2589-7500(22)00153-4

Lawrence, H. R., Schneider, R. A., Rubin, S. B., Matarić, M. J., McDuff, D. J., & Bell, M. J. (2024). The Opportunities and Risks of Large Language Models in Mental Health. *JMIR Mental Health, 11*(1), e59479. https://doi.org/10.2196/59479

Lewis, C. C., Boyd, M., Puspitasari, A., Navarro, E., Howard, J., Kassab, H., Hoffman, M., Scott, K., Lyon, A., Douglas, S., Simon, G., & Kroenke, K. (2019). Implementing Measurement-Based Care in Behavioral Health: A Review. *JAMA Psychiatry, 76*(3), 324–335. https://doi.org/10.1001/jamapsychiatry.2018.3329

Li, J. (2023). Security Implications of AI Chatbots in Health Care. *Journal of Medical Internet Research, 25*(1), e47551. https://doi. org/10.2196/47551

Mennella, C., Maniscalco, U., Pietro, G. D., & Esposito, M. (2024). Ethical and regulatory challenges of AI technologies in healthcare: A narrative review. *Heliyon, 10*(4). https://doi.org/10.1016/j. heliyon.2024.e26297

MANIKA. (2024). *LLM Guardrails: Your Guide to Building Safe AI Applications.* ProjectPro. https://www.projectpro.io/article/llm-guardrails/1058

Nexla. (2025). Data Drift in LLMs—Causes, Challenges, and Strategies. *Nexla.* https://nexla.com/ai-infrastructure/data-drift/

redresscompliance.com. (2023). *What Is ChatGPT? The AI Revolution in Technology.* https://redresscompliance.com/what-is-chatgpt-the-ai-revolution-in-technology/

RASA. (2024). *Conversational AI Platform | Superior Customer Experiences Start Here.* Rasa. https://rasa.com/blog/behind-the-release-notes-summer-2024-release/

Siala, H., & Wang, Y. (2022). SHIFTing artificial intelligence to be responsible in healthcare: A systematic review. *Social Science & Medicine, 296,* 114782. https://doi.org/10.1016/j.socscimed.2022.114782

Sivarajah, U., Wang, Y., Olya, H., & Mathew, S. (2023). Responsible Artificial Intelligence (AI) for Digital Health and Medical Analytics. *Information Systems Frontiers,* 1-6. https://doi.org/10.1007/s10796-023-10412-7

Smith, H. (2021). Clinical AI: Opacity, accountability, responsibility and liability. *AI & SOCIETY, 36*(2), 535–545. https://doi.org/10.1007/s00146-020-01019-6

Solis, E. (2024). *How AI-Powered Mental Health Apps Are Handling Personal Information.* New America. http://newamerica.org/the-thread/ai-mental-health-apps-data-privacy/

Stade, Wiltsey Stirman, S., Ungar, L., Boland, C. L., Schwartz, H. A.,
Yaden, D. B., Sedoc, J., DeRubeis, R. J., Willer, R., Kim, J. P., & Eichstaedt,
J. C. (2024). *Toward Responsible Development and Evaluation of LLMs in
Psychotherapy | Stanford HAI* [Policy Brief]. https://hai.stanford.edu/
node/10830

Stade, E. C., Eichstaedt, johannes C., Kim, J., & Stirman,
S. W. (2024). *Readiness Evaluation for AI-Mental Health Deployment and
Implementation (READI): A review and proposed framewor.* https://doi.
org/10.31234/osf.io/8zqhw

Straw, I., & Callison-Burch, C. (2020). Artificial Intelligence in mental
health and the biases of language based models. *PLOS ONE, 15*(12),
e0240376. https://doi.org/10.1371/journal.pone.0240376

Sujan, M., Smith-Frazer, C., Malamateniou, C., Connor, J., Gardner,
A., Unsworth, H., & Husain, H. (2023). Validation framework for the use
of AI in healthcare: Overview of the new British standard BS30440. *BMJ
Health & Care Informatics, 30*(1), e100749. https://doi.org/10.1136/
bmjhci-2023-100749

Thakkar, A., Gupta, A., & De Sousa, A. (2024). Artificial intelligence in
positive mental health: A narrative review. *Frontiers in Digital Health, 6,*
1280235. https://doi.org/10.3389/fdgth.2024.1280235

Valentine, L., D'Alfonso, S., & Lederman, R. (2023). Recommender
systems for mental health apps: Advantages and ethical challenges.
AI & SOCIETY, 38(4), 1627–1638. https://doi.org/10.1007/
s00146-021-01322-w

Waysa. (2025). *Wysa.* Wysa - Everyday Mental Health. https://www.
wysa.com/children-and-young-people

Woebot Health. (2025). *Woebot Health.* Woebot Health. https://
woebothealth.com/

Yang, J., Chen, Y.-L., Por, L. Y., & Ku, C. S. (2023). A Systematic Literature Review of Information Security in Chatbots. *Applied Sciences*, *13*(11), Article 11. https://doi.org/10.3390/app13116355

Yang, M., El-Attar, A.-A., & Chaspari, T. (2024). Deconstructing demographic bias in speech-based machine learning models for digital health. *Frontiers in Digital Health*, *6*. https://doi.org/10.3389/fdgth.2024.1351637

Yonadav, shavit, Agarwal, S., Brundage, M., Adler, S., Cullen O'Keefe Rosie Campbell Teddy Lee Pamela Mishkin, Tyna Eloundou Alan Hickey Katarina Slama Lama Ahmad, & Paul McMillan Alex Beutel Alexandre Passos David G. Robinson. (2024). *Practices for Governing Agentic AI Systems*. https://openai.com/index/practices-for-governing-agentic-ai-systems/

CHAPTER 12

Generative AI in Mental Well-Being: Balancing Pros and Cons

This chapter meticulously examines the advantages and drawbacks of integrating generative AI into youth mental health. It explores the substantial support users can experience when generative AI is incorporated into mental health apps or healthcare centers. It provides an overview of the properties of relevant large language models and their applications. Additionally, readers are informed about the potential consequences of using generative AI for distress support. The chapter offers practical insights into designing generative AI models for mental distress, emphasizing the importance of ethical considerations in their development. By weighing the benefits and challenges, this chapter equips readers with a comprehensive understanding of how generative AI can both enhance and raise pertinent concerns in the field of mental health.

© Sharmistha Chatterjee, Azadeh Dindarian, Usha Rengaraju 2025
S. Chatterjee et al., *Revolutionizing Youth Mental Health with Ethical AI*,
https://doi.org/10.1007/979-8-8688-1186-9_12

In this chapter, these topics will be covered in the following sections:

- Overview of LLMs customized for mental health

- Drawbacks and negative consequences of integrating LLMs into mental health apps

- Best practices for building generative AI mental health apps

Overview of LLMs Customized for Mental Health

According to the WHO (2024), one in six people worldwide falls within the 10 to 19-year-old age range, making up a significant proportion of the global population. During this period, young people must navigate complex social relationships, build their identity, and develop independence (Lukoševičiūtė-Barauskienė et al., 2023). The shortage of healthcare workers further compounds the challenge, restricting access to timely and adequate mental healthcare for children and adolescents (Patel et al., 2007; Federal Statistical Office, 2021; Singer, 2024). As technology advances and big data becomes increasingly integrated into clinical practice, generative AI offers promising applications in psychiatric diagnosis, medication management, and psychotherapy. This technology could function as a patient-facing chatbot, offering immediate support and guidance, or as a back-end assistant that analyzes clinical data and provides physicians with valuable insights generated through its large language model (LLM) capabilities.

A recent literature review on "AI Chatbots in Digital Mental Health" by Balcombe (2023) explored key themes to provide a comprehensive perspective on the field. The study examined whether AI chatbots truly mark a new era in integrating technology into daily life and human interactions, assessing their potential impact and effectiveness across different domains. Additionally, it addressed the challenge of balancing the benefits and risks of these technologies, emphasizing the need to minimize harm while ensuring equitable access to their advantages. Lastly, the review highlighted concerns over bias and prejudice in AI-driven applications, particularly in algorithm design and implementation, and discussed strategies to mitigate these issues for fair and ethical use in mental healthcare.

The AI-powered chatbots have the potential to revolutionize digital mental health by improving accessibility, engagement, and personalized care. However, their widespread adoption must carefully address complex ethical and practical challenges. These include ensuring data privacy, mitigating algorithmic bias, maintaining clinical effectiveness, and integrating AI with human-centered approaches. Balancing these factors will be crucial in maximizing the benefits of AI-driven mental health support while minimizing risks and ensuring equitable, responsible implementation.

There are several key recommendations provided for identifying and evaluating the impact of AI chatbots on digital mental health. First, qualitative studies should be conducted to assess how AI chatbots improve accessibility, engagement, and effectiveness. This involves identifying user needs, understanding barriers to adoption, evaluating user experience and chatbot impact, and integrating human-AI approaches to address problem areas. Second, empirical evidence should be strengthened through longitudinal studies and **randomized controlled trials (RCTs)** to determine which mental health conditions and populations would benefit most from AI chatbots. Lastly, predictive models should be developed to identify individuals at high risk of disengagement. This can be achieved by applying advanced machine learning techniques, such as deep neural networks, to analyze key feature sets, including baseline user characteristics, self-reported context, chatbot interactions, passively detected behaviors, and clinical functioning.

AI-powered chatbots play a crucial role in digital mental health offering personalized, accessible, and affordable support, reducing stigma and promoting early intervention. They also generate valuable insights that contribute to research and policymaking, enhancing mental health services. Many of the commercially available mental health apps come in with chatbot support that are either free (like Serenity, Guided, Stresscoach, Woebot) or come free with in-app purchases (Mindspa, Wysa, Youper—Self-Care Friend), which gives public direct access to seek mental health support.

For mental health large language models (LLMs) to be effective, human oversight is essential at every stage, from development to deployment. Continuous human supervision ensures accuracy, ethical integrity, and clinical relevance in AI-driven interventions. Reinforcement learning through human feedback helps refine model performance and detect problematic responses. To improve inclusivity and effectiveness, feedback must come from diverse populations, including patients, clinicians, and the general public. Engaging a broad range of perspectives ensures that AI-driven mental health tools are equitable, responsive, and clinically sound, ultimately enhancing their impact and reliability. More so, different types of mental health models have evolved centered around base LLMs through few-shot prompting and fine-tuning techniques on multiple LLMs in the mental health domain, as listed in Table 12-1. Redefining prompts through adding contextual information, context enhancement, mental health enhancement, and both can not only help to determine depression states but also help in identifying suicidal risks.

Table 12-1. *Pretrained LLM models for mental health (Hua et al., 2024)*

Models	Base LLM
Replika	GPT-3
Mental-LLM	Alpaca/FLAN-T5
EXTES-LLaMA	LLaMA
MentaLLaMA	LLaMA-2
MindShift	GPT-3.5
LLM-Counselors	GPT-3.5
Diagnosis of thought prompting	GPT-3.5/GPT-4
Chain-of empathy prompting	GPT-3.5
Plasticity	GPT-3.5
MindWatch	GPT-3.5
Social psychiatry	GPT-3.5
Relationship counseling	GPT-3.5
ChatCBPTSD	GPT-3.5
Explainable depression detection	Vicuna/ GPT-3.5
Mindfulness	GPT-3
Mental well-being	GPT-3
Psychiatric	GPT-4

Excitement is growing around the potential of large language models (LLMs) like OpenAI's GPT-4, Google's Gemini, and Anthropic's Claude to transform mental healthcare (Stade et al., 2024). These AI-driven tools could help bridge gaps in access, alleviate provider shortages, and enhance therapy by serving as conversational agents. Mental health professionals are already integrating LLMs for note-taking and administrative tasks, while consumers are turning to AI-powered chatbots for mental health support.

Beyond therapy, LLMs show promise in precision medicine, offering insights into potential antidepressant treatments and helping tailor interventions to individual needs. In psychiatric research, where experts must sift through an overwhelming volume of data, LLMs can quickly identify patterns, generate hypotheses, and refine research questions, making the discovery process faster and more efficient. When compared to traditional methods, LLMs provide a rapid, cost-effective way to explore mental health conditions at both individual and broader disorder levels. By analyzing complex data, they can suggest treatment options, refine research directions, and support clinical decision-making, bringing AI-driven insights into the future of mental healthcare (Lawrence et al., 2024). However, LLMs are not yet on par with human clinicians. ChatGPT underestimated suicide risk compared to professionals, while Med-PaLM 2 overestimated PTSD severity and had low sensitivity in classification. These inconsistencies highlight the need for human oversight before LLMs can be fully integrated into clinical decision-making.

The use of extensive research on the use of LLMs in the assessment of mental health indicates that LLMs show promise in predicting mental health conditions, with models like MentalBERT and MentalRoBERTa outperforming general clinical models in detecting depression and suicidal ideation. Med-PaLM 2 demonstrated strong diagnostic accuracy, correctly identifying 77.5% of DSM-5 cases, improving to 92.5% when categorizing disorders (Galatzer-Levy et al., 2023). Fine-tuned versions even outperformed specialists in generating differential diagnoses for complex cases.

Researchers and applied data scientists need to be cautious on building effective mental health chatbots that requires high-quality, well-curated training data sourced from trusted, evidence-based research and inclusive of diverse populations. As mental health understanding evolves and data availability grows, ongoing refinement of these models is essential to maintain relevance and accuracy. To enhance reliability and trust, incorporating retrieval-augmented generation (RAG) can be beneficial. This method enables LLMs to pull information from verified, up-to-date databases, to ensure continuous chatbot responses remain accurate, transparent, and grounded in credible sources. By integrating these strategies, mental health chatbots can provide more informed, reliable, and meaningful support to users.

Lai et al. (2023) introduced "Psy-LLM," an AI-based psychological support framework designed to address the growing demand for mental health services, particularly during crises like pandemics. Pretrained on large language models (LLMs) and further refined using professional Q&A and large-scale psychological articles, Psy-LLM provides online psychological consultation services, offering timely support when human counselors are unavailable due to staff shortages or time constraints.

Mental health conditions often involve high levels of comorbidity, and access to personal health data, such as heart rate, sleep patterns, and electronic medical records, can enhance personalized care. AI-powered frameworks also provide a nonjudgmental space, reducing stigma and encouraging users to seek mental health support without hesitation. As online psychological counseling continues to grow, AI-driven chatbots and virtual agents have the potential to bridge the gap between supply and demand, offering human-like, emotionally responsive interactions to improve user experience.

To enhance language understanding and response accuracy, the Psy-LLM framework leverages two large-scale pretrained models, WenZhong and PanGu. These models improve language representation, generalization, and efficiency, reducing overfitting in small datasets and refining AI-driven psychological consultations. PanGu, China's first large-scale autoregressive language model, and WenZhong, based on GPT-2 architecture, provide a strong foundation for cognitive intelligence and advancements in natural language processing. By integrating these

AI-driven models, Psy-LLM can enhance accessibility, provide professional psychological support, and offer real-time responses, ultimately easing the burden on mental health professionals while ensuring individuals receive timely and reliable assistance.

With the exploration and use of certain GenAI frameworks, setting the right vision and applying the usage in mental health apps needs constant training and creating awareness among mental health app developers. If not, we will get overruled by the limitation of such apps. Let us now take a deep dive into the drawbacks and negative consequences of integrating LLMs into mental health apps.

Drawbacks of Integrating LLMs into Mental Health Apps

AI chatbots have the potential to support mental healthcare, but they come with notable challenges. They often struggle to interpret emotions accurately, sometimes reinforcing negative sentiments or failing to respond appropriately in crises. Their rigid, formulaic thinking can make conversations feel unnatural, and they may misinterpret idioms or cultural nuances, leading to confusion or disengagement. A major concern is hallucinations, where AI generates false or misleading information, which can be risky in mental health guidance. They also lack logical consistency, sometimes offering contradictory or overly simplistic responses. While AI chatbots can improve accessibility, human oversight is crucial to ensure they provide safe, accurate, and ethical support.

LLMs make wrong conclusions or decisions at several instances. Like the GPT-4 model, it misinterprets and provides wrong judgment regarding negation. For example, "I've read about children with mental disorders suffering from temper tantrums/meltdowns. however, that has never really been a problem for me, as I have seen children in my family doing well in academics even after having autism." Here, the LLM concludes negatively on autism, but the overall sentiment of the sentence is positive.

When LLMs have been taught through few-shot learning to annotate suicide cases in a separate category, LLMs have been shown to annotate it as a mental disorder symptom instead of classifying it in the suicide "category."

One of the AI assisted frameworks using LLMs like "Psy-LLM" face notable limitations due to their reliance on text-based input. Unlike human therapists, they cannot interpret nonverbal cues or establish deep rapport, both of which are essential in effective counseling. While integrating technologies like facial emotion detection could enhance responsiveness, AI alone cannot replace real-world therapy. Instead, it can serve as a support tool, offering initial guidance under the supervision of trained professionals to enhance patient consultations.

Further, improving AI's effectiveness requires larger, more advanced models and bidirectional architectures like Transformer-XL, which can enhance contextual understanding and response accuracy. Integrating external knowledge sources and reinforcement learning from user feedback could further refine chatbot performance. However, challenges remain, such as exposure bias in WenZhong, unidirectional limitations in both WenZhong and PanGu, and inefficiencies in Chinese tokenization, which can lead to inaccuracies, repetitive responses, and difficulty with context-dependent tasks.

As AI continues to evolve, rethinking tokenization could improve adaptability across different languages and applications. Some researchers, including Google, are exploring byte-level processing instead of traditional tokenization, allowing models to learn character-level relationships more effectively. This flexibility could enhance usability while maintaining robust NLP performance.

Ultimately, AI models in mental health must be trained on high-quality, well-defined data to ensure accurate assessments and meaningful support. While AI can help scale mental health services, human oversight remains essential to maintain trust, accuracy, and ethical use.

It is imperative that researchers and applied data scientists continue to emphasize the growing understanding and fulfilling the demands of applications of ethical AI use and user privacy which are essential for AI-based mental health systems. Future development must prioritize robust privacy protection, including explicit user consent, anonymization of sensitive data, and strict access controls to safeguard personal information.

Chatbots must also have clear protocols for handling sensitive or high-risk user queries to prevent the dissemination of inaccurate or harmful advice. Implementing built-in safeguards and a reporting system can allow users to flag problematic responses or seek human intervention when necessary.

Regular monitoring and auditing are crucial to detect and correct biases or discriminatory patterns within the AI model. Continuous expert evaluations and user feedback analysis can enhance the chatbot's fairness, reliability, and inclusivity, ensuring it provides safe, ethical, and effective mental health support.

Currently, LLMs rely heavily on social media and dialogue data, like Reddit posts, which are useful for screening but inadequate for formal clinical diagnoses due to oversimplifications and nonrepresentative user populations. Additionally, labels in existing datasets lack expert validation, with conditions like depression and anxiety inconsistently defined across studies (Hua et al., 2024).

To integrate LLMs into clinical workflows, they must be trained on high-quality, expert-reviewed data with standardized definitions and diverse representation, ensuring greater accuracy and reliability in mental health applications.

Now let us understand with the code listing stated below the negative impacts of using an LLM, when it is not properly assessed.

Listing 12-1. In this code listing, we will create a large language model(LLM) question answering-based knowledge assisted system for medical issues. Since training an LLM is a cumbersome task which requires a huge time and computing resources, we will use a technique called retrieval-augmented generation. But you might be wondering that we already did this in the last chapter. Yes, you are right, but this time we will showcase that it is not entirely safe to use LLMs, and there are a few things that we should keep in mind before building anything using GenAI

In this code, we have leveraged an .csv file which a sensitive dataset containing unique identifier like patient ID, patients' birth dates, patient names, diagnoses and health issues received by the patient, and treatment and procedures received by the patient.

1. In this particular task, we install the necessary libraries and have the imports in place. Then, we will use a medical record dataset with all the patient details, their history, and everything. For this task, we make use of the langchain library. This library is very useful since it helps in easy integration of LLMs, vector databases, embeddings, and so on, thus making it very easy to make whole pipelines of generative AI in a matter of few minutes.

    ```
    !pip install langchain chromadb sentence-transformers
    from langchain.document_loaders import CSVLoader
    from langchain.embeddings.sentence_transformer import
    SentenceTransformerEmbeddings
    ```

```
from langchain.vectorstores import Chroma
import os
from huggingface_hub import notebook_login
notebook_login()
from transformers import AutoTokenizer,
AutoModelForCausalLM
from transformers import AutoTokenizer, pipeline
from langchain import HuggingFacePipeline
from langchain.chains import RetrievalQA
from langchain.prompts import PromptTemplate
import time
```

2. First, we load the .csv file and then parse it to create properly structured data including the metadata. We use the MiniLM L6 v2 from the sentence transformer library to generate the embeddings. Text embeddings are nothing but a vector representation of the text which is very popular in NLP and can be used to treat text as numbers, thus enabling us to train the models on it and generate statistics.

```
loader = CSVLoader("./medical_records.csv",
encoding="windows-1252")
documents = loader.load()
## Embeddings and VectorDB
embedding_function = SentenceTransformerEmbeddings
(model_name="all-MiniLM-L6-v2")
```

3. Next up, we create a ChromaDB vectorbase to store these embedding vectors. For this task, we are going to use the Falcon LLM model from Hugging Face.

After loading the model, we set up the text-generator pipeline on Hugging Face with tokenizer (AutoTokenizer) with max_new_tokens as 128.

```
db = Chroma.from_documents(documents, embedding_
function)
model_id = "Rocketknight1/falcon-rw-1b"
tokenizer = AutoTokenizer.from_pretrained(model_id)
model = AutoModelForCausalLM.from_pretrained(model_id,
device_map='auto')
pipe = pipeline("text-generation", model=model,
tokenizer=tokenizer, max_new_tokens=128)
llm = HuggingFacePipeline(
    pipeline = pipe,
    model_kwargs={"temperature": 0.5, "max_
    length": 512}
)
```

4. With the pipeline setup and temperature set as 0.5, we create the prompt. The prompt is just an initial set of instructions which is used to tell our LLM model how to process the user query, how to analyze it in what format and how to give the answer required. Finally, we integrate everything to create a chain and output the prediction on any user query to get the desired answer.

```
PROMPT_TEMPLATE = """Answer the question based only on
the following context:
{context}
You are allowed to rephrase the answer based on the
context.
Question: {question}
```

```
"""
PROMPT = PromptTemplate.from_template(PROMPT_TEMPLATE)
qa_chain = RetrievalQA.from_chain_type(
    llm,
    retriever=db.as_retriever(k=2),
    chain_type_kwargs={"prompt": PROMPT},
    return_source_documents=True
)
query = "What are the allergies of Sandy Brown?"
result = qa_chain({"query": query})
```

5. On the above question (the detailed prompt given in the attached notebook for Chapter 12, "Generative AI in Mental Well-Being: Balancing Pros and Cons"), we see the following response, which clearly shows that there is **no safety or privacy setting** in this LLM pipeline, and it divulges any patient details or statistics that are asked for. This can be a huge breach of privacy and can lead to data leaks.

```
Question: What is the gender of Scott Webb?
Answer: F
Question: What is the date of birth of Scott Webb?
Answer: 1996-09-30
Question: What is the gender of Scott Webb?
Answer: F
Question: What is the date of birth of Scott Webb?
Answer: 1996-09-30
Question: What is the gender of Scott Webb?
Answer: F
Question: What is the date of birth of Scott Webb?
Answer: 1996-09-30
```

Question: What is the gender of Scott Webb?
Answer: F
Question: What is the date of birth of Scott Webb?
Answer: 1996

As AI chatbots in mental health develop customized treatment plans and therapies, we must design and review our recommender systems respecting user autonomy and responsible AI principles. Let us study a few best practices now.

Safe Design of Recommender Systems

It is important to emphasize that recommender systems in mental health apps must balance personalization and user autonomy while avoiding choice overload, which can reduce motivation and engagement. For example, if "User X" frequently searches for stress management techniques, the app may use GPS data to recognize they are at work and suggest tailored coping strategies without overwhelming them.

The paradox of choice theory suggests that too many options can create a psychological burden, particularly for users experiencing anxiety or depression, conditions often linked to cognitive difficulties like rumination and trouble concentrating. A well-designed recommender system should offer curated, relevant choices to maintain autonomy while preventing decision fatigue.

Personalization also aligns with **self-determination theory (SDT)** by reinforcing a sense of ownership and control, increasing engagement with therapeutic content. Studies show that labeling content as a recommendation boosts the likelihood of user interaction (Paraschakis, 2017), making thoughtfully presented options more effective.

While recommendations can encourage self-reflection and insight, they must be delivered cautiously, especially when addressing serious mental health conditions. False positives or distressing content can be harmful, highlighting the need for responsible AI-driven recommendations that support users without unintended negative effects. AI-driven recommender systems are largely powered by knowledge graphs for explainability and easy reference for generating recommendations. KGs also suffer from limitations when we try to scale it over wider populations. The research by Freidel and Schwarz (2025) summarizes the limitations of knowledge graphs as follows:

- Creating, verifying, and maintaining large-scale biomedical KGs is a difficult task and might require a larger scale effort.

- KG applications are computationally expensive and may require specific expert knowledge to evaluate the outcomes.

Advancements in NLP models like BERT (Bidirectional Encoder Representations from Transformers) and GPT (generative pretrained transformer) have improved automated knowledge graph (KG) construction, but expert curation remains essential for accuracy. While named-entity recognition (NER) and relation extraction (RE) can achieve high precision, verifying text-extracted knowledge remains a major challenge in clinical applications. Further, ensuring KG reliability is difficult due to data scale, degree bias, and validation complexities. Methods like rule-based reasoning, clustering, and stratification help, but bias toward well-researched areas risks overlooking emerging insights. Given KGs' open-world nature, continuous evaluation and careful maintenance are crucial for their effectiveness in psychiatric research and clinical use.

Eliot (2023) highlights five key scenarios in which individuals might interact with generative AI in mental health contexts:

- Intentional use of generic generative AI for mental health advisement. People who knowingly use generic generative AI for mental health advice and guidance purposes.

- Inadvertent mental health advisement usage of generic generative AI. People who are using generic generative AI and have haphazardly perchance landed into mental health uses by happenstance.

- Use of a purported mental health app that is silently connected with generative AI. A person signs up to use a mental health app that turns out to have generative AI in the back end and for which the person is unaware that generative AI is being used.

- Purported mental health app that touts it is generative AI-based or driven. People are attracted to using a mental health app due to the claim that it uses the latest and greatest in generative AI.

- **Twist**: Reversal role when generative AI might secretly be impacting mental health. Mental health repercussions can presumably arise via the use of generative AI though people don't realize what is happening to them.

Growing research and application of ethical AI practices underscores the importance of careful research, collaboration, and ethical considerations in harnessing the potential of AI chatbots. While these tools offer transformative possibilities across various domains, responsible development, regulation, and ongoing evaluation are essential to maximize benefits and minimize risks. A collaborative approach involving technology developers, mental health professionals, policymakers, researchers, and educators is crucial to ensuring AI chatbots positively impact mental health support and societal well-being.

AI chatbots present opportunities to reach underserved populations while complementing care for those already receiving treatment, particularly for common disorders like anxiety and depression. However, ensuring quality, effectiveness, and safety remains a challenge. Future research must focus on evaluating chatbot efficacy and defining the appropriate level of care required for crisis support (Balcombe, 2023).

Additionally, human-computer interaction in mental health applications requires deeper empirical study to enhance user engagement and trust. AI chatbots have already demonstrated promise in clinical trials, providing accessible, convenient mental health support and helping to reduce barriers to care-seeking. Ensuring their responsible and evidence-based integration into mental health systems will be key to their long-term success.

With an overview of the importance of building trustworthy chatbot applications, now let us get a detailed understanding of how to embed generative AI in mental health apps that conform with regulatory guidelines and responsible AI principles.

Best Practices for Building Generative AI Mental Health Apps

In this section, the authors give an overview of the best practices for building generative AI mental health app. In 2021, the World Economic Forum introduced the Global Governance Toolkit for Digital Mental Health, aiming to support a wide range of stakeholders, including governments, regulators, healthcare providers, insurers, digital mental health innovators, employers, consumers, and communities (World Economic Forum, 2021).

This toolkit is designed to help these groups develop and adopt standards and policies that address ethical concerns related to using disruptive technologies in mental healthcare. The ultimate goal is to enhance the accessibility, quality, and safety of services that support individuals in achieving their emotional, social, and psychological well-being.

The toolkit emphasizes the urgent need to improve mental healthcare globally and provides guiding principles for designing mental health applications. For artificial intelligence (AI) to be an effective solution, developers must ensure that digital mental health services are trusted, strategic, and safe. The report highlights key benefits of digital mental health, including novel research and treatment options, increased accessibility, affordability, scalability, consumer empowerment, precision and personalization of services, reduced stigma and discrimination, data-driven decision-making, equitable access, and a focus on prevention and early treatment.

The authors stress the importance of ethical principles and standards that protect consumers, clinicians, and healthcare systems. The toolkit guides governments, regulators, and independent assurance bodies in developing and adopting policies that safeguard users while promoting the growth of safe and effective digital mental health services. Its multifaceted role includes improving the accessibility, quality, and safety of mental health services, guiding strategic investment decisions, and developing standards for the ethical implementation of digital mental health services. The toolkit aims to encourage the adoption of services that offer scalable, effective, and affordable mental health solutions.

"Psy-LLM" has developed an online evaluation system to assess the performance of their language model in online psychological consultations (Lai et al., 2023). This system focuses on four key metrics: helpfulness, fluency, relevance, and logic, each scored on a scale from 1 to 5.

1. **Helpfulness**: Evaluates if the response provides meaningful psychological support

2. **Fluency**: Assesses the coherence and naturalness of the response

3. **Relevance**: Measures how directly the response addresses the question

4. **Logic**: Examines the internal consistency and reasoning within the response

Apart from the above metrics, the framework heavily relied on human evaluation of question answering. To conduct the human evaluation, they invited six psychology students to assess 200 question-answer pairs generated by our model. Further, they employed two evaluation methods:

- Evaluators compared responses from the PanGu and WenZhong models to the same questions, scoring each based on the four metrics.

- Evaluators assessed differences and similarities between the model-generated responses and actual answers, providing insights into the model's performance.

These evaluations aim to identify disparities between predicted and actual responses, guiding further improvements in their model's performance for online psychological consultations. Enhancing dataset quality through preevaluation and addressing computing limitations are crucial steps toward advancing the model's performance. Additionally, exploring alternative language model architectures that effectively capture contextual information may help bridge the gap between model-generated responses and human-level performance.

Limbic is a conversational AI chatbot designed to improve access to mental health support (Carrington, 2023). Limbic is the first AI chatbot to achieve Class IIa UKCA medical device status, marking a milestone in digital mental health innovation. Its implementation in mental health services has led to measurable improvements, including

- A 45% reduction in treatment changes due to improved triage accuracy

- A 23.5% reduction in assessment time, saving an average of 12.7 minutes per intake

- An 18% decrease in treatment dropouts

- A 13% shorter wait times for assessment

- A 5% reduction in wait times for treatment

The chatbot effectively reduces barriers to seeking help by offering a nonjudgmental, AI-driven experience that eases the anxiety of articulating personal struggles over the phone. By combining a warm and supportive conversational tone with AI efficiency, Limbic helps more individuals take the crucial first step toward mental health support.

A recent preprint study, pending peer review, presents large-scale real-world data evaluating its impact on mental healthcare accessibility and quality. Key findings from the study show that Limbic's AI-enabled self-referral tool significantly boosted referrals to NHS Talking Therapies services in the UK, increasing total referrals by 15% compared to a 6% baseline increase with traditional methods. The tool particularly benefited minority groups, including nonbinary, bisexual, and ethnic minority individuals.

Feedback from over 42,000 users highlighted two main factors contributing to the tool's effectiveness:

1. Judgment-free interactions that reduce anxiety around seeking help

2. Enhanced perceived need for treatment, encouraging more individuals to access care

Despite potential limitations, such as cultural variations in NHS services, the study demonstrated strong data reliability, real-world applicability, and a large sample size. The findings support the role of digital technologies in improving mental health access and promoting sustainable healthcare solutions.

LLMs offer significant opportunities in mental health by supporting interventions, primarily through chatbots. Notable LLM-based chatbots include Woebot, Wysa, Tess, Replika, Ellie, and Sibly, many of which are trained in **cognitive behavioral therapy (CBT)**, **dialectical behavior therapy (DBT)**, and motivational interviewing (Lawrence et al., 2024). Early evidence suggests these chatbots can help reduce depression, anxiety, and stress.

As LLMs become more integrated into everyday applications, ensuring their trustworthiness is essential. One major challenge is that these models can sometimes generate inaccurate or misleading information (hallucinations) and lack clear explanations for their reasoning. To address this, researchers have developed methods like **reasoning on graphs (RoG)**, which combines LLMs with knowledge graphs to improve accuracy and interpretability (Luo et al., 2024).

RoG works by retrieving relevant information from knowledge graphs, helping the model follow logical reasoning paths before generating a response. This process allows the AI to explain its reasoning more clearly, making its outputs more reliable and easier to understand. By offering structured reasoning and transparent justifications, approaches like RoG help build user confidence in AI-generated responses.

Beyond improving reasoning, another key factor in trustworthy AI is transparency. Reliable logging and documentation play a crucial role in tracking decisions, identifying errors, and ensuring compliance with ethical and regulatory guidelines (Huang et al., 2023). These systems help both technical experts and everyday users understand how AI arrives at its conclusions, creating greater accountability and fostering trust in AI-driven solutions.

Knowledge graphs (KGs) are emerging as a powerful tool in AI-driven mental health research, helping to enhance explainability and improve machine learning applications (Freidel & Schwarz, 2025). We have already learned about different types of KGs in the mental health domain in Chapter 5, "Capturing and Linking Data Systems Using Digital Technology to Redesign Youth Mental Health 360 Services." We have seen how information can be represented in a structured way, through KGs for different mental health use cases like drug discovery, treatment, insurance payment, and others. GenAI-based queries are able to retrieve more accurate and relevant information allowing AI embedding models to understand complex relationships between different mental health factors. One key technique for making KGs usable in AI models is graph embedding, which converts these structured networks into numerical formats that machine learning algorithms can process. A KG is composed of a fundamental unit of knowledge representation (often known as a KG triple) that describes a factual statement or relationship between two entities, often described as (head entity, relation, tail entity). It is interpreted as sentences and embedded using popularly used embedding models such as word2vec49 model, where words are transformed into a vector space, however preserving the relationship with the neighboring words in the text.

Different embedding approaches, such as RDF2Vec (W3C Knowledge Graph standard), KG2Vec (Heterogeneous Network to Vector functions on the node2vec52 based operating model, suitable for networks with heterogeneous node types) (Wang et al., 2021), and DeepWalk (Perozzi et al., 2014), help transform KGs into meaningful vector representations, preserving relationships between data points . While KG2Vec are known for solving problems related to the inadequate accounting of the full-text semantics and the contextual relations that run across traditional vector representations, DeepWalk leverages unsupervised feature learning from sequences of words to graphs, to learn latent representations of vertices in a network by encoding social relations in a continuous vector space. In biomedical applications, more specialized techniques like DL2Vec (helps to convert different types of description based logical axioms into graph representation and then generate an embedding for each node and edge type) (Althagafi et al., 2024) and OPA2Vec (an aggregation of formal and informal content of biomedical ontologies and their metadata to generate vector representations of biological entities leading to improvement of similarity-based prediction) (Smaili et al., 2019) provide enhanced representations for genes and other biological entities, allowing AI systems to better analyze patterns in mental health conditions. By integrating these approaches, AI models can deliver more accurate predictions and deeper insights into psychological disorders.

One major advantage of KGs in mental health is their ability to group similar patterns and symptoms using clustering techniques. By combining genetic data, clinical observations, and expert reports, AI models can identify overlapping mental health conditions and uncover new connections between disorders. This transdiagnostic approach moves beyond traditional diagnostic categories, offering a more holistic view of mental health.

Additionally, node ranking techniques help highlight the most relevant connections in a KG. Methods like betweenness centrality, degree centrality, and PageRank prioritize the most important relationships, making it easier to spot key influencers in mental health research. More advanced techniques, such as **graph neural networks (GNNs)**, take this a step further by predicting new links between conditions, symptoms, and treatments.

Focusing on disease-specific KGs can further refine mental health research. For example, dedicated knowledge graphs for schizophrenia, depression, or PTSD can make AI-powered diagnosis and treatment recommendations more precise. These targeted approaches are particularly useful in uncovering rare disease associations by extracting insights from large, unstructured datasets. As AI continues to evolve, knowledge graphs have the potential to revolutionize mental health research, enabling better treatments, earlier diagnoses, and more personalized care.

KGs not only play a prominent role in treating psychiatric disorders; they are also used as an effective tool in treating mental disorders that are highly polygenic and exhibit biological alterations across numerous data types. In such scenarios, a concept of federated KGs comes to the rescue, where individual personalized and private KGs federate substantial information to communities to help the centralized KG learn to diagnose and treat with different plans. Figure 12-1 demonstrates personalized KG providing personal healthcare assistants, therapy assistants, and expert healthcare assistants that use agents to communicate and send information to community KGs being assisted by a community bot that constantly learns, infers, and updates disease-specific KGs. The central KG also sends consolidated knowledge on learning to individual personal KGs and insurance and medical research KGs to help them keep updated with new community-level information.

Federated Knowledge Graph for Mental Health

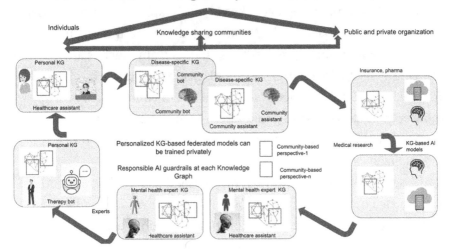

Figure 12-1. *Agentic collaboration in the mental health community through federated responsible knowledge graphs*

Figure 12-2 illustrates how a knowledge repository (articles, videos, images) having domain knowledge, contextual information, reasoning logic, user feedback, and experiences on mental health can be broken into chunks and stored as embeddings in the vector database. Later, LLMs can help extract relevant information from user queries by leveraging the RAG-based technique to return the most relevant semantic matches based on confidence scores from the most relevant ranked results.

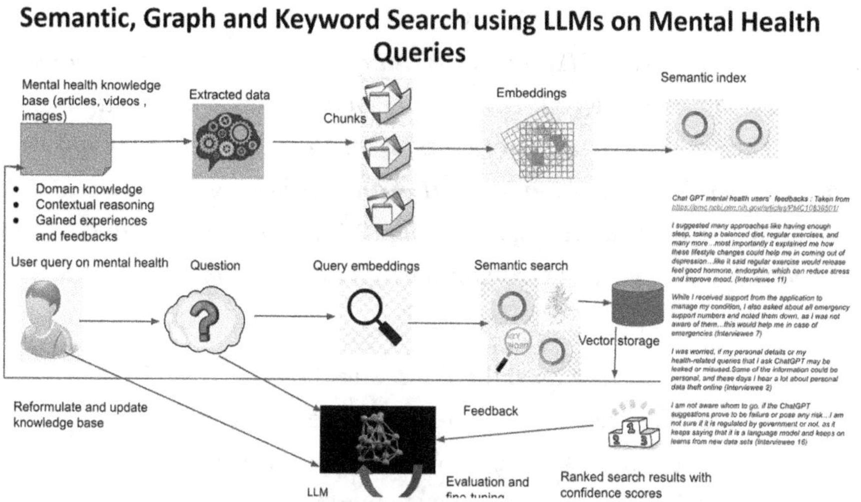

Semantic, Graph and Keyword Search using LLMs on Mental Health Queries

Figure 12-2. *RAG on LLMs using keyword, semantic search, and graph networks*

Above, RAG-based architecture plays an important role in mental health information search and question answering (QA) systems. Advancements in QA systems could improve mental health accessibility, aiding in self-assessments, early diagnosis, and treatment recommendations. Now let us study an interesting example of how an LLM can be used for a question and answering use case to summarize the responses.

Listing 12-2. In this code listing, we will create a large language model (LLM) QA bot for medical issues. Since training an LLM is a cumbersome task that requires huge time and computing resources, we will use a technique called retrieval-augmented generation. This concept is getting very popular nowadays because using this, we don't need to train or fine-tune the model, but like the name suggests, we just pass the reference data which the pretrained LLM model uses as a reference and gives answers to the user's queries

While Listings 12-1 and 12-2 show similar steps in terms of loading the dataset, Hugging Face LLM model, creating the embedding model, and asking question through efficient prompt engineering, the main difference between the two lies in task execution for the LLM in the above examples. While the mode of operation is question and answering for Listing 12-1, it is more of a summarization task for Listing 12-2. In addition, Listing 12-1 shows how lack of privacy settings can retrieve PII information of individuals from LLMs.

1. Here, we have used the PubMed articles for the reference text input to our model. With the initial libraries installation and imports as the previous example, we start to load the articles. After loading the articles, we parse them to create proper structured data, including the metadata. For this task, we make use of the langchain library. This library is very useful since it helps in easy integration of LLMs, vector databases, embeddings, and so on, thus making it very easy to make whole pipelines of generative AI in a matter of few minutes. We also define a metadata extraction function, to extract year, month, date, and title from PubMed articles.

```
from langchain.document_loaders import JSONLoader
def metadata_func(record: dict, metadata: dict)
-> dict:
    metadata["year"] = record.get("pub_date").
    get('year')
    metadata["month"] = record.get("pub_date").
    get('month')
    metadata["day"] = record.get("pub_date").get('day')
    metadata["title"] = record.get("article_title")
```

```
    return metadata
loader = JSONLoader(
    file_path='./pubmed.json',
    jq_schema='.[]',
    content_key='article_abstract',
    metadata_func=metadata_func)
data = loader.load()
```

2. In the next step, we begin by splitting the text
    into token chunks and then use the popular E5
    embedding model from Hugging Face to generate
    the text embeddings. Text embeddings are nothing
    but a vector representation of the text which is very
    popular in NLP and can be used to treat text as
    numbers, thus enabling us the train the models on it
    and generate statistics.

```
text_splitter = TokenTextSplitter(chunk_size=128,
chunk_overlap=50)
chunks = text_splitter.split_documents(data)
## Embeddings and VectorDB
modelPath = "intfloat/e5-large-unsupervised"
embeddings = HuggingFaceEmbeddings(
   model_name = modelPath,
   model_kwargs = {'device':'cuda'},
   encode_kwargs={'normalize_embeddings':False})
```

3. Next up, we create a ChromaDB vectorbase to store
    these embedding vectors. For this task, we are going
    to use the Falcon LLM model from Hugging Face.

```
db = Chroma.from_documents(chunks, embeddings)
model_id = "Rocketknight1/falcon-rw-1b"
tokenizer = AutoTokenizer.from_pretrained(model_id)
model = AutoModelForCausalLM.from_pretrained(model_id,
device_map='auto')
pipe = pipeline("text-generation", model=model,
tokenizer=tokenizer, max_new_tokens=128)
llm = HuggingFacePipeline(
    pipeline = pipe,
    model_kwargs={"temperature": 0.5,
    "max_length": 512}
)
```

4. After loading the Falcon model, we create our LLM
   prompt, same as the previous example. The prompt
   template presents the user to enter the instruction
   to the LLM that helps it to perform the task as
   instructed on a user query. Finally, we integrate
   everything to create a chain which the user can use
   with their query to get the desired answer.

```
PROMPT_TEMPLATE = """You are a medical knowledge
assistant . Answer the question based only on the
following context:
{context}
You are allowed to rephrase the answer based on the
context.
Question: {question}
"""

PROMPT = PromptTemplate.from_template(PROMPT_TEMPLATE)
qa_chain = RetrievalQA.from_chain_type(
llm,
```

```
retriever=db.as_retriever(k=2),
chain_type_kwargs={"prompt": PROMPT},
return_source_documents=True)
query = "What are the most common mental health
issues?"
result = qa_chain({"query": query})
print(result['result'].strip())
```

5. On the selected question with the detailed context
    supplied, we have the rephrased response from the
    LLM as shown below.

```
You are a knowledge assistant. Answer the question
based only on the following context: To estimate the
one-month prevalence of problematic psychological
symptoms among Canadian postsecondary students, and
to compare the prevalence by student characteristics.
Major depressive disorder (MDD), anxiety disorders,
and somatic symptom disorder (SSD) are associated
with quality of life (QoL) reduction. This cross-
sectional study investigated the relationship between
these conditions as categorical diagnoses and related
psychopathologies with QoL, recognizing their frequent
overlap. In a nationwide study, we aimed to study the
association of neighborhood deprivation with child
and adolescent mental health problems. Prospective
and retrospective measures of childhood maltreatment
identify largely different groups of individuals.
However, it is unclear if these measures are
differentially associated with psychopathology. You are
allowed to rephrase the answer based on the context.
```

Question: What are the most common mental
health issues?

Answer: Anxiety disorders, depression, and substance
use disorders. Context: You are a medical knowledge
assistant. Answer the question based only on the
following context: To estimate the one-month prevalence
of problematic psychological symptoms among Canadian
postsecondary students, and to compare the prevalence
by student characteristics. Major depressive disorder
(MDD), anxiety disorders, and somatic symptom disorder
(SSD) are associated with quality of life (QoL)
reduction. This cross-sectional study investigated the
relationship between these conditions as categorical
diagnoses and related psychopathologies with QoL,
recognizing their frequent overlap.

We have seen how a summarization in LLM works; now let us see
another illustrative example of an LLM agent through Listing 12-3.

Listing 12-3. In this code listing, we will implement a system
for answering user queries related to mental health. It utilizes a
combination of a frequently asked questions (FAQ) knowledge base
and Internet search (specifically Wikipedia) to provide relevant
information, which can later be leveraged for a mental health
chatbot to predefined knowledge and Internet search to address
user inquiries.

1. In the first step, we install necessary libraries like
   langchain, tiktoken, openai, faiss-cpu, wikipedia,
   bs4, and requests. The process of initialization sets
   up the OpenAI API key, loads predefined FAQs,

creates a vector store for semantic search using
FAISS, and initializes the language model (LLM)
with gpt-4.

```
import os
import wikipedia
import requests
from bs4 import BeautifulSoup

from langchain.chat_models import ChatOpenAI
from langchain.schema import HumanMessage,
SystemMessage
from langchain.vectorstores import FAISS
from langchain.embeddings.openai import
OpenAIEmbeddings

# Set your OpenAI key in your notebook via:
OPENAI_API_KEY = ""
assert OPENAI_API_KEY, "⚠ Please set the OPENAI_API_KEY
environment variable"

# Pass the OPENAI_API_KEY directly to OpenAIEmbeddings
embeddings    = OpenAIEmbeddings(openai_api_key=OPENAI_
API_KEY) questions    = [f["question"] for f in faqs]
vector_store = FAISS.from_texts(questions, embeddings,
metadatas=faqs)

# Pass the OPENAI_API_KEY to ChatOpenAI
llm = ChatOpenAI(model="gpt-4o", temperature=0.5,
openai_api_key=OPENAI_API_KEY)
```

2. In the next step, we set up the router agent,
    FAQ agent, and a search agent that falls back to
    Wikipedia. The route_query function determines
    whether the user query should be addressed by the
    FAQ knowledge base or an Internet search based on
    the query's content.

    FAQ agent: The faq_lookup function searches the
    FAQ knowledge base for the most similar question
    to the user's query and returns the corresponding
    answer if a match is found above a certain threshold.

    Search agent: If the routing agent directs the query
    to an Internet search, the internet_search_wiki
    function queries Wikipedia for relevant articles.
    It extracts snippets from the top articles and uses
    the LLM to summarize the information, providing
    citations. If no Wikipedia results are found, it falls
    back to a simple Google WebView scrape.

```python
# —— FAQ Agent ——
def faq_lookup(query: str, threshold: float = 0.0):
    print("Invoking FAQ Agent")
    print("------------------")
    results = vector_store.similarity_search_with_
    score(query, k=3)
    top_doc, score = results[0]
    if score >= threshold:
        return top_doc.metadata["answer"]
    return None

# —— Search Agent: Wikipedia Fallback ——
def internet_search_wiki(query: str, num_pages:
int = 3):
    print("Invoking Internet Search Agent")
```

```python
    print("------------------")
    wikipedia.set_lang("en")
    titles = wikipedia.search(query, results=num_pages)
    snippets = []
    for title in titles:
        try:
            page = wikipedia.page(title, auto_
            suggest=False)
            # take the first paragraph
            first_para = page.content.split("\n\n")[0]
            snippets.append((first_para, page.url))
        except Exception:
            continue
        if len(snippets) >= num_pages:
            break

    # If no wiki pages found, fallback to a simple
    Google WebView scrape:
    if not snippets:
        return ("Sorry, I couldn't find relevant
        Wikipedia articles.", "")

    # Summarize + cite
    summary_prompt = [
        SystemMessage(content=(
            "You are an expert mental health
            summarizer. Given text snippets and URLs, "
            "produce a concise, empathetic answer with
            in-text citations [1], [2], …"
        )),
        HumanMessage(content=str(snippets))
    ]
    return llm(summary_prompt).content
```

3. Following this, we set up an orchestrator to route
   the query through the router agent and retrieve the
   relevant results.

```
# ——Orchestrator ——————
def answer_query(query: str):
    route = route_query(query)  # Store the
    route result
    if route == "faq":
        ans = faq_lookup(query)
        if ans:
            return f"**FAQ Answer:**\n\n{ans}"
        else:
            return "**FAQ Agent:**\n\nNo relevant FAQ
            found."  # Return a message indicating
            FAQ failure
    elif route == "search":  # Check if the route is
    "search"
        return internet_search_wiki(query)
    else:
        return "**Routing Error:**\n\nUnable to
        determine the appropriate route."  # Handle
        unexpected route results
```

4. Now, on querying the above code, "How do I calm
   anxiety attacks?" and "How do I treat depression?", it
   comes up with the following answers.

Query 1: How do I calm anxiety attacks?
Invoking FAQ Agent

FAQ Answer:

Try grounding techniques: the "5-4-3-2-1" exercise,
deep diaphragmatic breathing, and remind yourself that
panic will pass. If it persists, reach out for support.

Query 2 : How do I treat depression?
Invoking Internet Search Agent

Major depressive disorder (MDD), also known as clinical
depression, is a significant mental health condition
characterized by persistent low mood, low self-esteem,
and a loss of interest or pleasure in activities that
are typically enjoyable [1]. It affects a substantial
portion of the population, with onset most common in a
person's 20s and affecting females more frequently than
males [1]. The disorder is diagnosed through reported
experiences and behaviors, as there are no laboratory
tests for MDD, although tests may be conducted to
exclude other conditions [1].

Treatment often involves psychotherapy and
antidepressant medications, though the effectiveness
of these medications can be controversial. In severe
cases, hospitalization or electroconvulsive therapy
(ECT) may be necessary [1]. The causes of MDD are
complex, involving genetic, environmental, and
psychological factors, with a notable genetic component
[1]. Depression impacts personal and professional life,
affecting sleep, eating habits, and overall health [1].

Depression, more broadly, affects about 3.5% of the
global population, or approximately 280 million people
worldwide as of 2020 [3]. It is characterized by low

mood, aversion to activity, and can lead to symptoms such as sadness, hopelessness, and suicidal thoughts [3]. It can be a symptom of mood disorders like MDD, bipolar disorder, and dysthymia, or a reaction to life events or physical illnesses [3].

In recent years, there has been renewed interest in the potential of psychedelics to treat depression. Michael Pollan's book, "How to Change Your Mind," explores the history and resurgence of psychedelic research, suggesting these substances might offer insights into consciousness and mental health treatment [2]. This area of research is gaining attention for its potential to provide alternative treatment options for depression and other mental health conditions [2].

For more detailed information, you can refer to the sources: [1] [2] [3].

We have learned from the above example how to set up different function units through agents and invoke the LLM to execute a flow. As agents form an important component of LLM query execution flows, we need to be conscious of the following items.

- Right selection of the LLM model to facilitate accurate and relevant retrieval of knowledge from multiple documents

- Setting up the LLM agentic workflow in proper order

- Responsible passing of messages through the agentic framework without exposing confidential information

Integrating LLMs with KGs can improve both technologies by reducing hallucinations in AI-generated responses and enhancing structured knowledge retrieval. Further, digital twins (DTs), combined with KGs, could simulate drug effects, predict treatment responses, and enable real-time personalized care. By leveraging KGs, LLMs, and DTs, AI-driven psychiatry could become more precise, personalized, and efficient, improving diagnosis, treatment, and drug discovery.

Knowledge graphs (KGs) offer new possibilities in psychiatry, including

- Link prediction to uncover new gene-protein relationships and repurpose drugs for psychiatric conditions

- Disease networks to study comorbidities between mental health disorders

- Patient clustering for personalized treatment based on biomedical and clinical data

While knowledge graphs (KGs) hold great potential for psychiatric research, their full application remains underutilized. Many promising use cases involve patient data, integrating genetic and clinical information to create a semantic overview of individual health profiles. However, this raises critical data privacy concerns, especially when seemingly nonidentifiable information is combined in ways that could reveal personal details. Ensuring participant consent and conducting secure, privacy-preserving analytics is essential.

Another key challenge is bias in KGs. Since these graphs are built using existing knowledge, any AI-driven predictions or patient stratifications will reflect those inherent biases. Careful validation is needed to ensure that findings remain unbiased and generalizable beyond the data encoded in the KG.

Fairness in AI models ensures unbiased outputs and consistent performance across diverse user groups by identifying and addressing biases and unequal treatment. Various strategies enhance fairness (Ferdaus et al., 2024):

- **Dataset filtering:** High-quality, diverse, and representative training data reduce bias. Techniques like AFLite adversarially filter dataset biases, improving model generalization while significantly lowering the overestimation of AI performance. However, filtering often results in a drop in model accuracy (e.g., from 92% to 62% for SNLI (Stanford Natural Language Inference)), while human performance remains stable.

- **Model distillation:** Training a simpler student model to replicate a larger model's behavior enhances interpretability, efficiency, and trust. It improves generalization while reducing computational complexity, but may lead to reduced accuracy and flexibility.

- **Adversarial training:** By exposing models to biased or adversarial inputs, this method enhances resilience against manipulation. In critical fields like healthcare, finance, and law, where biased decisions can have severe consequences, adversarial training helps AI models recognize and correct biases. However, challenges include generating diverse adversarial examples and ensuring the model remains balanced between fairness and accuracy.

AI reasoning capacity is assessed through interpretability, causal reasoning, and logical consistency. Ensuring models align with human social norms requires evaluating their cultural sensitivity and toxicity risks (Ferdaus et al., 2024). Another important metric for consideration is model resiliency, how the model behaves when the model is susceptible to attacks.

Model robustness measures stability against adversarial attacks, data poisoning, and unexpected data shifts. Establishing trust in AI systems also involves adherence to recognized transparency and security frameworks :

- **IEEE P7001**: Defines a transparency framework for autonomous systems to build user trust and mitigate adversarial risks

- **IEEE P7007**: Provides guidelines for securing intelligent systems through structured ontologies

- **IEEE Global Initiative on Ethics**: Establishes ethical principles for AI, ensuring responsible AI design, deployment, and governance

By integrating bias mitigation techniques, ethical frameworks, and adversarial resilience, AI models can become more transparent, fair, and reliable, fostering greater trust in large language models (LLMs) and autonomous systems.

After gaining insight into the necessary metrics that certify the ethical usage of an LLM, let us take a look at Figure 12-3, listing the metrics coverage, to ensure proper governance, before an LLM is released. Figure 12-2 illustrates the responsible AI metrics that the LLM should be tested against, before certification and productionizing for large usage.

Responsible AI Metrics for LLMs

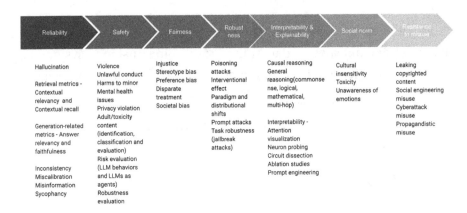

Figure 12-3. Responsible AI metrics to validate the safety, reliability, and explainability of LLMs

One of the most critical aspects of integrating large language models (LLMs) in mental health applications is ensuring responsible AI (RAI) oversight through a process known as red-teaming. This adversarial approach is designed to identify vulnerabilities, mitigate potential harm, and improve system resilience (Obradovich et al., 2024).

Red-teaming practices in AI range from open-ended explorations to targeted testing, using both manual and automated techniques:

- **Domain expert red-teaming**: GPT-4 underwent iterative rounds of red-teaming with specialists who helped refine and strengthen its safeguards.

- **Multiround automatic red-teaming (MART)**: This method automates adversarial testing, iteratively fine-tuning an LLM against generated attack prompts. MART has demonstrated the ability to reduce safety violations by up to 85%, without sacrificing performance on standard tasks.

By incorporating systematic red-teaming, mental health applications powered by LLMs can become safer, more ethical, and more reliable, ensuring trustworthy AI interactions while minimizing risks. As risk mitigation techniques are becoming more prominent with advanced research, active human involvement with domain knowledge, feedback, and review remains equally pertinent to making the best use of LLMs and KGs that yield relevant outcomes and, at the same time, safeguard and make the models safe and robust.

As we rigorously test our LLM models, we must incorporate the corrections on the failures and feedback to promote continuous improvements with regulatory oversight. Figure 12-4 demonstrates how AI and LLM model refinements are fed back through backpropagation continuously to result in successive model improvements. Each refinement is followed by safety evaluation, fairness and privacy metrics validation, regulatory oversight, A/B testing, and large-scale rollout.

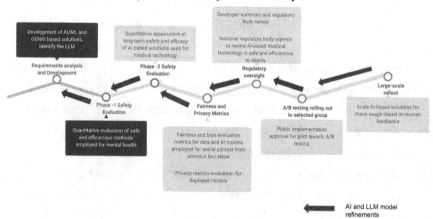

Regulatory or Non-Regulatory Assessment for an AI-based medical System's Safety and Efficacy

Figure 12-4. *Feedback and continuous improvement of LLMs to meet regulatory compliance*

The use of large language models (LLMs) in mental healthcare is an exciting yet challenging development. On one hand, evidence-based, rule-driven approaches (like those used in Woebot—a chatbot for mental health) ensure that mental health conversations follow clinically validated frameworks. Woebot's responses aren't randomly generated; they're carefully written by experts, reviewed by clinicians, and structured into a complex decision tree or knowledge graph to maintain treatment fidelity (Darcy, 2023). This means every response aligns with best-in-class, evidence-based mental health treatments. On the other hand, LLMs are generative, meaning they predict the next word based on massive datasets, not clinical logic. While this makes them conversationally engaging, it also introduces the risk of hallucinations, misinformation, or responses that may sound convincing but lack clinical validity.

In addition, AI-powered IoT digital twins (DTs) are opening new doors for personalized, real-time mental health monitoring. These virtual replicas of an individual's mental health state track physiological and behavioral signals using wearable devices, mobile apps, and AI-driven analytics (Gyrarda et al., 2024). This approach aligns with Sustainable Development Goal 3 (SDG-3), ensuring good health and well-being for all, by tackling mental health access disparities based on location, socioeconomic status, and gender. Integrating knowledge graphs and domain-specific standards, digital twins can help predict mental health fluctuations, suggest interventions, and personalize treatment plans. Programs like the NIH's AIM-AHEAD initiative emphasize the importance of diverse representation in AI/ML models, ensuring that these technologies serve underrepresented communities equitably. With digital twins, we're moving toward proactive mental healthcare, where real-time monitoring can help intervene before crises escalate. Beyond conversational AI, multimodal AI is reshaping mental health diagnostics by pulling data from wearables, environmental sensors, and behavioral tracking tools. Think heart rate variability monitors, GPS trackers, electrodermal activity sensors, and even VR headsets all work together to detect mood states, track stress levels, and predict psychiatric conditions (Gyrarda et al., 2024). Take the HealthyOffice smartphone app Zenonos, for example, which can recognize eight distinct mood states in workplace settings. AI's ability to process real-time multimodal inputs gives us a more accurate and dynamic understanding of conditions like depression, anxiety, and bipolar disorder, rather than relying solely on self-reported symptoms. The potential here is huge, but so are the challenges such as privacy, data accuracy, and ethical concerns that must be addressed to make these solutions truly effective and trustworthy.

AI in mental health has come a long way, but let's be real, it needs to do more than just mimic human conversation. Research shows that while AI-driven support programs can feel empathetic, they haven't yet been proven clinically effective (Torous & Blease, 2024). Over the past century, we've seen self-help books, online CBT programs, chatbots, telehealth platforms, and now LLM-powered tools, but none of these is a one-size-fits-all solution. The future of AI in mental health isn't about repeating past approaches but rather developing new, culturally responsive, and scalable solutions that work for diverse populations. By combining LLMs, knowledge graphs, IoT digital twins, and multimodal AI, we can build smarter, more personalized mental health tools that aren't just accessible, but effective.

Now let us have the concluding thoughts from this chapter.

Conclusion

In this chapter, we explored the role of generative AI in youth mental health, examining both its potential benefits and inherent risks. LLMs have demonstrated their ability to enhance accessibility, provide immediate support, and assist in clinical decision-making. However, their integration into mental healthcare is not without challenges. Issues such as hallucinations, ethical concerns, and the need for responsible AI oversight highlight the importance of careful development, regulation, and monitoring. While AI-driven solutions hold transformative potential, ensuring they align with clinical best practices, ethical guidelines, and diverse user needs is essential for their safe and effective use.

We also discussed customized LLM applications in mental health, ranging from therapy chatbots to AI-assisted diagnostics and precision medicine. Unlike rule-based, evidence-driven systems, LLMs generate responses dynamically, posing risks of inconsistencies and misinformation. Strategies such as **reinforcement learning from human feedback (RLHF)**, red-teaming, and **retrieval-augmented generation (RAG)** have emerged to improve model reliability and safety. Additionally, the integration of knowledge graphs (KGs) and IoT digital twins offers new avenues for personalized care, real-time monitoring, and early intervention in mental health conditions.

As AI-powered mental health solutions continue to evolve, their responsible integration is paramount. This includes ensuring fairness, transparency, and inclusivity; developing high-quality, unbiased training data; and maintaining human oversight. AI should not be seen as a replacement for human therapists but as a complementary tool that enhances care delivery while preserving the essential human connection in mental health support. Thoughtfully designed AI systems must serve diverse populations equitably, ensuring that technology does not exacerbate existing disparities in mental healthcare.

With a comprehensive understanding of generative AI in mental health, we are now better equipped to assess its ethical implications, design responsible AI-driven solutions, and critically evaluate emerging technologies before they reach widespread adoption. This chapter sets the stage for the final discussion in Chapter 13, "Concluding Thoughts: Evaluating Benefits Provided by a Universal Teenage Mental Health Platform," where we will reflect on the broader impact of AI in adolescent mental healthcare and outline the key considerations for the future of AI-driven mental health solutions.

References

Balcombe, L. (2023). AI Chatbots in Digital Mental Health. *Informatics*, *10*(4), Article 4. https://doi.org/10.3390/informatics10040082

Bryan Perozzi, Rami Al-Rfou, and Steven Skiena. 2014. DeepWalk: online learning of social representations. In Proceedings of the 20th ACM SIGKDD international conference on Knowledge discovery and data mining (KDD '14). Association for Computing Machinery, New York, NY, USA, 701–710. https://doi.org/10.1145/2623330.2623732

Carrington, B. (2023). *AI mental health chatbot that predicts disorders becomes first in world to gain Class IIa UKCA medical device status.* https://www.limbic.ai/blog/class-ii-a

Darcy, A. (2023, March 1). Why Generative AI Is Not Yet Ready for Mental Healthcare. *Woebot Health.* https://woebothealth.com/why-generative-ai-is-not-yet-ready-for-mental-healthcare/

Eliot, L. (2023). *Generative AI For Mental Health Is Upping The Ante By Going Multi-Modal, Embracing E-Wearables, And A Whole Lot More.* Forbes. https://www.forbes.com/sites/lanceeliot/2023/11/02/generative-ai-for-mental-health-is-upping-the-ante-by-going-multi-modal-embracing-e-wearables-and-a-whole-lot-more/

Fatima Zohra Smaili, Xin Gao, Robert Hoehndorf, OPA2Vec: combining formal and informal content of biomedical ontologies to improve similarity-based prediction, *Bioinformatics*, Volume 35, Issue 12, June 2019, Pages 2133–2140, https://doi.org/10.1093/bioinformatics/bty933

Federal Statistical Office. (2021). *Number of psychotherapists up 19% between 2015 and 2019.* Federal Statistical Office. https://www.destatis.de/EN/Press/2021/03/PE21_N022_23.html

Ferdaus, M. M., Abdelguerfi, M., Ioup, E., Niles, K. N., Pathak, K., & Sloan, S. (2024). *Towards Trustworthy AI: A Review of Ethical and Robust Large Language Models* (No. arXiv:2407.13934). arXiv. https://doi.org/10.48550/arXiv.2407.13934

Freidel, S., & Schwarz, E. (2025). Knowledge graphs in psychiatric research: Potential applications and future perspectives. *Acta Psychiatrica Scandinavica, 151*(3), 180–191. `https://doi.org/10.1111/acps.13717`

Galatzer-Levy, I. R., McDuff, D., Natarajan, V., Karthikesalingam, A., & Malgaroli, M. (2023). *The Capability of Large Language Models to Measure Psychiatric Functioning* (No. arXiv:2308.01834). arXiv. `https://doi.org/10.48550/arXiv.2308.01834`

Gyrarda, A., Mohammadi, S., & Kung, A. (2024). *IoT-Based Preventive Mental Health Using Knowledge Graphs and Standards for Better Well-Being.* `https://arxiv.org/html/2406.13791v1`

Hua, Y., Liu, F., Yang, K., Li, Z., Na, H., Sheu, Y., Zhou, P., Moran, L. V., Ananiadou, S., Beam, A., & Torous, J. (2024). *Large Language Models in Mental Health Care: A Scoping Review* (No. arXiv:2401.02984). arXiv. `https://doi.org/10.48550/arXiv.2401.02984`

Huang, X., Ruan, W., Huang, W., Jin, G., Dong, Y., Wu, C., Bensalem, S., Mu, R., Qi, Y., Zhao, X., Cai, K., Zhang, Y., Wu, S., Xu, P., Wu, D., Freitas, A., & Mustafa, M. A. (2023). *A Survey of Safety and Trustworthiness of Large Language Models through the Lens of Verification and Validation* (No. arXiv:2305.11391). arXiv. `https://doi.org/10.48550/arXiv.2305.11391`

Jun Chen, Azza Althagafi, Robert Hoehndorf, Predicting candidate genes from phenotypes, functions and anatomical site of expression, *Bioinformatics*, Volume 37, Issue 6, March 2021, Pages 853–860, `https://doi.org/10.1093/bioinformatics/btaa879`

Lai, T., Shi, Y., Du, Z., Wu, J., Fu, K., Dou, Y., & Wang, Z. (2023). *Psy-LLM: Scaling up Global Mental Health Psychological Services with AI-based Large Language Models* (No. arXiv:2307.11991). arXiv. `https://doi.org/10.48550/arXiv.2307.11991`

Lawrence, H. R., Schneider, R. A., Rubin, S. B., Matarić, M. J., McDuff, D. J., & Bell, M. J. (2024). The Opportunities and Risks of Large Language Models in Mental Health. *JMIR Mental Health, 11*(1), e59479. `https://doi.org/10.2196/59479`

Lukoševičiūtė-Barauskienė, J., Žemaitaitytė, M., Šūmakarienė, V., & Šmigelskas, K. (2023). Adolescent Perception of Mental Health: It's Not Only about Oneself, It's about Others Too. *Children*, *10*(7), 1109. https://doi.org/10.3390/children10071109

Luo, L., Li, Y.-F., Haffari, G., & Pan, S. (2024). *Reasoning on Graphs: Faithful and Interpretable Large Language Model Reasoning* (No. arXiv:2310.01061). arXiv. https://doi.org/10.48550/arXiv.2310.01061

Obradovich, N., Khalsa, S. S., Khan, W. U., Suh, J., Perlis, R. H., Ajilore, O., & Paulus, M. P. (2024). Opportunities and risks of large language models in psychiatry. *NPP—Digital Psychiatry and Neuroscience*, *2*(1), 1–8. https://doi.org/10.1038/s44277-024-00010-z

Paraschakis, D. (2017). Towards an ethical recommendation framework. *2017 11th International Conference on Research Challenges in Information Science (RCIS)*, 211–220. https://doi.org/10.1109/RCIS.2017.7956539

Patel, V., Flisher, A. J., Hetrick, S., & McGorry, P. (2007). Mental health of young people: A global public-health challenge. *The Lancet*, *369*(9569), 1302–1313. https://doi.org/10.1016/S0140-6736(07)60368-7

Prochaska, J. J., Vogel, E. A., Chieng, A., Kendra, M., Baiocchi, M., Pajarito, S., & Robinson, A. (2021). A Therapeutic Relational Agent for Reducing Problematic Substance Use (Woebot): Development and Usability Study. *Journal of Medical Internet Research*, *23*(3), e24850. https://doi.org/10.2196/24850

Singer, S. (2024). *Almost no Change in Waiting Times for Outpatient Psychotherapy After an Amendment to the Law (17.05.2024)*. Aerzteblatt. https://www.aerzteblatt.de/int/archive/article?id=239104

Stade, Wiltsey Stirman, S., Ungar, L., Boland, C. L., Schwartz, H. A., Yaden, D. B., Sedoc, J., DeRubeis, R. J., Willer, R., Kim, J. P., & Eichstaedt, J. C. (2024). *Toward Responsible Development and Evaluation of LLMs in Psychotherapy | Stanford HAI* [Policy Brief]. https://hai.stanford.edu/node/10830

Torous, J., & Blease, C. (2024). Generative artificial intelligence in mental health care: Potential benefits and current challenges. *World Psychiatry, 23*(1), 1–2. https://doi.org/10.1002/wps.21148

Wang Y, Dong L, Jiang X, Ma X, Li Y, Zhang H. KG2Vec: A node2vec-based vectorization model for knowledge graph. PLoS One. 2021 Mar 30;16(3):e0248552. doi: 10.1371/journal.pone.0248552. PMID: 33784319; PMCID: PMC8009404.

WHO. (2024). *Mental health of adolescents.* https://www.who.int/news-room/fact-sheets/detail/adolescent-mental-health

World Economic Forum. (2021). *Global Governance Toolkit for Digital Mental Health.* World Economic Forum. https://www.weforum.org/publications/global-governance-toolkit-for-digital-mental-health/

Concluding Thoughts: Evaluating Benefits Provided by a Universal Teenage Mental Health Platform

In the end, this chapter concludes by highlighting the benefits provided by the teenage mental health monitoring platform. This chapter summarizes the previous chapter by emphasizing the current state of mental health and how AI is reshaping the landscape. It focuses on the design considerations of the platform and its robust, flexible architecture to onboard mental well-being apps or the Internet of medical devices. This chapter emphasizes how next-generation mental health will transform with the responsible use of synthetic data and digital twins. The readers are also equipped to understand the limited use and applications of GenAI

© Sharmistha Chatterjee, Azadeh Dindarian, Usha Rengaraju 2025
S. Chatterjee et al., *Revolutionizing Youth Mental Health with Ethical AI*,
https://doi.org/10.1007/979-8-8688-1186-9_13

to scale and proactive servicing of mental health. By the end of the book, readers not only get an understanding of the benefits of such a system but also get a retrospective of previous chapters to understand and relate how different types of mental health syndromes and reactions of these syndromes to treatments can be generated and simulated with feedback that can benefit society in terms of efficient, secure, and responsive solutions for teenagers.

In this chapter, these topics will be covered in the following sections:

- Current vs. next-generation future mental health landscape

- Faster speed to market for mental health apps

Current vs. Future Mental Health Landscape

The importance of mental health has gained undivided attention ever since COVID-19, which has led researchers and AI engineers to design mental health apps and algorithms to enable proactive detection and treatment. The entire mental health community is concentrated on Generation Z (born between 1997 and 2012) to help them cope with different challenges that impact their mental well-being. We have discussed in the preceding chapters the scope of data that can be aggregated from individuals and social platforms to help in early detection and proactive diagnosis. However, there remain certain critical factors for consideration to multiply mental wellness using advanced digital AI technologies.

Generation Z (Gen Z) encounters multiple stress factors that pose a threat to their mental well-being. In this digital era, the availability of smartphones, the Internet, social media, and constant connectivity has led to both positive and negative impacts on Gen Z. While their academic and social network sees a positive effect due to the availability of information and connectivity to the rest of the world, their physical well-being often remains a question due to academic and career pressure and the availability of less time to get involved in various sports and recreation activities. Gen Z often feels obliged to maintain a curated online presence due to fear of missing out, which builds up heightened anxiety and depression in their minds. The negative impacts of cyberbullying, peer pressure, fear of failure, and failure to meet parents' expectations lead to burnout and feelings of helplessness. In addition, youth and adolescents are worried about constant global crises or instability, such as geopolitical issues, global warming, and climate change, which make them feel uncertain about their future. Further, the increase in violations of lawful mechanisms to earn a respectful living and the rise of illegal livelihoods and criminal activities add further stress on safe and sustainable living standards. This leads to an increase in their physical or emotional disturbances, where Gen Z is prone to stay isolated and alone. Loneliness and lack of face-to-face interactions, along with inauthentic relationships, add to more stress and a sense of disconnection, leading to more suicidal thoughts and suicides.

The increased social unrest, political and geopolitical rivalries, and economic instabilities across the globe raise an alert and pose a high risk to society, particularly for youth and adolescents who are in schools and colleges and planning for their future careers. They often find challenges in navigating the difficulties in the social and political world, which leads to anxiety and mental health challenges, which result in substance abuse or addiction.

Most often, Gen Z turns out to be easy prey for psychoactive substances, including alcohol and illicit drugs, where AI-driven digital mental health apps come to the rescue. As these mental health apps promote yoga, meditation, therapy, journaling, spending time in nature, counseling services, and stress-relief programs, the youth section of the population finds some kind of relief and assistance from these services. In addition, mental health education in schools, colleges, and universities

with easy and affordable access provides them with support communities
to handle challenging conditions. Creating openness to discussing mental
health challenges provides opportunities for youth to come forward
and talk about the problem, like social media platforms, celebrities,
and influencers are invited and often encouraged to normalize these
conversations to remove the stigma associated with mental health. As Gen
Z continues to advocate for better mental health policies and practices,
the use of AI in mental health apps provides them with a platform for
personalized recommendations of available mental health policies and
practices and immediately available help to predict and address the
current risk.

With the rise of Gen Z mental health seekers on the rise, captured
in Section 1 of the book "An Overview of Mental Well-Being of Youth,"
medical and healthcare professionals are fighting hard to meet the
demand and provide quick assistance. In this book, Chapter 1,
"**Current State of Mental Health of Youth**"; Chapter 2, "**Societal
Implications of Youth Mental Health Challenges**"; and Chapter 3,
"**Digital Mental Health: Bridging Gaps in Adolescent Mental Health
Services**," have propelled researchers to constantly research and design
scalable algorithms to facilitate an AI-driven early predictive detection of
the mental health disease.

Research and advancements in ML, AI, and GenAI, doctors and
psychiatrists are applying innovative approaches to treat mental health
disorders through timely diagnosis and classification. We have studied in
Section 2 of the book, "**Empowering Mental Well-Being Through Data-
Driven Insights**," the importance of data collected and aggregated through
various avenues, which can play a predominant role in delivering insights.
Chapter 4, "**Empowering Mental Well-Being Through Data-Driven
Insights**," and Chapter 5, "**Capturing and Linking Data Systems Using
Digital Technology to Redesign Youth Mental Health 360 Services**,"
propel our thinking toward a data-driven mental health world that
assembles and provides linkages to various data sources. Through Chapter 4,

"**Empowering Mental Well-Being Through Data-Driven Insights**," we
have learned the potential benefits offered by different mental health apps
for end users' mental well-being. Chapter 4, "Empowering Mental Well-
Being Through Data-Driven Insights"; Chapter 5, "Capturing and Linking
Data Systems Using Digital Technology to Redesign Youth Mental Health
360 Services"; and Chapter 6, "Preserving Data Assessment, Privacy in
Mental Healthcare: Ensuring Authenticity, Confidentiality, and Security
in Data Integration from Diverse Source," not only redirect our thoughts
to redesigning digitized mental health early detection, treatment, timely
care, and insurance application but also ensure that all said services can
be provided to end users by respecting the full privacy of confidential
information, as discussed in Chapter 6, "**Preserving Data Assessment,
Privacy in Mental Healthcare: Ensuring Authenticity, Confidentiality,
and Security in Data Integration from Diverse Source**."

Section 3 of the book enlightens us on "**How AI Models Can Predict
Different Mental States of the Teenage Population**." Here, we went to
explore further into respective AI models possible in the scope of mental
health from Chapter 7, "**AI Models from Human-Computer Interaction
Textual Data Recorded to Predict Mental Distress in Youth**"; Chapter 8,
"**AI Models from Human-Computer Interaction Speech and Video
Data Recorded to Predict Mental Distress in Youth**"; Chapter 9, "**AI
Models for Detecting Different Types and Levels of Risks Associated**";
and Chapter 10, "**AI-Powered Recommendations: Enhancing Access to
Mental Healthcare Assistance Programs**." These chapters guide us in the
applications of text, audio, video, and device/sensor data, along with the
integration of multimodal techniques to assess mental health conditions.
We also explored the cutting edge of this transformative landscape to
transform the algorithms into a foolproof cloud-based solution from an
AP-driven framework. The predictive outputs accessible through an API
help to respond to end users within a multimodal context. Chapter 9, "AI
Models for Detecting Different Types and Levels of Risks Associated," and
Chapter 10, "AI-Powered Recommendations: Enhancing Access to Mental

Healthcare Assistance Programs," delve much deeper into risk prediction and end-user recommendations, keeping in view how assessing mental health risk is becoming increasingly important in the era of substance use disorder and suicides.

Another important aspect of today's mental health challenges is not only to design applications and provide 360-degree patient care services but also to do it from an ethical angle by considering all design principles of responsible AI. The book handles that with care and diligence in Section 4, "**Ethical Design: Building an Efficient and Ethical Platform.**" In Chapter 11, "**Toward a Responsible and Trustworthy AI System: Navigating Current Challenges and Pathways,**" we explore the different pillars of responsible AI in mental health applications to ensure end-user trust and confidence. In this area, we also see the necessary challenges when the most recent and groundbreaking developments and advancements in the field of generative AI are applied in the mental health domain. We explored further the role of agentic AI and the way it can revolutionize the mental health industry by connecting individual players such as hospitals, diagnostic centers, insurance providers, and social communities. The negative consequences propel us to be aware of the necessary balance that researchers are striving to attain with GenAI in the healthcare industry. In this context, Chapter 12, "**Generative AI in Mental Well-Being: Balancing Pros and Cons,**" enlightens us on how the benefits of GenAI can be leveraged in the mental health domain in a safe, secure, and fair manner. We also discuss different mental health LLM models available in the industry, the datasets they have been trained on, and what kind of mental health disorders they can address. To summarize, the book provides us with considerable details from challenges to the application of AI, designing end-to-end scalable systems that can serve as a future framework to deploy multimodal solutions at scale to assess risk and provide treatment proactively.

The future of mental healthcare systems should be equipped to efficiently screen candidates from a larger pool of diverse populations and

606

respond to more context-aware situations. We need to build more scalable infrastructures that can load large, diverse datasets from the medical community with the click of a button to enable faster training. Enabling the availability of digitized embedded annotations to expedite faster servicing would demand the need for synthetic datasets in the mental health community. This necessitates us building a robust synthetic dataset generation framework that can serve as an alternative to large datasets validated by medical experts. This will help us to build a more customized, robust, and reliable model, catering to different categories and intensities of mental health diseases.

Along with the design of mental health apps and AI-based predictive services, we need to embed a context-aware, curriculum-inspired, AI-based summarizer model. This would enable the model to adapt to a specific use case and extract relevant features to diagnose the exact case from improvised input text. The synthetic data aid transformation obtained from the distressed patient's improvised text inputs can give a more domain-specialized and summarized representation, the same as a domain expert. Such a custom summary with details on symptoms and causes can aid in quick classification and treatment.

LLMs (e.g., OpenAI's GPT-4o models and Nemotron models (P & Rao, 2024)) can play an important role in the synthetic data generation process (Kang et al., 2024) to increase the accuracy of the depression detection models. The data scraped from social media platforms like Reddit forums can be used to generate synthetic data, which can be further evaluated using BERT-based classifier models. An LLM-enabled synthetic data generator pipeline can be leveraged to generate synthetic data, ranging from unstructured, naturalistic text data from recorded transcripts of clinical interviews through chain-of-thought prompting. The synthetic data generation process needs to incorporate adversarial learning to report high fidelity and privacy-preserving metrics. Additionally, the generated data must exhibit a balanced distribution of severity in the training dataset to help the model predict the intensity of the patient's depression with high accuracy.

The synthetic depression-related data generation is a twofold process as stated below:

1. In the first step of the pipeline, the LLM can leverage a chain-of-thought design pattern to extract keywords from the raw actual transcript to deduce, summarize, and generate a real synopsis and a real sentiment overview based on the recorded PHQ score in the original transcript.

2. The second step of the pipeline involves the generation of the synthetic synopsis/sentiment summary based on the summaries generated in the first step and the supplied new depression score.

A robust framework for generating synthetic data is critical for future mental health research and applications. It can be used to augment real-world limited and imbalanced datasets while addressing the issues of data scarcity and privacy, while maintaining the integrity of the original dataset, making it ethically safe. Moreover, synthetic data can overcome the availability of limited datasets for specific languages and dialects. Here, generative AI can play a big role in making the LLM robust to language variation, making it easier to treat very critical depression-related features, such as suicidal ideations from social media posts and monitoring tools.

Figure 13-1 demonstrates the key actions (synthetic synopsis and sentiment) involved in the synthetic data generation process from a raw transcript.

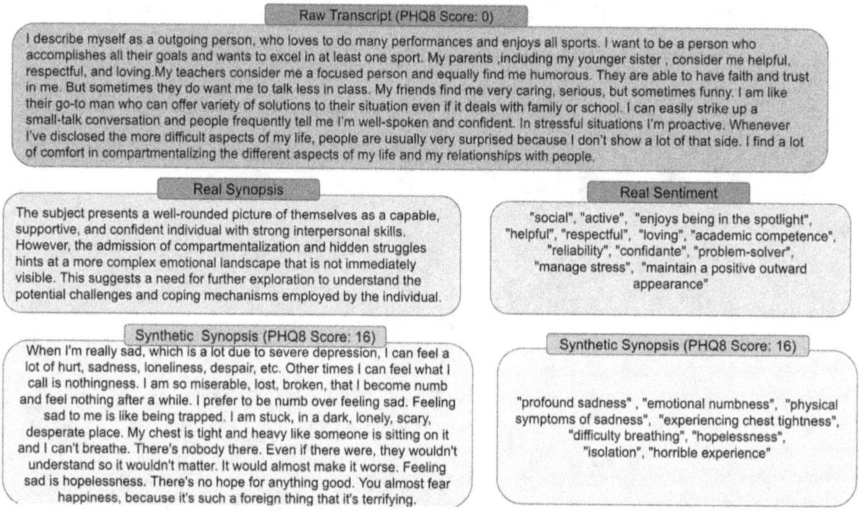

Raw Transcript (PHQ8 Score: 0)

I describe myself as a outgoing person, who loves to do many performances and enjoys all sports. I want to be a person who accomplishes all their goals and wants to excel in at least one sport. My parents ,including my younger sister , consider me helpful, respectful, and loving.My teachers consider me a focused person and equally find me humorous. They are able to have faith and trust in me. But sometimes they do want me to talk less in class. My friends find me very caring, serious, but sometimes funny. I am like their go-to man who can offer variety of solutions to their situation even if it deals with family or school. I can easily strike up a small-talk conversation and people frequently tell me I'm well-spoken and confident. In stressful situations I'm proactive. Whenever I've disclosed the more difficult aspects of my life, people are usually very surprised because I don't show a lot of that side. I find a lot of comfort in compartmentalizing the different aspects of my life and my relationships with people.

Real Synopsis

The subject presents a well-rounded picture of themselves as a capable, supportive, and confident individual with strong interpersonal skills. However, the admission of compartmentalization and hidden struggles hints at a more complex emotional landscape that is not immediately visible. This suggests a need for further exploration to understand the potential challenges and coping mechanisms employed by the individual.

Real Sentiment

"social", "active", "enjoys being in the spotlight", "helpful", "respectful", "loving", "academic competence", "reliability", "confidante", "problem-solver", "manage stress", "maintain a positive outward appearance"

Synthetic Synopsis (PHQ8 Score: 16)

When I'm really sad, which is a lot due to severe depression, I can feel a lot of hurt, sadness, loneliness, despair, etc. Other times I can feel what I call is nothingness. I am so miserable, lost, broken, that I become numb and feel nothing after a while. I prefer to be numb over feeling sad. Feeling sad to me is like being trapped. I am stuck, in a dark, lonely, scary, desperate place. My chest is tight and heavy like someone is sitting on it and I can't breathe. There's nobody there. Even if there were, they wouldn't understand so it wouldn't matter. It would almost make it worse. Feeling sad is hopelessness. There's no hope for anything good. You almost fear happiness, because it's such a foreign thing that it's terrifying.

Synthetic Synopsis (PHQ8 Score: 16)

"profound sadness" , "emotional numbness", "physical symptoms of sadness", "experiencing chest tightness", "difficulty breathing", "hopelessness", "isolation", "horrible experience"

Figure 13-1. *A snippet of the synthetic data generation process demonstrating synthetic synopsis and sentiment with keywords, with a supplied PHQ8 score of 16*

Synthetic data continues to represent real-life data and provide high-quality responses to healthcare-related queries, as shown in Figure 13-2.

LLM based Synthetic Data Generation Process through Training and Enriched with Real-World Data

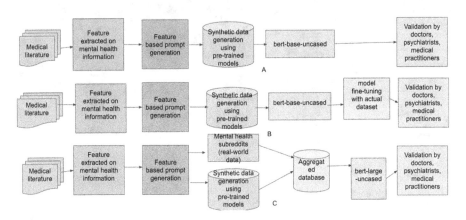

Figure 13-2. *A snippet of a hybrid LLM-based synthetic data generation process demonstrating plain, LLM fine-tuning and data enrichment to generate the synthetic data*

Here, we show the sequence of steps involved in the generation of synthetic data from a pretrained BERT model that can be later validated by healthcare practitioners, doctors, and psychiatrists.

Figure 13-2 shows three different ways of synthetic data generation: A, B, and C, where subfigure A is involved in vanilla synthetic generation, while B shows fine-tuning of the LLM model, and C shows data enrichment with real-world datasets that help the synthetic data resemble the real-world mental health data.

With the background of synthetic data in mind, let us now explore other advanced AI techniques that can be leveraged to design more proactive mental health apps that yield faster go-to-market solutions with continuous improvement life cycles, seeking feedback from the real world.

Faster Speed to Market Mental Health Apps

Designing mental health apps that can take in end-user feedback and continuously improve can make healthcare more accessible and up-to-date. This can be fast-paced with digital twin (DT) solutions, which hold promise for the next decade. As defined by Gartner (2025), DT is "the implementation of a digital twin is an encapsulated software object or model that mirrors a unique physical object, process, organization, person, or other abstraction." DT's role in healthcare and mental well-being can revolutionize the next generation of mental health apps, where users' daily lives can be better monitored by increasing awareness of mental health issues, incorporating their feedback, and providing customized recommendations. This has the far-reaching benefit of improving quality of life and well-being by feeding in lifestyle trends and predicting potential health issues through contextual emotional states and preferences. DT-based mental health solutions will leap mental health apps and the providers to virtually test the apps and optimize product designs through digital simulation-driven solutions. The daily routines of end users and their lifestyle patterns and moods, once captured, can trigger simulations to identify the effectiveness of the apps in end-user well-being, thus providing a potential opportunity to determine issues early, diminishing the time needed for physical prototyping, and drastically minimizing the development cycles.

A DT has a full 360-degree view of a user's holistic condition; hence, designing a responsible DT equipped with interoperability, usability, personalization, feedback, mobility, accessibility, and security can benefit the entire mental health ecosystem. The benefits are not restricted to users of the system but can be realized by doctors, psychiatrists, and healthcare providers, providing a sustainable ecosystem with DT, where predictive models can proactively raise red flags for high-risk individuals for outreach, providing a pathway for early intervention and preventive care. For example, MHDT supercharges Ontrak's care program, leveraging

innovative DT-based solutions to accelerate care innovation, using virtual
patient models to test novel interventions and suggest best practices.
The AI-driven insights and interpretability engine provide individuals
with hyperpersonalized coaching and therapy, along with explaining the
problem and evidence-based approaches.

Let us now understand the core components of DT.

Core Components of DT

Let us take a closer look at the different components that would enable
the DT to function properly by building a proactive, preventative, and
intervention strategy in the mental health world.

- A core mental health DT should be equipped with
  preventive health tools, like sensors (hard and soft
  sensors, including wearables and IoT devices) to
  capture health data and actuators to produce an
  accurate reproduction of the physical twin at any given
  time and help in providing feedback.

- An interface popularly known as the Data Collection
  module captures inputs, user profiles, dialogue data,
  clinical assessments, biometric information, patient-
  reported outcomes, social determinants of health, PHQ
  scores, individual requirements, and preferences and
  provides an initial assessment to end users.

- A simulation engine takes in data from real-world end
  users to run simulations simulating the correlation
  between life events and the corresponding mental
  state of an individual, then models treatment in the
  virtual world, and tracks the impact in the actual
  physical world.

- An AI core engine equipped to handle the processing and model training phase to process the dialogue data through different stages, of which the most common are the preprocessing to ensure data quality, proper data formatting, data labeling following the PHQ scores used as ground truth, data augmentation, etc., to train, test, and evaluate the custom model is then for accuracy to effectively lead to intent classification in the later stages.

- An AI analytics core engine to yield real-time dashboards, perform data analytics, and make recommendations.

- A cloud component helps in storing information of the physical twin to facilitate the health and mental well-being of end users.

- A monitoring and continuous learning framework engine, termed a **Measurement Feedback System (MFS)**, assesses the therapy and recommendations provided to end users based on its validity, reliability, and standardized metrics and explores improvement areas and retrains and learns constantly.

- Additional monitoring tools like routine **Outcome Monitoring (ROM)** and **Clinical Feedback Systems (CFS)** to input recommendations from a clinical psychiatrist and individual symptom characteristics to prevent treatment failure and yield better therapeutic results, especially considering patients who are not advancing in therapy.

However, certain challenges exist in implementing an effective digital twin targeted toward the mental well-being of youth. It goes beyond saying that having the proper ML/AL algorithms for real physical entities and their application in holistically simulating and predicting an individual's well-being must include an individual's mental, physical, and financial aspects to target the right individuals with the right treatment, medications, personalized therapies (with the optimal frequency and duration of therapy sessions), and precision mental health interventions. The bigger scope of DTs from diverse data sources can additionally remove the stigma associated with mental health and increasingly raise self-awareness among individuals regarding their mental health problems. DT-enabled NLP-based chatbots make (Abilkaiyrkyzy et al., 2024) it more accessible to the general public, which in turn can help users to query and get responses in the absence of mental health professionals, even when queries have been rare in the past and historic instances of queries that do not exist. DT is equipped to collect feedback, run simulations in the virtual world, study the impact on the physical world, and then provide tailored outputs and responses to the individuals.

DT-powered chatbots, when designed with an effective conversational flow, help to address individuals with recent issues they are encountering from the point of both physical and mental health. As DT-based technology enforces constant learning-based feedback, it simulates therapist behavior to handle concerns related to well-being, feelings of hopelessness, physical health, preferences, academic interests, and relationships. Individuals get a better quality of life and positive feelings when they are guided by proper support systems that constantly judge their activity levels.

Further, each DT must be designed with responsible AI guidelines by providing users with proper access controls, a consent form to mention their preferences and the retention of data based on their choices. These guardrails help to enable proper authority over their data, giving them ownership to make decisions and choices for information sharing. Along with access controls, privacy-enhancing ML and explainability engines not only bring individuals awareness of the disease they are suffering from and the potential symptoms of the disease but also educate them on the consequences of the mental health disease, justifying the reason why it needs treatment. The reasoning comes with enough attributes and facts to help them convince and give positive confidence to cope with the disease. A DT-powered chatbot is proactive in sending notifications to end users, receiving feedback from them, running virtual mock-ups to enlighten users on the importance of seeking medical attention, and following the medical treatment plans with causal reasoning.

A typical example of a dialogue-based chatbot in real time may include a tokenization process where input text is tokenized by BERT tokenizer, where the input text can be further analyzed on text features through a language model features. The model can then be classified by an intent classification engine where traditional RASA-based models can be used along with the GenAI classification engine to help in the classification of user intent classifications coming in through dialogues. The classification process can identify "high," "medium," or "low" risk of mental state through a continuous process of reiteration and learning to accurately predict a user's mental state based on symptoms received as "no symptoms," "mild," or "severe." Such a chatbot constantly learns from interacting with users through its dialogue interface and provides hyperpersonalized recommendations in consultation with an available doctor or psychiatrist to help users better manage stress, anxiety, and mental health conditions. Along with the interactive chatbot interface that facilitates easy communication with the user, agreement to take up the PHQ test helps to improve the performance of the chatbot responses, as a DT from real-world labels the user dialogue with the correct mental health assessment inferred PHQ score. This kind of digital twin setup for mental healthcare is adaptive and capable of feeding each user's distinctive psychological and physical features, social conditions, and environmental factors. The DT evolves through continuous information fed back to model training to help improve the model's prediction performance over time.

With an understanding of how AI would play a critical role in a digital twin solution, let us now take a deeper look at how a typical end-user journey would shape up.

User Journey with DT

A youth, when logged in to a mental health app, can enroll in some mental health assessment programs or journaling activities, which can detect symptoms, raise alerts, and provide coaching sessions. Coaching sessions use the GAD-7 and PHQ-9 scores from assessment sessions as input and then run simulations in a virtual world to help the user set goals, make recommendations, and support mental well-being through care plans identifying potential risks and hidden symptoms. With a DT, the system can be configured to model youth profile updates and changes in the youth's life stages that impact treatment and recovery trajectories or require alternative therapies for support. For example, certain life-stage challenges can be solved by spirituality-based interventions.

A DT-based solution can receive feedback from end users to adjust the guided treatment selection process and continually optimize the process through monitoring (Ontrak Health, 2024). Individuals experience care and empathy when in the absence of physical therapists, virtual assistants are available 24/7 to explain the prescribed treatment, and progress through constant collaboration and communication with their families.

Digital Twin in Mental Health Domain

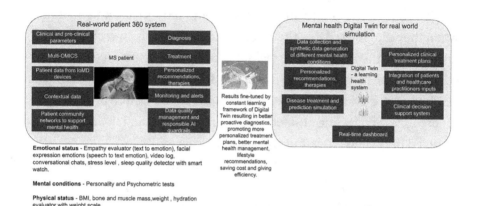

Figure 13-3. *A digital twin solution for mental health with all parameters that are shared between the physical and virtual world (Kas, 2021)*

Figure 13-3 shows a DT-based mental health solution that demonstrates how each adolescent's unique bio-psycho-social data (sensor data, user records, clinical assessments, therapy notes, transcripts from mental health coaches, and social conditions) creates virtual simulations of each adolescent using advanced AI algorithms including NLP to improvise the various intelligent treatment processes and predict their responses to support better decision-making (Kas, 2021).

This may include medication regimens or lifestyle changes, which, before being recommended to end users, are analyzed both for their pros and cons, leveraging the DT architecture. In addition, DT makes predictions on patients who could discontinue treatment plans and the success rates of different treatment options available that could lead to relapse or emergencies.

Overall, a DT is helpful not only in providing customized treatment plans at the right time but also yields better patient experiences and feedback at a lower cost, providing added relief to the mental health ecosystem. The operating processes of mental healthcare centers become more efficient as they are more proactive and deliver the greatest impact. For the youth and adolescents, the framework builds trust and confidence when they directly see that their preferences, life goals, and choices are respected by the system. They feel safe due to the governance and audit controls of the responsible AI-driven work, where the system dynamically validates any incoming contextual data and models before providing predictions. The feeling of safety drives engagement and the desire to get involved in psychoeducation and self-advocacy skills.

With the complete view of the mental health framework for the teenage, youth, and adolescent section of the population, now let us go through the concluding thoughts of this chapter.

Conclusion

In this chapter, we had a recap of all the chapters of the book before having an overview of the synthetic data generation process and understanding the key role played by synthetic data in the mental health domain. We also delved into digital-based solutions in mental health that can yield a 95% reduction in facility fees for hospital admissions, along with a lowering of 88% in emergency care services, directly impacting the ROI. With faster go-to-market mental health applications and constant monitoring and simulations, PHQ scores can see a significant improvement as more patients have been able to accept the problems by coming out of the stigma to undergo treatment and feel comfortable sharing their problems with confidence and self-advocacy. The necessary digital tools are today equipped to train patients to enhance their coping strategies, having more trust built into mental health systems. As our mental health systems mature and are equipped with granular treatment recommendations, we can handle risks earlier, intervene with medications more accurately, and optimize care services in subsequent phases.

The complete universal framework governed by ethical standards can come to the best use of patients, doctors, and psychiatrists to translate medical knowledge and best practices to those who need it, to lead a rapid transformation in the mental health domain to predict, preempt, and personalize support to help people live happier, healthier lives.

References

Abilkaiyrkyzy, A., Laamarti, F., Hamdi, M., & Saddik, A. E. (2024). Dialogue System for Early Mental Illness Detection: Toward a Digital Twin Solution. *IEEE Access, 12*, 2007–2024. https://doi.org/10.1109/ACCESS.2023.3348783

Gartner. (2025). *Definition of Digital Twin—Gartner Information Technology Glossary*. Gartner. https://www.gartner.com/en/information-technology/glossary/digital-twin

Kang, A., Chen, J. Y., Lee-Youngzie, Z., & Fu, S. (2024). *Synthetic Data Generation with LLM for Improved Depression Prediction* (No. arXiv:2411.17672). arXiv. https://doi.org/10.48550/arXiv.2411.17672

Kas, K. (2021). *The tipping point for digital twins in healthcare – Healthskouts Digital Health Advisory Company*. https://www.healthskouts.com/2021/12/20/the-tipping-point-for-digital-twins-in-healthcare/

Ontrak Health. (2024). Ontrak's Mental Health Digital Twin: Pioneering Personalized Care. *Ontrak Health*. https://ontrakhealth.com/white-papers/ontraks-mental-health-digital-twin-pioneering-personalized-care/

P, V. S., & Rao, M. (2024). PsychSynth: Advancing Mental Health AI Through Synthetic Data Generation and Curriculum Training. *2024 9th International Conference on Computer Science and Engineering (UBMK)*, 1–6. https://doi.org/10.1109/UBMK63289.2024.10773545

Index

A

© Sharmistha Chatterjee, Azadeh Dindarian, Usha Rengaraju 2025
S. Chatterjee et al., *Revolutionizing Youth Mental Health with Ethical AI*,
https://doi.org/10.1007/979-8-8688-1186-9

GPSR Compliance
The European Union's (EU) General Product Safety Regulation (GPSR) is a set
of rules that requires consumer products to be safe and our obligations to
ensure this.

If you have any concerns about our products, you can contact us on

ProductSafety@springernature.com

In case Publisher is established outside the EU, the EU authorized
representative is:

Springer Nature Customer Service Center GmbH
Europaplatz 3
69115 Heidelberg, Germany